Protein–Ligand Interactions: structure and spectroscopy

A Practical Approach

Edited by

Stephen E. Harding

NCMH Physical Biochemistry Laboratory,
University of Nottingham, School of Biosciences,
Sutton Bonington, Leicestershire LE12 5RD,
U.K.

and

Babur Z. Chowdhry

School of Chemical and Life Sciences,
University of Greenwich, Wellington Street,
Woolwich, London SE18 6PF, U.K.

OXFORD

UNIVERSITY PRESS

OXFORD
UNIVERSITY PRESS

Great Clarendon Street, Oxford OX2 6DP

Oxford University Press is a department of the University of Oxford.
It furthers the University's objective of excellence in research,
scholarship, and education by publishing worldwide in

Oxford New York

Athens Auckland Bangkok Bogotá Buenos Aires Calcutta Cape Town
Chennai Dar es Salaam Delhi Florence Hong Kong Istanbul Karachi
Kuala Lumpur Madrid Melbourne Mexico City Mumbai Nairobi Paris
São Paulo Singapore Taipei Tokyo Toronto Warsaw

with associated companies in Berlin Ibadan

Oxford is a registered trade mark of Oxford University Press in the UK
and in certain other countries

Published in the United States by Oxford University Press Inc., New York

British Library Cataloguing in Publication Data
Data available

Library of Congress Cataloguing in Publication Data

1 3 5 7 9 10 8 6 4 2

ISBN 0 19 963750 4 (Hbk.)
ISBN 0 19 963747 4 (Pbk.)

Typeset in Swift by Footnote Graphics, Warminster, Wilts
Printed in Great Britain on acid-free paper
by The Bath Press, Bath, Avon

Preface

Interactions involving proteins with other molecules (termed as 'ligands'), whether they be nucleic acid, carbohydrate, vitamin, hormone, steroid, cofactor, metal ion, peptide, or even other proteins, underpin the whole of Biological Science. Nature, when it manufactures or engineers a protein does not generally do so with a single polypeptide chain as the intended end-point but more often than not with a composite structure involving other molecules of life in view. A thorough practical grounding in the experimental and computational methodologies used to investigate such interactions is therefore highly appropriate to the modern day protein scientist, and these two volumes in the Internationally renowned 'Practical Approach' series have been constructed with this intention. The volumes consider the range of principal techniques for studying interactions of proteins with ligands and from a practical standpoint: 22 chapters, produced by experts in the various areas cover: equilibrium dialysis, chromatography, analytical ultracentrifugation, surface plasmon resonance, capillary electrophoresis, electro-optics, X-ray and neutron scattering, isothermal titration calorimetry, differential scanning microcalorimetry, X-ray crystallography, molecular modelling, circular dichroism, fluorescence, stopped flow Fourier transform infrared, Raman and mass spectroscopy, paramagnetic probes, nuclear magnetic resonance, and atomic force microscopy. The volumes are designed for the Academic and Industrialist alike, intelligible by young (Masters, PhD, Postdoctoral) scientists and yet penetrating enough for established scientists.

The whole subject of Protein Science is one of the most intellectually attractive, diverse, and exciting subjects in contemporary science. It is theoretically as well as experimentally demanding and its scope is truly enormous. In fact it encompasses and impacts upon virtually every branch of science, from fundamental theoretical blue skies research at the subatomic level through chemical synthesis, molecular characterization/biology, clinical medicine, biotechnology, and the existing and emerging biophysical technologies (e.g. proteonomics) at the other end of the spectrum.

Proteins are so numerous and diverse in e.g. size, shape, geometry, topology, and other physico-chemical properties that there are very few methodologies/ techniques which have, at some stage or another, not been applied in order to parse their structure–function relationships. Moreover the *in vivo* function of proteins is superbly—and obviously optimally—conditioned by the inter-relationship between structure, short- and long-range interactions, together with environmental properties toward accomplishment of evolutionary directed function. In addition their microscopic and macroscopic structural/electronic behaviour is precisely modulated, both spatially and temporally, by interactions with other macromolecules (nucleic acids/proteins/carbohydrates/lipids or lipid assemblies) and small molecular weight intra- and extracellular (metabolite) molecules. The latter are often referred to as ligands, although this term can encompass virtually any kind of molecule, and is not well defined. Understanding such interactions, at a detailed level (micro- and macroscopic) is essential for gaining an intrinsic, curiosity driven knowledge base relating to cellular behaviour. In addition the applied clinical, agricultural, pharmaceutical, and biotechnology knowledge base has and continues (be it more slowly than often envisaged) to be extended by the need to design specific ligands which will alter/modify intrinsic protein function.

The many scientific and technical disciplines embodied in the subject of protein science and, more specifically, protein–ligand interactions, are making significant contributions to the goal of elucidating structure–function relationships in proteins. For example spectroscopists define electronic structure, enzymologists determine equilibrium/rate parameters, and molecular/structural biologists perturb structure and observe the effect on reactivity and structure at different levels. The theoretical/physical organic chemist can attempt to synthesize and assemble minimal reactivity site representations and determine and/or model intrinsic geometric, electronic, and reactivity characteristics involved in proteins and protein–ligand interactions. All this, and more, is directed towards what is undoubtedly the ultimate goal of protein research: to define function in terms of the detailed inter-relationship between structure/dynamics/kinetics and thermodynamics of a system under known physico-chemical conditions and systematic variations therein.

The field of protein–ligand interaction studies is at a propitious stage of development as well as being in a very active growth phase. It is hoped that both of these volumes will be used by both professional and inexperienced research workers wishing to characterize protein–ligand systems by using a multidisciplinary technique approach: such an approach is more or less inevitable given that the research ethos is now (rightly) more problem solving orientated rather than technique orientated. Both volumes are meant to complement and enhance the information content contained within the other titles in the Practical Approach series.

The techniques used to probe protein–ligand interactions have and continue to improve dramatically over recent years due to the development of new (or continued evolution of) reagents, protocols, strategies, and instrumental techniques/theoretical concepts. The methods presented here are not a comprehensive compilation of all the numerous techniques, which can be used to probe protein–ligand interactions. That would require an encyclopaedia! They are, instead, a selection of what we believe are the most useful and easily applied methods—many of which are now routinely used in laboratories world-wide for research into protein–ligand interactions. It is hoped that the two volumes will act as an introduction for investigating the basic principles and practical application of the subject before embarking on more detailed and/or other techniques, which are not, as yet, so commonly used. Towards this end the co-operation of all contributing authors together with the quality of their presentations are greatly appreciated.

Nottingham and Greenwich   S. E. H. and B. Z. C.
October 2000

Contents

CONTENTS

Protocol list

Abbreviations

a,b,c	semi-axial dimensions of an ellipsoidal protein
a,b,c	unit cell dimensions (Å)
a,b,c	unit cell parameters (Vol. 2, Chap. 1)
a,b,c	unit cell vectors/axes
AC	alternating current
AFM	atomic force microscopy
A_x	absorption (UV/visible) of component x or at wavelength x
CARS	coherent anti-Stokes Raman spectroscopy
CAT	chloramphenicol acetyl transferase
Cc	correlation coefficient
CCD	charge-coupled device
CCD	charge-coupled device area detector
CCDB	Cambridge Crystallographic Data Base
CCDC	Cambridge Crystallographic Data Centre
CD	circular dichroism
CE	capillary electrophoresis
CID	collision-induced dissociation
CMP	cytidine monophosphate
CoA	coenzyme A
Con A	concanavalin A
C_p	excess molar heat capacity ($J.K^{-1}mol^{-1}$; $erg.K^{-1}mol^{-1}$)
CRT	
CSLM	confocal laser scanning microscopy
C_x	weight (mass) concentration ($g.ml^{-1}$) of species x
c_x or [x]	molar concentration (M or $mol.l^{-1}$) of species x
d(hkl)	crystal plane interplanar spacing (Å)
d^*	reciprocal lattice vector
DB3	antibody DB3 (and its Fab′ fragment)
Dc	crystal density
DC	direct current
DM	density modification
D_{max}	maximum dimension of a protein

dmin	resolution of X-ray data (Å)
DMS	polydimethylsiloxane
DMSO	dimethyl sulfoxide
DSC	differential scanning calorimeter
DTE	dithioerythritol
DTNB	5,5′-dithiobis (2-nitrobenzoic acid)
DTT	dithiothreitol
D_x	translational diffusion coefficient of component x ($cm^2.s^{-1}$)
E	electric field strength ($V.cm^{-1}$)
EDTA	ethylenediamine tetraacetic acid
EPR	electron paramagnetic resonance
ESI	electrospray ionization
ESRF	European Synchrotron Radiation Facility
EtOH	ethanol
f	atomic scattering factor for x-rays
$\mathbf{F}(hkl)$	X-ray structure factor
f/f_o	frictional ratio of macromolecule to that of a spherical particle of the same anhydrous mass and volume
Fab′	antibody Fab′ fragment
FAB	fast-atom bombardment
FCS	fetal calf serum
FF	force field
FFT	fast Fourier transform
fL	fluorescent ligand
FMS	polytrifluoropropylmethylsiloxane
FPE	fluoresceinphosphatidylethanolamine
FPLC	fast protein liquid chromatography
FSD	Fourier self-deconvolution
FT	Fourier transform
FTICR	Fourier transform ion cyclotron resonance
G	molar Gibbs free energy ($J.mol^{-1}$, $erg.mol^{-1}$)
GA	genetic algorithm
GdnCl	guanidinium chloride
H	molar enthalpy ($J.mol^{-1}$, $erg.mol^{-1}$)
h	Planck constant
Hb	haemoglobin
Hkl	indices of a crystal plane
HSA	human serum albumin
HSQC	heteronuclear single quantum coherence
HUVECS	human umbilical vein endothelial cells
I	intensity
I	ionic strength ($mol. l^{-1}$)
IL	interleukin
ILL	Institut Laue Langevin (Grenoble, France)
IPDA	intensified photodiode array

IR	isomorphous replacement
IRS	internal reflection spectroscopy
ITC	isothermal titration calorimetry
J_x	absolute concentration (fringe units) of component x
j_x	relative concentration to the meniscus (fringe units) of component x
k	dialysis rate constant (s^{-1})
K, K_a	association constant (M^{-1} ($l.mol^{-1}$) or $ml.mol^{-1}$)
k_B	Boltzmann's constant
K_D, K_d	dissociation constant (M ($mol.l^{-1}$) or $mol.ml^{-1}$)
k_{off}	dissociation rate constant (s^{-1})
k_{off}	first-order dissociation rate constant (s^{-1})
k_{on}	association rate constant (s^{-1})
k_{on}	first-order association rate constant (s^{-1})
k_s	Gralén coefficient, $ml.g^{-1}$
K_{xy}	intrinsic binding constant between acceptor x and ligand y
L	ligand
l, L	litre
mAb	monoclonal antibody
MALDI	matrix-assisted laser desorption/ionization
MC	Monte Carlo
MD	molecular dynamics
MetJ	methionine repressor protein
mH	mass of H atom
MIR	multiple isomorphous replacement
MLA	mistletoe lectin A-chain
MLB	mistletoe lectin B-chain
MLI	mistletoe lectin
MLVs	multilamellar vesicular
MPD	methylpentanediol
MR	molecular replacement
MS-MS	tandem mass spectrometry
$M_{w,x}$	weight average molecular weight (molar mass) of component x
MWPC	multiwire proportional counter
M_x $M_{r,x}$	molecular weight (Da) or molar mass ($g.mol^{-1}$) of component x
N	Newton
N	native folded polypeptide
N, N_A	Avogadro's number (6.02252×10^{23} mol^{-1})
NA	numerical aperture
NAG	N-acetyl glucosamine
NCS	non-crystallographic symmetry
NDSB	non-detergent sulfobetaines
nfL	non-fluorescent ligand
NMR	nuclear magnetic resonance
NOEs	nuclear Overhauser enhancement effects

NOESY	nuclear Overhauser enhancement effects spectroscopy
P	protein
P	Patterson function
P	Perrin shape function
PADS	peroxylamine disulfonate or Fremy's salt
PBS	phosphate-buffered saline
PC	phosphatidylcholine
PDB	Protein Data Bank
PEG	polyethylene glycol
δ	phase shift or optical retardation (Chap. 8)
PLVs	phospholipid vesicles
PMT	photomultiplier tube
PROXYL	(2,2,5,5-tetramethyl-1-pyrrolidinyloxy)
PS	phosphatidylserine
PSCARS	polarization-sensitive coherent anti-Stokes Raman spectroscopy
PSI	photosystem I
Q	scattering vector (nm^{-1})
QSAR	quantitative structure–activity relationship
R	crystallographic R-factor
R	aggregate radius (Vol. 2, Chap. 1)
r	fluorescence anisotropy
R	gas constant ($8.3143 \ J.K^{-1} \ mol^{-1}$)
R	rotation function
R_{int}	combined structure factor index (Vol. 2, Chap. 1)
RAF	Rabbit Anti-Fluorescein antibody
RCA	ricin agglutiin A-chain
REPs	resonance enhancement profiles
R_f or Rfree	free R-factor
R_G, R_g	radius of gyration (Å, nm, cm)
RIDS	reaction-induced difference spectroscopy
Rint	internal consistency R factor
Rmerge	merging R factor
σ	RMS density (Patterson or electron density)
RMS	root mean square
rmsd	root mean square deviation
RNase	ribonuclease
RNase A	ribonuclease A
ROA	Raman optical activity
ROESY	rotating frame nuclear Overhauser enhancement effects spectroscopy
RTA	ricin A-chain
RTB	ricin B-chain
RU	response units or resonance units
s	equilibrium solubility
S	supersaturation

S	electron spin
S	molar entropy ($J.mol^{-1}K^{-1}$, $erg.mol^{-1}K^{-1}$)
s	reciprocal of the Bragg spacing ($Å^{-1}$, nm^{-1}, cm^{-1})
s	solvent fraction of the cell mass (Vol. 2, Chap. 1)
SAP	human serum amyloid P component
SCRs	structurally conserved regions
SDS	sodium dodecyl sulfate
SERRS	surface-enhanced resonance Raman spectroscopy
SERS	surface-enhanced Raman spectroscopy
SIR	single isomorphous replacement
SPR	surface plasmon resonance
SR	synchrotron radiation
SRS	synchrotron radiation source (Daresbury, UK)
ST-EPR	saturation-transfer electron paramagnetic resonance
SVD	singular value decomposition
s_x	sedimentation coefficient of component x (sec or Svedberg units, $S = 1 \times 10^{-13}$ sec)
T	translation function (in MR)
T_1	spin–lattice relaxation time
T	temperature
TEMPO	(2,2,6,6-tetramethyl-1-piperidinyloxy)
TF	tissue factor
TFE	trifluoroethanol
TOF	time-of-flight
TR^3	time-resolved resonance Raman
trNOE	transferred nuclear Overhauser enhancement effects
U	unfolded polypeptide
U_i	energy of interaction (induced dipole)
U_p	energy of interaction (permanent dipole)
UV	ultra-violet
UVRR	ultra-violet resonance Raman
uvw	a point in Patterson space
Vx	elution volume (ml) of species x
$\bar{v}_x$	partial specific volume ($ml.g^{-1}$) of component x
Web	World Wide Web
X	interaction constant (equilibrium constant) in weight concentration units $ml.g^{-1}$ or $l.g^{-1}$
x; y; z;	fractional coordinate of atom j
Z	atomic number
z_x, Z_x	net charge (valency) on a species, x
λ	wavelength (Å, nm, cm)
θ (hkl)	Bragg angle for X-ray reflection from hkl planes
φ (hkl)	phase angle of structure factor $\mathbf{F}(hkl)$
ρ (xyz)	electron density at point (x,y,z)
α, β, γ	unit cell inter-axial angles

ABBREVIATIONS

ψ_x	psi (sedimentation equilibrium) function for component x
$[\Theta]$	molar ellipticity
$|F|$	structure factor amplitude
$1/T_2$	transverse relaxation rate
^2H_2O	deuterium oxide
3D	three-dimensional
$\Delta\delta$	chemical shift separation
$\Delta\varepsilon$	molar differential extinction coefficient ($M^{-1}cm^{-1}$, $l.mol^{-1}cm^{-1}$), sometimes abbreviated as ε
α	polarizability (Chap. 8)
δ	hydration
ε_λ	extinction coefficient (either $ml.g^{-1} cm^{-1}$, or $M^{-1}.cm^{-1}$) at a wavelength λ
ϕ_p	maximum packing fraction
$\lambda/4$ plate	quarter-wave plate
μ_p	dipole moment (permanent dipole)
τ_c	rotational correlation time (s)
ω	angular velocity of centrifuge rotor (radians.s^{-1})

Chapter 1

X-ray crystallographic studies of protein–ligand interactions

Rex A. Palmer

Department of Crystallography, Birkbeck College, University of London, Malet Street, London WC1E 7HX, UK.

1 Introduction

This chapter is concerned with current methods employed for the X-ray analysis of protein structures complexed with ligands. It deals with the methods used for growing crystals, the background to the use of X-ray intensity data, collection and measurement of the X-ray data, methods for analysing and interpretation of the structures. Whilst emphasis is given to the practical aspect of this type of work, in view of the highly specialized nature of X-ray crystallography it has been felt necessary to provide the reader with a background to the underlying theory. X-ray crystallography is highly computational and a great variety of software is available for performing the various stages of the calculations and graphics display. Some of the software used is specific to local research groups whilst other programmes are widely available and employed world-wide. Those programmes mentioned in the text are readily available and prospective users should have no difficulties obtaining copies complete with instructions. Important Internet addresses are included and major suppliers of specialised equipment are listed. The flow chart in *Figure 1* indicates the various stages covered in the chapter.

2 Crystallization of proteins and complexes for X-ray analysis

2.1 Introduction

Haemoglobin crystals are known to have been observed in dried blood samples in the 1800s, but the first reported crystallization of a protein was urease in 1926, with other important experiments being carried out in the 1920s and 30s (1). The use of crystals in structural biology dates from 1934, when Bernal and Crowfoot (2) produced the first X-ray diffraction pattern of a protein (pepsin). Since then hundreds of biological macromolecules have been crystallized, repre-

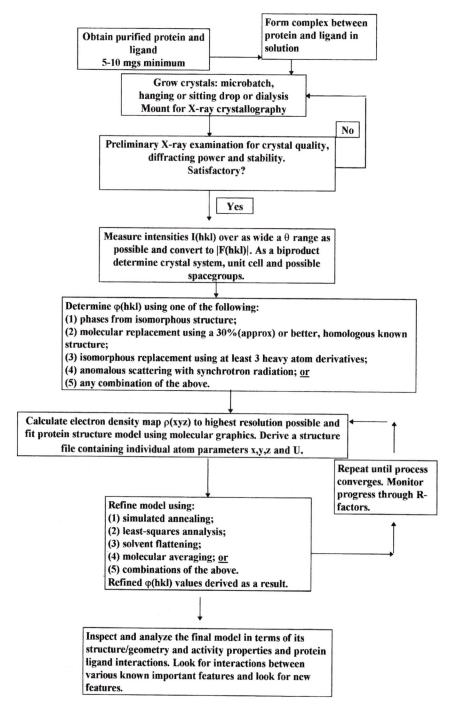

Figure 1 Flow diagram for X-ray analysis of protein–ligand interactions.

sentative of the major families including proteins, enzymes, nucleic acids, chromatins, and viruses, almost always through an empirical approach and with little or no control over their growth. Recently however it has been possible to rationalize the procedures, at least to some extent, through the use of hanging and sitting drops which require minimal amounts of protein, and multiple sampling techniques, which enable a wide variety of conditions to be tried.

Once established, crystallization protocols are generally reproducible. However there are well known instances where once-successful crystallizations have failed to be repeated, or subsequently yielded only inferior crystals. Such problems may be due to one of several factors: minor differences in protein composition or purity, for example, through extraction from different sources; or failure to reproduce exact crystallization conditions, perhaps due to poor reporting or recording protocols. Even the chance presence of a trace of dirt or grease can make all the difference, one way or the other!

Special properties of protein crystals include the following:

(a) They are small compared to common crystals, rarely > 1 mm on any edge.

(b) Because in the vast majority of cases only one stereoisomer of any biological macromolecule exists in nature, their crystals contain only symmetry elements with *rotation* and/or *translation* (*inversion* and *reflection* are excluded as these symmetry operations involve a change of hand). As a consequence the crystals themselves tend to exhibit fairly simple shapes compared to many minerals and other naturally occurring crystals (snowflakes etc.).

(c) Protein crystals are extremely fragile and require ultra gentle handling (or none at all). This is due to their high solvent content, which can be as much as 70%, or even more, and is rarely less than 40% (3).

(d) Protein crystals are extremely sensitive to pH, ionic strength, and temperature. Stability to low temperature can be improved through the use of *cryoprotectants* (see below) and is a useful technique for improving the extent and quality of X-ray intensity data.

(e) They exhibit weak optical properties (e.g. birefringence), and diffract X-rays both weakly and to resolutions far short of the theoretical limit (see section 6.2, p. 48). Again these effects are associated with the presence of large amounts of disordered or partially disordered solvent.

2.1.1 Crystallization conditions for macromolecules

Almost every quantitative aspect of crystal growth is a direct function of supersaturation. Macromolecular crystals usually form by *nucleation* at extremely high levels of supersaturation (100 to 1000%). In comparison, small molecule crystals usually nucleate at only a few per cent supersaturation.

Although high levels of supersaturation may be essential for promoting nucleation of macromolecular crystals, in general this is far from ideal in terms of the formation of good quality crystals.

Supersaturated macromolecular solutions tend to produce *alternative solid states* (*amorphous precipitates*). Consequently there is *competition* between crystals and precipitates at both nucleation and growth stages, and this competition is particularly acute because it is promoted at high levels of supersaturation. Because amorphous precipitates are kinetically favoured, even though they are associated with higher energy states, they tend to dominate the solid phase and inhibit or even preclude crystal formation.

2.1.2 Properties of protein crystals

Protein crystals may contain one or more of the following factors, any or all of which may influence their size, quality, and X-ray diffraction characteristics:

(a) Ordered layers.

(b) Other, more complex layers.

(c) Air pockets.

(d) Disordered molecular deposits or clusters.

(e) Inclusions.

(f) Ordered solvent.

(g) Disordered solvent.

(h) Precipitant ions.

(i) Impurities (e.g. bound carbohydrate, which may lack homogeneity).

(j) Various defects and dislocations.

(k) Inhibitors, non-covalently bound sugars, prosthetic groups, or other ligands.

(l) Covalent or non-covalently bound heavy atoms.

2.2 Crystallization of proteins

Crystallization from solution, like the formation of ice from water, represents a change of phase. Macromolecular crystallization depends on three factors:

(a) Perturbing the relationship between the macromolecules and the solution components (water molecules and ions).

(b) Altering the structure of the solvent so that the molecules are less well accommodated, thus promoting phase separation.

(c) Enhancing the number and strength of favourable interactions between macromolecules.

In the non-equilibrium state of supersaturation the supersaturation is defined as:

$$S = \ln (c/s)$$

where c = concentration and s = equilibrium solubility. In this state, molecules are continuously associating to form clusters and aggregates of unknown order, whose size may be defined by a radius R. Molecules that are free in solution are continuously recruited into a potential nucleus, while others dissociate. If at some point R exceeds a critical value, R_c, then it will be energetically favourable for the aggregate to accumulate new molecules more rapidly than it loses old. A crystal nucleus will then be born and growth will proceed.

The higher the value of s, the greater the probability that molecules will be gained rather than lost to the aggregate, and the smaller is R_c. The energy barrier to achievement of critical nucleus size R_c is therefore supersaturation dependent.

The nature of the nucleus and the process by which R_c is attained is largely a mystery. It is not known whether the critical nucleus is initially ordered or assumes order through restructuring, nor whether it forms by coalescence of arbitrary sub-nuclear clusters, or by strict monomer or oligomer addition, or by all of these factors simultaneously.

2.2.1 Molecular purity

To form crystals, the macromolecules have to be ordered in regular 3D arrays. All forms of interference with regular packing will hinder crystallization. Lack of purity and homogeneity are major factors in causing unsuccessful and irreproducible crystallization experiments. It is advisable to conduct purification on well-defined material, and to carry out crystallization assays on fresh samples without mixing batches of macromolecules. Note however that some micro-heterogeneities can be tolerated provided they do not occur in parts of the molecule involved in packing contacts in the crystal structure—a factor that is unpredictable prior to the structure being solved!

Types of impurity that can occur are as follows:

(a) Molecular contaminants of any type, large or small.

(b) Small sequence heterogeneities.

(c) Conformational heterogeneities especially flexible domains.

(d) Batch effects (no two batches of protein are absolutely identical).

2.3 Practical considerations

The *high supersaturation* of the molecules required for crystal nucleation can be achieved using a variety of precipitants. Widely used precipitants include:

• ammonium sulfate

• polyethylene glycol (PEG)

• methylpentanediol (MPD)

• sodium chloride

When starting from non-saturated solutions, supersaturation can be reached by varying parameters such as *temperature or pH*. It should be noted that the conditions for optimal nucleation are not the same as those for optimal growth. If possible these different stages should be uncoupled but this is nearly always serendipitous! *Nucleation* may be homogeneous or heterogeneous, occurring in the latter case on solid particles. This can lead to *epitaxial growth* (growth of crystals on other crystals). Interface or wall effects and the shape and volume of drops can affect nucleation or growth. Therefore the geometry of crystallization chambers or drops have to be defined.

2.3.1 Batch crystallization

This method is mentioned here mainly for historical interest. It was used for the early crystallization attempts in the 1940s, on proteins such as ribonuclease or pepsin, which were available in relatively large quantities. Protein solutions are prepared in millilitre quantities and crystallization is carried out in ordinary test-tubes, typically of 5–10 ml capacity. The crystals grow on the walls of the test-tubes and are harvested for storage when of suitable size.

Protocol 1

Batch crystallization of monoclinic ribonuclease II (4)

Equipment and reagents

- Ribonuclease A (1 g from Sigma)
- Standard test-tubes (1 cm diameter, 5 cm in length)
- Mechanical stirrer
- pH meter
- Absolute ethanol
- Distilled water

Method

1 Put 100 mg protein into a test-tube and dissolve in 2 ml water, cooled to 0 °C, and gradually add 1 ml ice-cold absolute ethanol with stirring.

2 Measure the pH and adjust to 5.0 by addition of a suitable buffer.

3 Keep the tube at room temperature for three days.

4 Re-cool and add a further 0.4 ml alcohol with stirring. The composition is now approximately 40% (v/v).

5 Small crystals should eventually appear on the walls of the tube, say within ten days.[a]

Seeding

1 Larger crystals may be obtained by seeding. Use crystals grown as above for seeds. This may be carried out three days after step 4.

2 Add a few crystal seeds (0.1–0.2 mm) to the tube. New crystals should be left to grow slowly.

3 To accelerate growth add 0.2 ml alcohol at seven day intervals.

Harvesting

1 Harvest the crystals when the alcohol concentration is 50–60%.

2 Individual crystals may be carefully detached from the tube wall using a glass fibre or spatula and transferred to another tube, washed with 50% aqueous ethanol, and then stored in 60% (v/v) ethanol–water at room temperature.

Protocol 1 continued

Controlling the rate of crystallization

1   Rapid increase in the alcohol concentration will promote fast crystallization with many very small crystals being formed. If solutions are left for several months at alcohol concentrations lower than 40%, very large crystals 2–3 mm in size may be obtained.

Amorphous precipitate

1   An amorphous precipitate of ribonuclease is obtained when the alcohol is first added at 0 °C. This can be partially redissolved by warming in a water-bath at 40 °C. This prevents excessive amorphous background on the tube walls where the crystals form.

ᵃ The crystals obtained should be the monoclinic $P2_1$ form (5). This form grows at pH 5.0.

2.3.2 Microbatch screening

In order to establish crystallization conditions initial experiments are usually conducted at a microscale level with sample volumes in the range of a few microlitres (i.e. approx. 1 μl) and protein concentrations around 5–10 mg/ml. Since it is difficult and expensive to prepare large quantities of highly purified protein this strategy allows many different crystallization conditions to be screened.

The set-up of a well for microbatch screening is illustrated in *Figure 2*. The wells are in the form of a matrix on a microbatch plate (Hampton) (*Figure 3*).

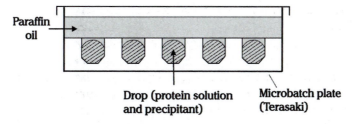

Figure 2 The set-up of a well for microbatch screening.

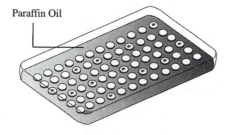

Figure 3 Microbatch plate showing the well matrix (Hampton, Terasaki).

Drops are pipetted using a fine Hampton or Gilson syringe under a layer of silicone (polydimethylsiloxane) or paraffin oil (or 1:1 mixture). The drops contain protein and precipitant solution. All of the reagents involved in the crystallization are present at a specific concentration and no significant concentration of the protein nor the reagents can occur in the drop. Reagent sampling kits are available from Hampton (Optimize™) or can be suitably made up in the laboratory. Diffusion of water from the drop takes place through the oil thus changing the concentration in the drop, hopefully towards the required crystallization conditions.

Using this method many different conditions and precipitants can be tried. When crystals do appear it will usually be in only a small number of wells. These localized conditions can be refined in order to optimize crystal size and quality for X-ray diffraction using one of the techniques described below. Microbatch droplets usually dry up completely within a few weeks and consequently require to be carefully monitored.

2.3.3 Vapour diffusion techniques

The most popular and successful techniques for establishing crystallization conditions rapidly and efficiently, with subsequent production of diffraction quality protein crystals, are based on vapour diffusion. There are several practical variations in use.

i. Sitting drop

The set-up for this method is illustrated in *Figure 4*. Wells are again in the form of a matrix, typically 4 × 6. A great variety of plates are available (e.g. Hampton, Qplate™), many of which are also adaptable for the hanging drop technique. In this method a droplet (5–20 µl) containing the macromolecule, a buffer and a precipitating agent is equilibrated against a reservoir (1–25 ml) containing a solution of the same precipitant at a higher concentration (say by a factor of two) than the droplet. Equilibration proceeds by evaporation of the volatile species until the vapour pressure of the droplet equals that of the reservoir. Crystals form in the droplet.

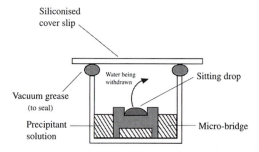

Figure 4 Sitting drop method of crystallization.

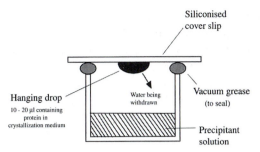

Figure 5 Set-up for hanging drop crystallization.

ii. Hanging drop

The principle here is essentially the same as that in the sitting drop technique except that the drop, in this case, is rapidly inverted and hangs by surface tension over the precipitant reservoir (*Figure 5*). Again if successful, crystals grow in the droplet. Similar drop size and conditions of precipitants, buffers etc., to those used for sitting drops apply.

iii. Vapour diffusion rate of control

As in the microbatch technique a layer of oil can be used with hanging or sitting drop techniques. This limits the rate of vapour diffusion. 200 μl of paraffin or silicone oil or a mixture of the two is applied over the reservoir solution. Varying the composition of the mixture provides additional control over the vapour diffusion rate. *Figure 6* illustrates the method for this hanging drop technique.

iv. Screening crystallization conditions

The above methods are all used with multiple screening protocols. A plate containing typically 4 × 6 wells, or for microbatch tests 6 × 12 (*Figure 3*) provides a matrix of conditions. The method allows a broad range of salts, polymers, and organic solvents over a wide range of pH to be sampled. When crystals are obtained a second, finer screening around the relevant conditions can be used in order to optimize crystallization conditions to produce X-ray diffraction quality crystals. Ready to use reagents formulated from highly pure salts, buffers, and precipitants at various concentrations are commercially available (Hampton).

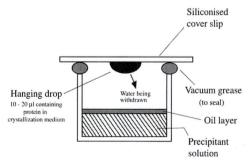

Figure 6 Set-up for under oil vapour diffusion in the hanging drop technique. The rate of vapour diffusion may be controlled by varying the composition of the oil mixture.

Home-made kits are employed in many laboratories where the appropriate expertise is available.

v. Liquid/liquid diffusion techniques

Two variants of this method are illustrated. *Figure 7* shows the use of a dialysis membrane cell. Precipitant diffuses slowly through the dialysis sac and crystals form when the conditions are optimal. In the interface diffusion method (*Figure 8*), equilibration proceeds by diffusion of the small precipitant molecules into the macromolecular compartment. To avoid too rapid a mixing of the precipitant and macromolecular phases, the equipment usually consists of cylindrical chambers of small diameter in which convection is reduced.

vi. Containerless crystallization

This method also employs a crystallization plate. A reservoir of two non-miscible oils, e.g. polytrifluoropropylmethylsiloxane (FMS) and polydimethylsiloxane (DMS) is formed in each well (*Figure 9*). Neither of the two oils is water miscible. Drops of sample combined with crystallization reagent are pipetted at the oil interface. Additional protein or reagent can be added to the drop if required. Large perfect single crystals have been reported using this method. Crystals typically nucleate at the oil/water interface and grow into the drop.

vii. Gel crystallization using silica hydrogel

Gels provide very efficient media for growing protein crystals. Silica gels in particular are stable over a wide range of conditions, and compatible with additives

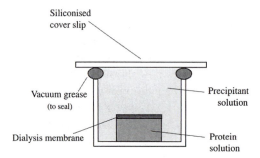

Figure 7 Use of a dialysis membrane cell. Precipitant diffuses slowly through the dialysis sac and crystals form when the conditions are optimal.

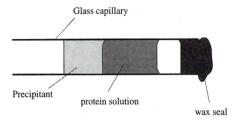

Figure 8 Interface diffusion method for macromolecular crystallization.

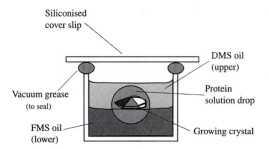

Figure 9 Containerless crystallization. This method employs a crystallization plate. A reservoir of two non-miscible oils, polytrifluoropropylmethylsiloxane (FMS) and polydimethylsiloxane (DMS) is formed in each well. Neither of the two oils is water miscible.

and precipitants commonly used in macromolecular crystal growing. The gel forms a porous network in which the crystals can grow, minimizes convection, sedimentation, and nucleation and therefore promotes the growth of large crystals without the strain usually imposed by the presence of a container. Silica hydrogel can be used for liquid-gel, liquid-gel-liquid, and vapour diffusion as well as dialysis techniques.

2.3.4 Practical hints

i. The importance of solubility data

Knowledge of macromolecular solubilities is a highly desirable factor for designing or varying crystallization conditions. Although it is known that the solubility of proteins is related to temperature, pH, solvation, and the nature of the salt used for salting-in or salting-out, its theoretical background is still controversial and solubility data always originate from experimental determinations. This requires copious supplies of protein, which may not be available. For example measurement of solubility data for the lectin concanavalin A (Con A) showed how the soluble and crystalline phases are significantly altered by pH, temperature, and concentration of the precipitating agent ammonium sulfate. This was carried out by plotting the data for constant ammonium sulfate concentrations. Reference to the phase diagram enabled crystallization of the protein to be achieved by shifts in pH. (It should be noted that solubilities cannot be extrapolated from just a few measurements and detailed phase diagrams should be established to find appropriate crystallization conditions. Unfortunately most proteins are not available in the sort of quantities required for this type of experiment.)

ii. The effect of anions on solubility

Using lysozyme as a model protein it has been shown that the major effect on solubility is due to anion concentration, the relative effect of anions corresponding to the reverse order of the Hofmeister series:

- Anions: $SO_4^{2-} > HPO_4^{2-} > CH_3COO^- > Cl^- > Br^- > NO_3^- > I^- > ClO_4^- > SCN^-$
- Cations: $NH_4^+ > K^+ > Na^+ > Li^+ > Mg^{2+} > Ca^{2+} > Ba^{2+}$

Ions at the left end of either series are usually the most effective for salting-out proteins. SCN$^-$ (thiocyanate) causes the greatest reduction in solubility. As a result, solubility maxima leading to crystallization can be obtained in the presence of thiocyanate and protein concentrations as low as 100 mM and about 2.5 mg/ml respectively. Supersaturation can be reached (or changed) over a large concentration range of protein and salt, provided the appropriate salts are used for a given protein.

iii. Use of non-detergents for solubilizing

Non-detergent sulfobetaines (NDSB) are zwitterionic ammonium propane sulfonic derivatives such as dimethyl ethyl ammonium propane sulfonate (NDSB195) which exhibit an unusual combination of both protein solubilizing and stabilizing properties. Using NDSBs it is possible to:

(a) Reduce protein aggregation.

(b) Simultaneously effect stabilization and consequently,

(c) Undertake protein crystallization over a wider range of conditions, for example pH through the use of otherwise ineffectual precipitants.

Protocol 2

Microbatch crystallization (selection of protein not specified)

Equipment and reagents

- Terasaki plate (Hampton Research): 72 wells, 10 μl capacity
- Colourless light paraffin oil (e.g. from BDH)
- Precipitant solutions
- Protein solution, centrifuged to remove particulates

Method

1 Fill the Terasaki plate with oil to cover the wells.

2 Add 1–5 μl of a selected precipitant to each well by pipetting just under the oil surface. The precipitant drop will sink to the bottom.

3 Add 1–5 μl of the protein solution to each well using the same method.

4 If crystals form at all it will be in only a few of the wells. Therefore make a careful note of which precipitant is in which well (a chart numbered in the same way as the wells is useful).

5 Store the plate in a temperature controlled spot and inspect under a polarizing microscope every two days, taking care not to jar the plate and its contents.

6 Crystals can be harvested by draining excess oil from the top of the wells and filling wells containing crystals with 15 μl of the appropriate precipitant to displace the oil. Crystals are ready to be mounted for X-ray work.

Protocol 2 continued

7   Having established crystallization conditions, optimization of the crystals can be attempted using one of the other methods described above.

Variations

1   Protein/precipitant solution can be pipetted into the well first and then topped up with oil.

2   The evaporation rate can be controlled by the thickness of the oil layer (see above). This in turn will affect the formation of crystals either for better or worse.

3   Further control of the evaporation rate from the well can be achieved as mentioned above by mixing the paraffin with silicone oil (e.g. Dow Corning silicone fluid 200/l cS from BDH). The silicone oil accelerates the evaporation rate.

Protocol 3

Hanging drop vapour diffusion

Equipment and reagents

- Linbro tissue tray (Hampton Research)
- Silicone vacuum grease in a 20 ml syringe
- Siliconized coverslips, 22 mm square (Hampton Research)
- Precipitant solutions
- Protein solution, centrifuged to remove particulates

Method

1   Apply silicone grease around the rim of each well. This will subsequently provide a seal with a coverslip placed over the well.

2   Pipette 1 ml of selected precipitant into each well.

3   Place 2–8 μl of protein solution into the centre of a coverslip.

4   Carefully layer 2–8 μl of precipitant solution from a well on top of the protein solution drop.

5   Invert the coverslip over the same well and press into the silicone grease to seal.

6   Store in a suitable location.

Variations

1   The equilibration rate may be controlled using one of the following methods:
    (a) Vary the drop size (larger drops take longer).
    (b) Add a layer of paraffin/silicone oil mixture to the top of the reservoir to slow down equilibration.
    (c) Use sitting drops on a bridge. This method slows down equilibration and provides for larger drop sizes, which enable crystals to grow undistorted by the drop surface. The Hampton Research Micro-Bridge sits neatly inside the wells of a Linbro tray.

Protocol 4

Batch crystallization of lysozyme at acid pH employing NSDB195 non-detergent

Reagents (see ref. 6)
- Lysozyme
- Acetate buffer
- NaCl
- NDSB195

Method

1 Prepare lysozyme in 50 mM acetate buffer pH 4.6, 20 °C, protein concentration 75 mg/ml, NaCl precipitant 0.6–1.8 M, depending on NDSB concentration. Use the three conditions given below:

(a) 0.0 M NDSB195, 1.17 M NaCl.

(b) 0.1 M NDSB195, 1.04 M NaCl.

(c) 0.25 M NDSB195, 0.91 M NaCl.

2 The batch method is carried out in bulk in small test-tubes or other suitable containers. Large crystals should appear within 4–5 h.

3 In relation to kinetics of crystallization-drop size; theoretical and experimental understanding of the kinetics of biomolecular crystal growth is required.

4 Intuition suggests that reaching supersaturation is related to the rate of evaporation of the solvent and when achieved can affect nucleation or crystal growth. A study by Fowlis *et al.* (7) describes parameters that determine the kinetics for hanging drops. Experiments by Gioge in Strasbourg have shown that drop size (for sitting drops) must be regarded as a major parameter in macromolecular crystallization.

5 Convection (density-dependent flow) plays an important role in crystal growth. Such flow may disrupt concentration gradients. As noted previously, convection can be minimized under particular experimental conditions, e.g. in gels, and almost completely abolished in a microgravity environment in space. In microgravity conditions, only diffusive transport of molecules takes place. Furthermore, sedimentation and wall effects, which also interfere with crystal growth, do not occur in space. Consequently, more regular growth and larger crystals are expected under reduced gravity.

2.3.5 Examples of protein crystallization

(a) Lentil (*Lens culinaris*) lectin. Two high diffraction quality crystal forms grow together from one set of conditions:

(i) Monoclinic, space group $P2_1$, unit cell: a = 58.0 Å, b = 56.0 Å, c = 82.1 Å, β = 104.4°.

(ii) Orthorhombic, space group $P2_12_12_1$, unit cell: a = 56.4 Å, b = 74.6 Å, c = 124.9 Å.

Conditions: Hanging drop; crystals grow from a 3 mg/ml solution in a 20 mM combined phosphate/acetate buffer (pH = 5.9) combining 1 mM $CaCl_2$ and 0.1 mM $MnCl_2$, which was brought to supersaturation with 40% (v/v) 2-methyl-2, 4-pentanediol. Platelet shaped crystals (2 mm × 1 mm × 0.3 mm) appeared after two to three weeks. Although the crystals belong to different crystal systems they were both present in the same solution.

(b) MLI from mistletoe (*Viscum album*). Crystal properties: hexagonal, space group $P6_522$, unit cell: a = b = 110.0 Å, c = 309.3 Å. Conditions: Hanging drop; 20 mg/ml, 0.1 M phosphate buffer (pH 6.7) plus 0.5 M galactose. 5 ml protein solution equilibrated against an equal volume of reservoir solution containing 50% saturated $(NH_4)_2SO_4$ precipitant plus 10 ml of 20% glacial acetic acid. Hexagonal bipyramidal crystals, 0.4 × 0.3 × 0.3 mm³ in dimension can be obtained after two weeks (8).

2.4 Crystal mounting for X-ray data collection

For X-ray data collection at room temperature protein crystals are usually mounted in sealed thin-walled glass capillary tubes as shown in *Figure 10*. The presence of a drop of mother liquor inside the tube allows a stable equilibrium to be established which prevents the crystal from drying out and becoming denatured. Most proteins rapidly deteriorate once exposed to X-rays, through the formation of free radicals. Crystal lifetime can be anything from a few minutes to several days under normal laboratory conditions. Exposure to the highly intense beam from a synchrotron source usually decreases crystal lifetime dramatically. This however is compensated by the increased intensity, which permits shorter individual exposures and consequent increase in the quantity of data recorded.

2.4.1 Cryo crystallography

When the temperature of a crystal is lowered the so-called thermal motion of the constituent atoms becomes less marked and the X-ray diffraction pattern can be improved both in intensity and resolution. These two factors are highly desirable in protein crystallography, as protein crystals are notoriously poor diffractors. It is not surprising therefore that since the innovation of more efficient crystal cryo systems in the past ten years [a popular and efficient model is the Oxford Cryosystems Cryostream Cooler] there has been a marked increase

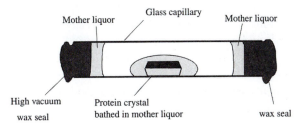

Figure 10 Capillary tube mounting of a protein crystal for X-ray data collection at room temperature.

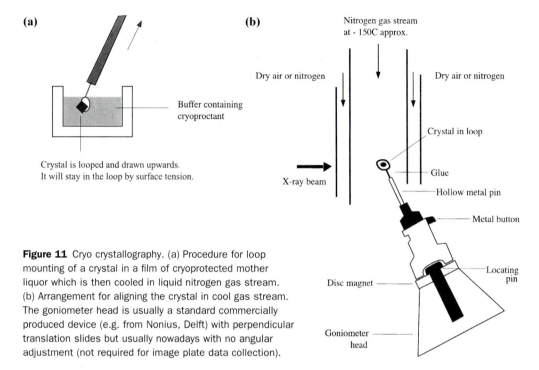

Figure 11 Cryo crystallography. (a) Procedure for loop mounting of a crystal in a film of cryoprotected mother liquor which is then cooled in liquid nitrogen gas stream. (b) Arrangement for aligning the crystal in cool gas stream. The goniometer head is usually a standard commercially produced device (e.g. from Nonius, Delft) with perpendicular translation slides but usually nowadays with no angular adjustment (not required for image plate data collection).

in the number of protein structures being determined at low temperature. However for each new protein crystal the method does require to be carefully set up empirically in order to establish the best conditions. To achieve success with the method, the crystal has to be mounted and flash-cooled rapidly, usually to 100 K, without damage, free from ice and set up in the X-ray beam. Damage from freezing is prevented through the use of a cryoprotectant liquid. The cryoprotectant may be incorporated in the mother liquor after crystallization or, less commonly, as a component of the crystallization reagents. The procedure involves supporting the crystal in a film of cryoprotected mother liquor in a small fibre loop (*Figure 11*) which is then cooled in liquid nitrogen.

The liquid surrounding the crystal must freeze as an amorphous glass to avoid crystal damage and diffraction from ordered ice crystals. Glycerol is the most frequently used cryoprotectant. The required glycerol concentration to achieve this must be carefully established and unfortunately this may involve loss of several crystals in the early stages of the experiment. Once frozen, protein crystals are usually extremely stable, transportable, and can be stored and kept ready for subsequent X-ray diffraction experiments.

2.5 Trouble shooting

In the following discussion it is assumed that initial trials have been conducted using sitting or hanging drop techniques (as this is the most usual approach).

(a) No crystals have appeared. Try varying screening conditions—buffer, pH (possibly finer intervals), precipitant, protein concentration, drop size, method used, temperature (crystals will sometimes form in the fridge or cold room). If limits of solubility or other special requirements exist (e.g. addition of a stabilizing cofactor or other ligand), careful attention should be paid to ensure that these conditions are satisfied.

(b) Crystals are obtained in one or two wells but are poorly formed and small. Using a finer matrix vary the conditions around the wells with crystals. Try another method. Try seeding—this works by transferring finely crushed crystal particles in the wells using a cat's whisker or hair.

(c) Depending on the individual protein, one or more of the following additives may help the crystallization process: Cu^{2+}, Zn^{2+}, Ca^{2+}, Co^{2+}; EDTA; acetone, dioxane, phenol; for membrane proteins, n-octyl-β-D-glucopyranoside (up to critical micelle concentrations) and substrates, cofactors, inhibitors, or binding sugars.

(d) With limited amounts of protein, concentrate on a few parameters known to be important. pH: initially in 0.5 intervals of pH unit, decreasing the interval as conditions are established. Temperature: 4, 22, and 37 °C are most commonly useful. Precipitants: $(NH_4)_2SO_4$, NaCl, PEG, EtOH, MPD are highly favoured.

(e) Crystals are obtained but their X-ray diffraction pattern looks like that of a small molecule (see section on X-ray diffraction). Try the 'click' test on one of the crystals. Small molecule crystals such as ammonium sulfate are usually physically hard and difficult to crush. When crushed on a glass slide with a fine needle a clicking noise may be heard and the crystal may even flick off the slide.

(f) Large crystals are obtained but the diffraction pattern disappears after 30 minutes exposure to the X-ray beam before a full set of data can be collected. Try freezing the crystals; book a session on a synchrotron facility, or both!

(g) Crystals are very soft and disintegrate when mounted in a glass capillary tube. Try being more careful when mounting; transfer through a larger pipette; use a larger diameter glass capillary.

2.6 Protein–ligand complex crystals

Preparation of protein complex crystals falls into many categories, including the following.

2.6.1 Protein–small molecule complexes

(a) Protein–heavy atom (covalent or non-covalent) for isomorphous replacement phasing (MIR, Section 8).

(b) Enzyme–inhibitor (non-covalent).

(c) Protein–carbohydrate (e.g. lectin–carbohydrate, non-covalent).

(d) Protein–nucleic acid fragments.

(e) Protein–peptide (e.g. cyclophilin/cyclosporin).

(f) Antibody–antigen complexes.

2.6.2 Protein–macromolecule complexes

(a) Protein–nucleic acid (DNA/RNA).

(b) Protein–protein (both covalent and non-covalent e.g. ribosome inactivating proteins, RIP's).

(c) Protein–protein (e.g. protein–antibody).

2.6.3 Small molecule complexes

This includes protein–heavy atom complex crystals for MIR (Section 8).

A very wide range of heavy atom reagents has been compiled for this purpose, mainly organic or inorganic compounds of Hg, Au, Pt, or U. The initial choice of possible compounds may be influenced by a knowledge of the protein's amino acid sequence (9). Lists of known useful compounds may also be found for example in the article by Blundell and Jenkins (10) or in the book 'Practical protein crystallography' (11).

If good quality native crystals are already available, soaking them in heavy atom solutions may be the best way to prepare derivatives. This is carried out under controlled conditions of pH, temperature, time, and concentration. Heavy atom concentrations in the range 0.5–20 M may be tried for anything between a few minutes to several days. The crystals may change colour but should not undergo any other significant physical change. Detailed screening by X-ray techniques is necessary in order to establish isomorphism and to ensure that the heavy atoms have been incorporated into useful locations (see Section 8). Alternatively, co-crystallization can be tried. This involves setting up crystallization experiments using one of the methods described in the previous section, but with the addition of heavy atom reagent into the solution. It is essential to maintain the integrity of the solution prior to crystallization attempts. This may require careful adjustment of the conditions and a good deal of patience! If crystals form they will again require careful monitoring by X-ray means to establish the hoped for incorporation of heavy atoms. Change of crystal unit cell and space group is more likely to occur in this method than with soaking but is sometimes preferable.

Covalently bound heavy atoms may be incorporated if the protein contains exposed Cys residues with free SH groups, although this is a rare luxury. Avey and Shall (12) showed that in bovine ribonuclease it is possible to modify Lys 41 with homocysteine thiolactone and covalently bind Hg, in the form of PCMBS to produce a useful heavy atom derivative (13, 14) but this type of chemical modification is rarely undertaken. It is now possible to mutate S atoms (on Met residues) to the heavier Se, which can be useful in some advanced X-ray techniques such as MAD phasing. A recent of example of this technique can be found in the structure of the protein ALAD (15). The use of heavy atom derivative crystals for phasing in the MIR method is described in Section 8.

2.6.4 Protein complex crystals with other small molecules

Many proteins associate with small molecules as part of their biological function, for example: enzyme–inhibitor complexes (usually non-covalent); protein–

carbohydrate complexes (e.g. lectin–carbohydrate, non-covalent interactions); protein–nucleic acid fragments; protein–peptide complexes (e.g. cyclophilin/ cyclosporin); antibody–antigen complexes (e.g. progesterone).

Complex crystals may again be formed either by soaking or co-crystallization under controlled conditions. The methodology is similar to that used for the preparation of heavy atom derivatives, requiring similar precautions and screening to establish incorporation of the adduct molecules. There is a slight preference for co-crystallization, which is more likely to lead to full occupation of the ligand binding sites. This, again, may lead to non-isomorphous derivatives but this problem should be easily overcome using molecular replacement to determine the structure, assuming the native structure is known (see Section 8). Examples:

1 Crystallization of the monoclonal anti-progesterone Fab′ antibody DB3 with progesterone and other steroid derivatives. Arevelo *et al.* (16) describe the culture of the unliganded DB3 mAb molecule and subsequent inhibition studies to establish the binding characteristics of eight progesterone derivatives with respect to progesterone. Large X-ray quality crystals were grown at 22°C by vapour diffusion methods from 1.7 M ammonium sulfate (pH 7.4) using 0.002% (w/v) PEG 20 000 and 1% (v/v) ethanol as additives, in 1–5 μl drops containing 5–25 mg of protein and added steroid.

2 Crystals of RNase A complexed with the mononucleotide inhibitor 2′-cytidine monophosphate (2′-CMP) were produced (17) by soaking native RNase crystals (13) in a 20 mmol solution of 2′-CMP for 12 h and then mounted for X-ray data collection. Native RNase also provides an excellent example of a protein with a non-covalent inorganic ligand (SO_4^{2-}) bound in its active site (see Section 10). Native bovine RNase crystals diffract to 0.9 Å, which is exceptional for proteins, enabling the protein–SO_4^{2-} ion interactions to be studied in detail (18).

2.6.5 Crystalline complexes involving macromolecules

MLI is a type II RIP protein comprising two covalently linked chains, A and B. Crystallization of MLI (8) has already been mentioned. It is of particular interest here because the A and B chains form a cleft, thus providing an interesting example of a protein–protein interaction (19) as well as involving protein– carbohydrate interactions through the co-crystallized β-D galactose (20).

The above examples are, amongst the structures, discussed in detail in Section 10.

3 Crystal geometry and symmetry

3.1 The crystal lattice

3.1.1 Crystal faces and their indices

Crystals are characterized by their regular internal three-dimensional molecular arrangements. *Figure 12a* shows a simulated two-dimensional example in which

(a)

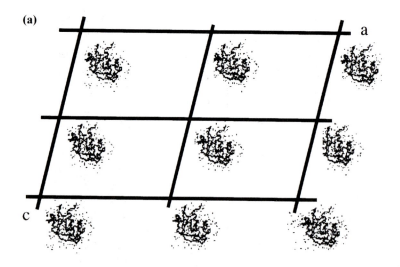

(b)

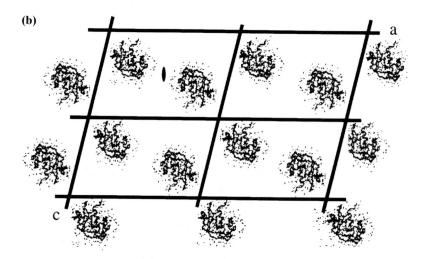

Figure 12 Unit cell of a lattice arrangement in 2D showing molecules related: (a) by simple translations; (b) with an added 2-fold rotation. The extended structures are formed by repeating the patterns shown according to the unit cell translations.

the protein ribonuclease is incorporated into an inclined lattice. *Figure 12b* shows a development of this arrangement involving 2-fold symmetry. Crystals used in single crystal X-ray analysis are regular and periodic in three dimensions over distances which are large compared with the unit of periodicity. Typical crystals used in macromolecular studies have overall dimensions of around 0.1–0.5 mm and unit cell dimensions in the range 20–400 Å (1 Å = 10^{-8} cm). To develop a feel for the magnitude of dimensions such as this, the reader should estimate the number of unit cells with unit cell edges 30 × 40 × 50 Å in a crystal of

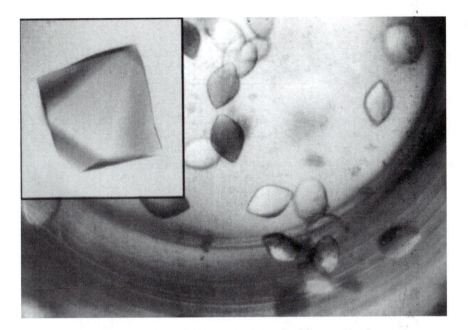

Figure 13 Crystals of *Pleurotus ostreatus* lectin (115). The crystals shown in the main picture are not fully mature being very rounded in shape. They do however diffract well enough to provide preliminary data for the crystals. A mature crystal with hexagonal symmetry is also shown (inset).

dimensions $0.2 \times 0.3 \times 0.4$ mm. Real crystals may be composed of a large number of ordered mosaic blocks, which are crystallographically perfect, slightly disordered, or displaced relative to one another by a few minutes of arc. To a first approximation such specimens can be considered to be perfectly imperfect single crystals, and are an assumed prerequisite in the theory of X-ray diffraction analysis. Protein crystals tend to be highly mosaic, an effect that is related to disorder associated with their high solvent content. At room temperature their Bragg reflections (section 4.1) tend to occur over an angular range of about 1°.

Crystallographers look, ideally, for crystal specimens with large well-formed faces and definite geometrical shapes. Whilst this is no guarantee of their quality as single crystals, it is a good first indication. Examination under a polarizing microscope (see for example ref. 21) is a rapid and invaluable technique for providing confirmation of crystal quality. Sometimes however incompletely developed protein crystals, *Figure 13*, can provide useful X-ray diffraction.

The faces, which are a fundamental property of well-formed crystals, grow in zones which have a common axial direction, frequently parallel to a principal crystallographic axis, *Figure 14a*. Corresponding interfacial angles on different crystals of the same crystalline form of a substance are invariable. However it should be noted that many substances, including proteins, occur in polymorphic or pseudo-polymorphic forms which have different symmetries. Bovine ribo-

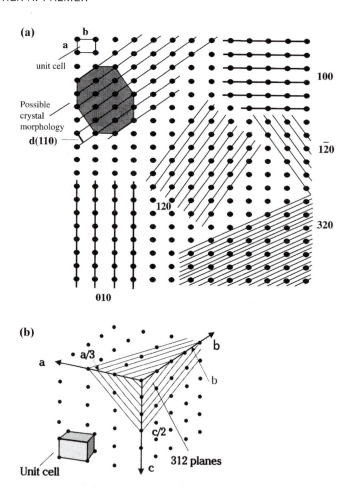

Figure 14 Families of lattice planes and their indices: (a) 2D examples showing also how such planes can act as crystal faces [in a 3D crystal all of the faces shown would be parallel to the c-axis if assumed perpendicular to **a** and **b**]. The meaning of the interplanar spacing for d(110) is also indicated. (b) A 3D example showing the 312 set of planes.

nuclease A, for example was reported in 1959 by King *et al.* (22), to occur in several different space groups. Crystal faces occur at definite angles to each other because they bear a precise, simple relationship to the internal periodic structure of the crystal. The planes that occur as external faces are related to the internal crystal structure. *Figure 14a* illustrates this principle in a two-dimensional example, indicating also the selected unit cell. The solid lines represent possible crystal faces seen in projection (see above). Lines labelled 100 and 010 are principle planes seen in projection and are related to the unit cell axes of the crystal. In this example the unit cell has been defined by the two vectors **a** and **b**. The resulting repeat mechanism can be considered to fill the whole of this two-dimensional space. In terms of measurable quantities, lengths *a* and *b*

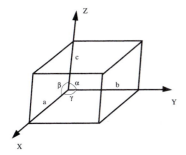

Figure 15 Conventional unit cell axes used in structure analysis forming a right-handed system. Note: there are six parameters defining the unit cell—three lengths a, b, and c, and three angles α, β, and γ. These six parameters must be determined as accurately as possible as part of the structure analysis.

define, together with the angle γ, the unit cell dimensions. The angle γ between **a** and **b** is 90° in this example.

Figure 15 shows a unit cell in three dimensions, defined by three non-colinear vectors **a**, **b**, and **c** which define the axial lengths a, b, and c and inter-axial angles α, β, and γ [α is the angle between **b** and **c** etc]. The axial vectors are always selected so as to define a right-handed system as illustrated.

The concept of three integers defining crystal planes is highly relevant to X-ray analysis. These numbers are in general written as *hkl* and known as the indices of a given plane. They are prime to each other when referred to an external crystal face (i.e. contain no common factors), and usually less than five. Negative values and 0 are allowed (see *Figure 14*).

When referred to X-ray diffraction spots *h*, *k*, and *l* are allowed all values in-cluding common factors and will be written (*hkl*). For example the crystal plane 111 has associated X-ray reflections (111), (222) etc., considered by Bragg as different 'orders' of diffraction. *Figure 14a* gives examples of various *hk0* planes in a two-dimensional crystal and *Figure 14b* is a 3D example. These examples illus-trate the fact that the crystal plane *hkl* makes intercepts of a/h, b/k, and c/l Å on the unit cell axes **a**, **b**, and **c** respectively. A crystal plane parallel to a given axis will have a corresponding zero index corresponding. Thus the plane 110 is parallel to **c**; and 100 is parallel to both **b** and **c** and simply intercepts the **a**-axis by $a/1 = a$. The advantage of this system of indexing planes is that it is in-dependent of the shape or symmetry of the unit cell and is therefore applicable to all crystals. Unit cell shape and symmetry are however important in X-ray crystallography. The reader may try drawing in sets of planes in the inclined ribo-nuclease examples, *Figure 12*, after extending the range by applying translations.

Figure 15 illustrates the most general type of unit cell having three unequal axial lengths *a*, *b*, and *c*, with three unequal inter-axial angles α, β, and γ having no special values. Other possible crystallographic unit cells can be derived from this by introducing special relationships between *a*, *b*, and *c*, and special values for α, β, and γ. There are thus seven crystal systems each characterized by unit cell shape, *Table 1*.

In addition, some unit cells are 'simple' or primitive (P); others are centred in one of several forms as defined in *Table 2*. *Figure 16* shows a selection of unit cells from the 14 standard or Bravais lattices listed in *Table 3*. Note it is possible to define a non-standard lattice for any crystal and this is sometimes done for

Table 1 The seven crystal systems: unit cell shapes, principal axial symmetry, and diffraction symmetry

Crystal system	Restrictions on unit cell axes and angles imposed by symmetry[b]	Principle axis of symmetry (for protein crystals) and location	Symmetry of diffraction pattern, assuming Friedel's Law holds,[a] i.e. patterns are all centrosymmeric
Triclinic	No restrictions: $a \neq b \neq c$ $\alpha \neq \beta \neq \gamma$	1-fold Any direction	Centre of symmetry
Monoclinic	a, b, c all different $\alpha = \gamma = 90° \beta > 90°$	2-fold Parallel to **b**	2/m (2-fold parallel, m perpendicular to b)
Orthorhombic	a, b, c all different $\alpha = \beta = \gamma = 90°$	Three 2-folds Parallel to **a, b, c**	2/m 2/m 2/m (mmm) 2-folds parallel, m's perpendicular to **a, b, c**
Tetragonal	$a = b \neq c$ $\alpha = \beta = \gamma = 90°$	4-fold Parallel to **c**	4/m (4-fold parallel, m perpendicular to **c**) **or** 4/mmm (as above with m's perpendicular and at 45° to **a or b**)
Cubic	$a = b = c$ $\alpha = \beta = \gamma = 90°$	Four 3-folds Along unit cell diagonals	m3 (four 3-folds along unit cell diagonals, 2-folds along **a, b, c** plus centre of symmetry **or** m3m (four 3-folds along unit cell diagonals, 4-folds parallel to **a, b, c** plus centre of symmetry
Hexagonal	$a = b \neq c$ $\alpha = \beta = 90° \neq \gamma = 120°$	6-fold Parallel to **c**	6/m (6-fold parallel plus m perpendicular to c) **or** 6/mmm (as above with m's parallel and perpendicular to **a**)
Trigonal	$a = b = c$ $\alpha = \beta = \gamma \neq 90°, < 120°$	3-fold Along body diagonal	3-fold parallel to **c** plus centre of symmetry **or** 3-fold parallel to **c** plus 2-fold perpendicular to **c** plus centre of symmetry

[a] Known as Laue symmetry. All Laue groups are centrosymmetric if Friedel's Law holds, which it always does to a first approximation. Descriptions given are abbreviated. Full details can be found in references.

[b] General notes: (1) = implies 'required by symmetry to equal exactly'. (2) $\neq$ implies 'not required by symmetry to equal, but may be equal within limits of measurement'. In the monoclinic system it is conventional to select $\beta > 90°$; if a value $\beta < 90°$ is selected by software it should be transformed to the complimentary angle $(180° - \beta)$.

crystals with monoclinic or orthorhombic symmetry. It should also be borne in mind that in practice such settings may also arise by chance, and may easily be transformed to a standard form. *Table 2*, indicates unit cell types, standard symbols and centring points. *Table 1* lists the seven crystal systems, unit cell requirements, principal symmetry axes, and diffraction symmetry (see below). *Table 3* indicates the 14 standard or Bravais lattices, symmetry requirements for **protein crystals** and cell volume formulae required for preliminary structure analysis (see below).

Table 2 Unit cell types, standard symbols, centring points, and limiting conditions

Centring site(s)	Indices of centred faces of unit cell	Symbol	Fractional co-ordinates[a] of centred sites	Limiting condition(s)[b]
None		P	–	None
bc faces	100	A	0, ½, ½	hkl: k + l = 2n
ca faces	010	B	½, 0, ½,	hkl: h + l = 2n
ab faces	001	C	½, ½, 0	hkl: h + k = 2n
Body centre		I	½, ½, ½	hkl: h + k + l = 2n
All faces	100		0, ½, ½	hkl: k + l = 2n
	010	F	½, 0, ½	hkl: h + l = 2n
	001		½, ½, 0	(hkl: h + k = 2n)

[a] A fractional co-ordinate x is given by X/a where X is the co-ordinate in absolute measure (Å) and a is the unit cell repeat distance in the same direction and in the same units.

[b] Systematic absences correspond to those *not* listed for each category, e.g. for hkl: k + l = 2n + 1 etc.

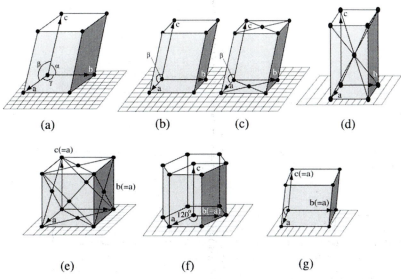

(a) (b) (c) (d)

(e) (f) (g)

Figure 16 Selected cells from the 14 standard or Bravais lattices: (a) triclinic P; (b) monoclinic P; (c) monoclinic C; (d) tetragonal I (see also orthorhombic I); (e) cubic F; (f) trigonal P (see also hexagonal, P); (g) rhombohedral R (trigonal system).

As shown in *Figure 14* each *hkl* actually represents a series or family of parallel planes in the crystal lattice. It was this idea that led to Bragg's formulation of X-ray diffraction which is employed universally by X-ray crystallographers. In practical applications this concept requires quantification of *d(hkl)* the inter-planar spacing. Referring to the two-dimensional case of *Figure 14a* where the angle γ between **a** and **b** is 90°, it is easily shown that:

$$1/d(hk)^2 = h^2/a^2 + k^2/b^2 \qquad [1]$$

Table 3 The 14 standard Bravais lattices, symmetry requirements for protein crystals, and cell volume formulae; see Table 1 for diffraction (Laue symmetries)

System	Standard unit cell(s)	Principal axes of symmetry[a] (protein structures)	Unit cell colume V_c
Triclinic	P	1-fold axis any direction	abcK, where $K = [1 - \cos^2\alpha - \cos^2\beta - \cos^2\gamma + 2\cos\alpha\,\cos\beta\,\cos\gamma]^{1/2}$
Monoclinic	P C	2-fold axis parallel to **b**	abc sinβ
Orthorhombic	P C I F	Three 2-fold axes parallel to **a, b, c**	abc
Tetragonal	P I	4-fold axis parallel to **c** **or** 4-fold axis parallel to **c** plus 2-fold parallel to **a**	a^2c
Cubic	P I F	Four 3-fold axes parallel to cell body diagonals **plus either** three 2-folds or three 4-folds parallel to **a, b, c**	a^3
Hexagonal	P	6-fold axis parallel to **c or** 6-fold axis parallel to **c** plus 2-fold parallel to **a**	a^2c sin120°
Trigonal (rhombohedral)	R	3-fold along cell diagonal **or** 3-fold along cell diagonal plus perpendicular 2-fold (see below for directions)	abcK, where $K = [1 - 3\cos^2\alpha + 2\cos^3\alpha]^{1/2}$
Trigonal[b]	P (using hexagonal cell)	3-fold parallel to **c** **or** 3-fold parallel to **c** **plus** 2-fold parallel to **a** **or** 3-fold parallel to **c** **plus** 2-fold perpendicular to **a**	a^2c sin120°

[a] Not all symmetry axes are listed. Other dependent ones may arise (see *International Tables for X-ray Crystallography*, Volume I or Volume A).

[b] In this setting trigonal crystals 'borrow' the hexagonal primitive lattice.

This formula can be extended to the special case of a three-dimensional lattice where the angles α, β, and γ are all 90° (orthorhombic, tetragonal, and cubic) as follows:

$$1/d(hkl)^2 = h^2/a^2 + k^2/b^2 + l^2/c^2 \qquad [2]$$

For the other crystal systems expressions for *d(hkl)* are best derived from the reciprocal lattice spacing *d*(hkl)* (see *Table 5* and Section 4.2).

It is easily shown in all cases that: *d(hkl)/n = d(nh,nk,nl)* where *n* is any integer and this relationship is used below (Section 4.1).

3.2 Symmetry elements, crystal systems, and space groups

3.2.1 Plane groups

Translation is considered to be a symmetry operation because it provides a repeat mechanism whereby molecular units fill the three-dimensional space of a crystal. Various other types of symmetry operations can occur in crystals. Further consideration of *Figure 12* will remind readers that unlike *Figure 12a* the two-dimensional structure of *Figure 12b* incorporates an axis of 2-fold symmetry into the pattern. The 2-fold axis is seen in projection in this diagram and is represented by the symbol (●). This operation clearly relates the two molecules either side of the top left-hand origin. Careful inspection of this diagram will reveal a total of nine such points, four of which belong to a single unit cell, the other five, being one translation unit away in each direction, are technically in neighbouring cells. Other rotation axes possible in crystals are 1-, 3-, 4-, and 6-fold. The arrangement in *Figure 12a* contains only 1-fold axes. In crystallographic notation this type of two-dimensional arrangement has the symbol p1, whereas *Figure 12b* is p2. In this notation the symbol p (primitive) corresponds to the lattice type (shown in lower case for 2D groups). These patterns are examples of 2D plane groups.

3.2.2 Space groups in protein crystals

A more complete investigation of crystal symmetry (see for example ref. 21) reveals that crystals in general can accommodate, in addition to rotational symmetry axes, the symmetry elements of inversion centres (centres of symmetry) and mirror planes (reflection planes) and various combinations of translation, rotation, inversion, and reflection. Because proteins are polymers of L-amino acids, the Cα atoms being specifically left-handed chiral centres, only axes of symmetry occur in protein crystals. This is because centres of symmetry and mirror planes will necessarily produce D-amino acids, which do not (normally) occur. This is good news for protein crystallographers because, as a consequence, protein crystals occur in only 65 space groups (*Table 4*) whereas crystals in general can have one of 230 space groups (see the *International Tables for X-ray Crystallography*). Space group determination, which must be accomplished in solving a crystal structure, is therefore, in the case of proteins, a relatively simple procedure (see data processing section 5).

3.2.3 Screw axes

In addition to the simple rotation axes 1-, 2-, 3-, 4-, and 6-, various combinations of rotation and translation called screw operations can occur in three-dimensional space groups. These symmetry operations work in the same way as a spiral staircase, i.e. there is a specific upwards movement (or translation) for a specific angular path travelled. The general notation for a screw axis is R_t where R represents rotation by R/360°, and t/R is the fractional translation along a unit cell

Table 4 The 65 enantiomorphic space groups applicable to protein crystals

Crystal system	Crystal class	Space group symbols[a] [Laue group symbol]	Number of asymmetric units per unit cell (Z)
Triclinic	1	P1 [$\bar{1}$]	1
Monoclinic	2	P2 P2$_1$ C2 [2/m]	2 4
Orthorhombic	222	P222 P222$_1$ P2$_1$2$_1$2, P2$_1$2$_1$2$_1$ C222 C222$_1$ I222 I2$_1$2$_1$2$_1$ F222 [mmm]	4 8 8 16
Tetragonal	4 422	P4 P4$_1$ P4$_2$ P4$_3$ I4 I4$_1$ [4/m] P422 P42$_1$2 P4$_1$22 P4$_1$2$_1$2 P4$_2$22 P4$_2$2$_1$2 P4$_3$2$_1$2 P4$_3$22 I422 I4$_1$22 [4/mmm]	4 8 8 16
Cubic	23 432	P23 P2$_1$3 I23 I2$_1$3 F23 [m3] P432 P4$_1$32 P4$_2$32 P4$_3$32 I432 I4$_1$32 F432 F4$_1$32 [m3m]	12 24 48 24 48 96
Hexagonal	6 622	P6 P6$_1$ P6$_5$ P6$_2$ P6$_4$ P6$_3$ [6/m] P622 P6$_1$22 P6$_5$22 P6$_2$22 P6$_4$22 P6$_3$22 [6/mmm]	6 12
Trigonal	3 321 312	P3 P3$_1$ P3$_2$[b] R3 [$\bar{3}$] P321 P3$_1$21 P3$_2$21[b] R32 P312 P3$_1$12 P3$_2$12 [$\bar{3}$m]	3 3 (9[b]) 6 6 (18[b])

[a] All protein crystals crystallize in one of these space groups. Corresponding Laue groups (diffraction pattern symmetry) are listed in [ ] brackets. For further details see references.

[b] Trigonal crystals are referred to hexagonal axes. The same applies to rhombohedral crystals (although this involves three times the number of molecules per unit cell, the cell shape is easier to handle).

direction. The symbol 2$_1$, in space group P2$_1$, thus represents rotation around **b**, of 180° combined with a translation of 1/2 = 0.5 along the **b**-axis. This operation, illustrated (artistically!) in *Figure 17* is very common in proteins. (For further information on the conventions involved in assignment of screw axes, see ref. 21/ Volume I or Volume A of *International Tables for X-ray Crystallography*. The relevant diagrams may also be found on the Internet (23) but it should be noted that the latter are based on a non-standard origin.)

a

2

b

2_1

Figure 17 Two-fold rotational symmetry. (a) Pure rotational symmetry 2. (b) Screw axis symmetry 2_1. Note: molecules can have symmetry 2 (or any other pure rotational symmetry) but not 2_1 (or any other screw symmetry) because a definite translation is involved.

4 X-ray diffraction from crystals

4.1 Bragg's treatment of X-ray diffraction: the location of diffraction maxima

When an X-ray beam strikes a crystal, X-rays are scattered or reflected in a characteristic manner. This process is also referred to as diffraction and produces an X-ray diffraction pattern, which is unique for any given crystal. When recorded on an X-ray film or image plate, a good quality diffraction pattern is composed of well-resolved small spots (see *Figure 18*). The following practical points are important.

Firstly the crystal to be used for measuring diffraction data should be well ordered and produce a single, clear diffraction pattern. Poor quality specimens, which are not single crystals, may suffer from a variety of faults including splitting, twinning, or the presence of slippage-disorder planes. Split or twinned crystals may diffract well but will produce multiple reflection spots, which are difficult to work with. Such specimens should be discarded. Badly disordered crystals are characterized by the spreading out of each diffraction spot, often in an irregular manner, and the poor overall resolution of the diffraction pattern, which rapidly fades away as a function of distance from the centre of the pattern (radially). Faults of this type are particularly associated with protein crystals which are relatively soft (containing up to 70% solvent), and can be

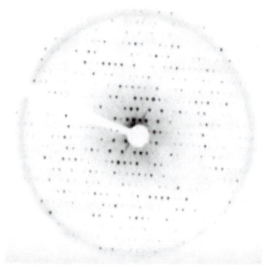

Figure 18 A b-axis zero-level precession photograph of monoclinic bovine RNase (13). The diffraction spots are clean and separate. The blank region in the centre is where a lead trap was positioned. The adjacent dark scattering is where the X-ray beam caught the edge of the trap! Note: (i) the spots are at definite relative positions; (ii) they have different relative intensity (blackness).

easily damaged through handling when being mounted for X-ray analysis. Even single crystals diffract poorly if heavily disordered.

If a crystal proves to be unsuitable through any of the above faults it should be discarded and another one selected either from the same crystallization batch or from a different one. Refinement of crystallization conditions is frequently necessary in order to produce better crystals.

Secondly the X-ray beam is well collimated and composed essentially of a single wavelength, λ; usually having a value in the range 0.7–1.6 Å. Methods for producing such a monochromatic beam are discussed in Section 5.1.

A given X-ray diffraction spot corresponds to a unique set of crystal lattice planes with indices *hkl* and as such depends on the size and shape of the crystal unit cell. Directional properties are defined in terms of the Bragg angle θ. This process is illustrated in *Figure 19*. Bragg's treatment accounts only for the spatial location of each diffraction maximum, in terms of θ, not for the variation in the relative intensities or energy associated with the diffraction spots, which depends on the positions and types of atoms in the structure. This important concept is dealt with in detail in Section 4.3.

Figure 19 shows that reinforcement, or constructive interference, of the X-ray wavelets reflected from the *hkl* set of planes at an angle θ occurs only when:

$$2d(hkl)\sin\theta = n\lambda \qquad [3]$$

This is the Bragg condition, corresponding to production of an X-ray reflection from the crystal in a definite direction given by the angle θ.

The whole number, n, was defined by Bragg as the 'spectral order'. In modern

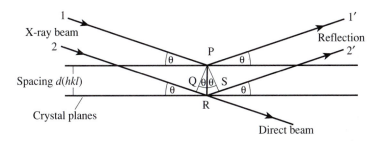

Figure 19 Geometry of Bragg's equation. Monochromatic X-rays of wavelength λ are incident at an angle θ to the general *hkl* planes. Constructive interference takes place by apparent reflection also at angle θ provided the path difference QR + RS = 2d(*hkl*)sinθ = nλ, where n is a whole number (order of diffraction in Bragg's notation). Note: most of the direct beam passes straight through the crystal and must be trapped to prevent ill effects to laboratory staff.

terminology however, *n* is incorporated into the equation using the relationship derived previously, i.e.:

$$d(hkl)/n = d(nh,nk,nl) \qquad [4]$$

Since *nh*, *nk*, and *nl* are simply other whole numbers, in terms of the diffraction pattern we replace them (for convenience) with respect to the diffraction pattern) by (different) *h,k,l*, where the indices can now have any integer value, including multiples. Bragg's equation then is absolutely specific for a given reflection with indices *hkl*. Under this definition θ is also specific for a given *hkl* and is designated θ (*hkl*). The indices can have any positive or negative value and Bragg's equation is thus written as:

$$2d(hkl) \sin\theta \ (hkl) = \lambda. \qquad [5]$$

X-ray cameras and other devices for recording diffraction patterns such as diffractometers and image plates may be considered as devices for satisfying Ewald's condition which is an ingenious concept, providing a clear visual image of X-ray diffraction.

4.2 The reciprocal lattice concept defining the location of diffraction maxima

Ewald devised an algorithm for Bragg's equation which provides a powerful conceptual interpretation of the physical processes which take place when X-ray diffraction occurs. Ewald's construction is represented in *Figure 20*. This shows the crystal located at C the centre of a hypothetical sphere (the Ewald sphere or sphere of reflection) of radius 1. When Bragg's equation is satisfied the geometry of the situation illustrated requires the point P labelled $\vec{d}^*$*(hkl)* to lie on the surface of the sphere. The point Q is the origin of the reciprocal lattice. This concept arises through the reciprocal relationship between the length d^**(hkl)* and crystal spacing *d(hkl)* thus:

$$OP = d^*(hkl) = 2\sin\theta \ (hkl) = \lambda/d(hkl) \qquad [6]$$

31

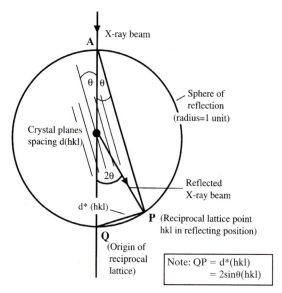

Figure 20 Ewald's construction leading to the reciprocal lattice analogue in X-ray diffraction. The crystal is at C and in the Bragg reflecting position. The point P, where the reflected beam cuts the Ewald sphere is the reciprocal lattice point (*hkl*). Q is the origin of the reciprocal lattice and QP = 2sinθ (*hkl*) = d*(*hkl*) the reciprocal lattice spacing.

Note that OP is the reciprocal lattice vector **d*(hkl)** of length $d*(hkl)$ reciprocal lattice units (RLU). The radius of Ewald's sphere by this definition is 1 RLU. It is also important to note that **d*(hkl)** is perpendicular to the *hkl* set of planes. We can thus consider the diffraction pattern to consist of reciprocal lattice points (*hkl*). The precession photograph in *Figure 18* is in fact a contact print of a reciprocal lattice section. Each spot has a unique *hkl* index. The process of assigning *hkl* values to diffraction spots (indexing) is vitally important and procedures for achieving this are discussed in the data collection section 5 below.

In summary: Ewald's construction predicts that an X-ray reflection occurs whenever a reciprocal lattice point P(*hkl*) passes through the Ewald sphere (Ewald's condition). X-ray diffraction recording devices (cameras, image plates, and other diffractometers) usually involve some motion of the crystal which will cause reciprocal lattice points to pass through the sphere and thus be energized.

As the terminology suggests, the points P when assembled for the whole crystal form a regular three-dimensional reciprocal array. This has unit cell lengths $a*$, $b*$, $c*$ and angles $\alpha*$, $\beta*$, $\gamma*$ (see ref. 21 for further details). The following important relationships apply to any crystal lattice for a diffraction experiment employing wavelength λ:

$$a* = \lambda/d(100); \quad b* = \lambda/d(010); \quad c* = \lambda/d(001) \text{ and}$$
$$d*(hkl)^2 = h^2a*^2 + k^2b*^2 + h^2c*^2 + 2hk\cos\gamma* + 2lh\cos\beta* + 2kl\cos\alpha* \qquad [7]$$

Table 5 shows how the formula for $d*^2(hkl)$ is crystal system dependent and

Table 5 Expressions for calculating $d*^2(hkl)$ in the seven crystal systems[a]

Crystal system	$d*^2$ (hkl)
Triclinic	$h^2a*^2 + k^2b*^2 + l^2c*^2 + 2klb*c*\cos \alpha* + 2lhc*a*\cos \beta* + 2hka*b*\cos \gamma*$
Monoclinic	$h^2a*^2 + k^2b*^2 + l^2c*^2 + 2lhc*a*\cos \beta*$
Orthorhombic	$h^2a*^2 + k^2b*^2 + l^2c*^2$
Tetragonal	$(h^2 + k^2) a*^2 + l^2 c*^2$
Cubic	$(h^2 + k^2 + l^2) a*^2$
Hexagonal and trigonal P	$(h^2 + k^2 + hk) a*^2 + l^2 c*^2$
Trigonal R rhomohedral	$[h^2 + k^2 + l^2 + 2(hk + kl + hl) \cos\alpha*] a*^2$

[a] These expressions can be used for calculation of $d(hkl)$, remembering that $d(hkl) = \lambda/d*(hkl)$. This assumes the definition of $d*(hkl)$ given in the text as a dimensionless quantity in reciprocal lattice units (RLU).

simplifies according to the special values which occur for $\alpha*$, $\beta*$, and $\gamma*$ (easily derived from *Table 1*).

4.3 The structure factor F(*hkl*): X-ray intensities I(*hkl*)

The concept and formulation of the structure factor enables the idea of Bragg reflection to be developed in order to account for the distribution of intensities associated with the diffraction spots (*Figure 18*). We have seen that Bragg's treatment accounts only for the positional features of spots recorded in the diffraction pattern, not for why some spots are very intense, graduating down to ones with practically no intensity at all. Bragg's treatment therefore ignores the actual arrangement of the individual atoms in a molecular structure. Since the object of X-ray analysis is to determine this structure we must now explain how the distribution of atoms in the unit cell affects the diffraction pattern.

Figure 21 indicates the situation for a simple (seven atom) structure in more detail. This shows how the seven atoms in the structure each contributes to an X-ray reflection. The scattering contribution of a given atom in the Bragg direction depends on:

(a) The atomic scattering power of the atom.

(b) The relative phase of the scattered wavelet.

The resultant or combined wave formed for the (*hkl*) reflection is called the structure factor. The important components of this wave are its amplitude |F(*hkl*)| and phase angle ϕ (*hkl*).

In an X-ray diffraction experiment X-rays are scattered by the electrons in the extra-nuclear clouds of the individual atoms. Obviously even a very small crystal contains many millions of atoms and consequently the X-ray scattering is amplified by the regular arrays of atoms, repeated by the crystal lattice and other symmetry operations which may be present. The total X-ray scattering

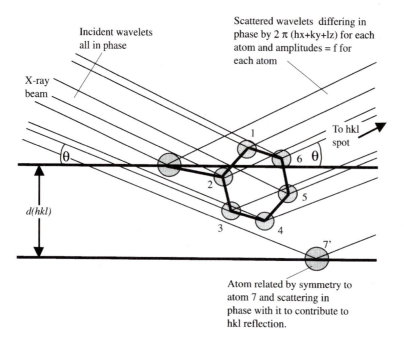

Figure 21 Simple seven atom structure showing X-ray wavelet scattering for the *hkl* planes and leading to I(*hkl*) in the diffraction pattern.

associated with a given *hkl* reflection is represented by the structure factor **F**(*hkl*) given by the complex equation:

$$\mathbf{F}(hkl) = A'(hkl) + iB'(hkl)$$

where

$$A'(hkl) = \sum_{j=1}^{N} f_j \cos 2\pi \left(hx_j + ky_j + lz_j\right)$$

and

$$B'(hkl) = \sum_{j=1}^{N} f_j \sin 2\pi \left(hx_j + ky_j + lz_j\right) \qquad [8]$$

As written above the summations are to be carried out over all atoms N in the unit cell. The use of symmetry, however, simplifies this process (*International Tables for X-ray Crystallography*, Vol. I; ref. 21). f_j is the theoretical atomic scattering factor for a given atom j, a measure of the X-ray scattering power, having a maximum value for a given atom of Z_j, the atomic number when $\theta = 0°$. For all other (actual) values of θ (*hkl*), f_j decreases slowly with $s = \sin\theta / \lambda$. The fall off is usually calculated from the formula:

$$f(s) = a_1 \exp(-b_1 s^2) + a_2 \exp(-b_2 s^2) + a_3 \exp(-b_3 s^2) + a_3 \exp(-b_3 s^2) + c \qquad [9]$$

where the coefficients a_1 to a_4, b_1 to b_4, and c are atom dependent with values listed in *International Tables for X-ray Crystallography*, Volume IV, pp. 99–101. Most software incorporates the values of these coefficients at least for all atoms likely to be encountered (i.e. for proteins C, H, O, N, and S). In addition to this natural fall off of f_j with $\sin\theta /\lambda$, for an actual crystal f_j has to be modified by a thermal displacement effect caused by the fact that the atoms in a crystal are subjected to (usually minor) variations in position which increase with increase of temperature and decrease with decrease in temperature. f_j thus becomes $f_j T_j$, where:

$$T_j = \exp(-B_j \sin^2\theta /\lambda^2) \text{ and } B_j = 8\pi^2 <U^2> \qquad [10]$$

$<U^2>$ is the mean square displacement of the atom from its average position (usually in Å^2 as the experimental wavelength λ is normally in Å).

B is designated colloquially as a 'temperature factor' although it may actually be associated with statistical or other type of disorder. Preference is currently given by the *International Union of Crystallography* to the use of the term 'thermal displacement factor'.

For a protein crystal, B_j typically has the following values: overall average, 20–30 Å^2; for main chain atoms, 10–20 Å^2; for side chain atoms 15–30 Å^2; for solvent and small ligand atoms, 25–45 Å^2. As a further rule of thumb it is common for an upper limit of around 80 Å^2 to be placed as a credibility indicator for any individual atom. Atoms with B_j greater than this value after refinement (Section 9) should be seriously re-examined. Values of T_j (for each atom) are usually derived during the course of structure refinement.

There are two important relationships associated with the structure factor:

(a) The structure factor amplitude:

$$|F(hkl)| = [A'^2(hkl) + B'^2(hkl)]^{1/2} \qquad [11]$$

and

(b) The phase angle:

$$\phi (hkl) = \tan^{-1} [B'(hkl)/A'(hkl)] \qquad [12]$$

$|F(hkl)|^2$ is directly related to the measured intensity $I_o(hkl)$ i.e.,

$$I_o(hkl) \rightarrow |F_o(hkl)|^2$$

where the subscript 'o' implies 'observed'.

It follows that observed values $|F_o(hkl)|$ can be obtained by direct measurement from the diffraction pattern (see Data processing, Section 5) at the completion of an X-ray diffraction experiment. For a successful analysis, as many values of $|F_o(hkl)|$ as possible must be measured (see Section 6.2 on resolution).

However it is very important to realize that the corresponding values of $\phi (hkl)$ cannot be measured but are essential for calculation of electron density (Section 7) and must be recovered somehow. Section 8 provides details of how this is carried out.

4.3.1 Friedel's Law

Consideration of the expressions for $d^{*2}(hkl)$ (*Table 4*) and $|F_o(hkl)|$ indicate that both are unchanged if (hkl) is replaced by $(-h, -k, -l)$. It follows that $I_o(hkl) =$

$I_o(-h, -k, -l)$, being located across a centre of symmetry in reciprocal space. This is known as Friedel's Law, and applies to all crystals to a first approximation, only breaking down significantly if atoms in the crystal scatter anomalously (see *International Tables for X-ray Crystallography*, Vol. 4; or ref. 21). In other words X-ray diffraction patterns are centrosymmetric, provided Friedel's Law holds, and it follows that they can be classified into the 11 centrosymmetric crystallographic point groups (Laue Groups, *Table 1*). The Laue group corresponding to each of the 65 possible protein space groups is listed in *Table 4*. Further details of crystal point groups and Laue groups can be found in *International Tables for X-ray Crystallography*, Volume I or Volume A, or in ref. 21.

4.4 Calculated structure factors and R-factors

Once a model of the structure has been proposed using one of the methods discussed in Section 8, values of the co-ordinates (x_j, y_j, z_j) for most of the atoms will be available, the exact number depending on the extent of the model. It is therefore possible to calculate values of the structure factor amplitudes and phases based on the model. These are designated $|F_c(hkl)|$ and $\phi_c(hkl)$ respectively and can be calculated for all values of (hkl) in the measured data set. The quality of the model can be tested by comparing $|F_o(hkl)|$ with $|F_c(hkl)|$ and since there are usually many thousands of (hkl) values in a data set, for convenience a combined index R is calculated, defined as:

$$R = [\Sigma \, ||F_o(hkl)| - |F_c(hkl)||] \, /[\Sigma \, |F_o(hkl)|] \qquad [13]$$

(the summations are carried out over all reflections in the data set).

For a protein structure R might be as high as 0.3 for the initial model, improving to 0.2 or better after refinement. The quality of any X-ray structure is restricted by the diffracting power of the crystal. Protein crystals, as explained previously, usually diffract X-rays rather poorly. This restricts the total number of data available compared to the large number of parameters involved in the calculation of the structure factors and is consistent with the correspondingly poor R values obtainable. By comparison the R-value for a good low molecular weight X-ray structure would be less than 7%.

4.4.1 Free R-factor

In the molecular replacement (MR) method of protein structure analysis described below (Section 8) new protein structures are derived from existing homologous structures by fitting the known structure into the unit cell and space group of the new one. The new structure is then refined by gradually transforming each amino acid in the sequence, in location and conformation, as necessary, in order to minimize the R-factor. Sometimes it is difficult to remove the initial bias which is built into the new structure as a consequence of using the co-ordinates of another structure. In order to monitor this process 5 or 10% of the $|F_o(hkl)|$ data are removed from the data set during the refinement process and do not contribute to the course of the analysis. This sub-set of data is used to calculate an R-factor called the free-R or R_f (24). As the refinement proceeds, if $|F_c(hkl)|$ truly approaches $|F_o(hkl)|$, R_f will drop together with R. If however the refined model fails to break away from the initial model, R will drop because its parameters are changing, but R_f will fail to improve because the model is not actually improving. Usually for a correct model, R_f will decrease but will stay a few per cent greater than R.

5 Recording and measurement of X-ray diffraction patterns

5.1 X-ray sources

5.1.1 Conventional X-ray laboratory sources

X-rays are a form of electromagnetic radiation in the wavelength range approximately 0.1–100 Å. For the purposes of practical crystallography the range used is around 0.6–3.0 Å.

With the exception of synchrotron installations, discussed below, in conventional crystallography laboratories (e.g. in university and industrial departments) in-house sealed vacuum X-ray tubes are employed to generate X-radiation. X-rays are produced through the impact of electrons on a metal target (anode). Electrons are emitted by a heated tungsten filament (cathode), and accelerated towards the anode through an extremely high voltage (40 kV or more) applied across the tube. In most laboratories the target used for protein crystallography is Cu (Mo may be used for some low molecular weight crystals). The rotating anode generator is a modification in which the anode is rotated rapidly in order to enable heat generated to dissipate more rapidly. As a consequence a higher accelerating voltage can be applied which provides a more powerful X-ray source. A metal filter (Ni in the case of a Cu tube) provides an essentially monochromatic X-ray beam, wavelength $\lambda = 1.5418$ Å. However about half the intensity is lost through absorption in the filter in this method. Alternatively a graphite monochromator can be used which selects a particular wavelength by employing a very intense Bragg reflection which then provides the radiation used for X-ray analysis, again usually, for proteins, with $\lambda = 1.5418$ Å.

5.1.2 Synchrotron X-radiation (SR)

X-rays are generated at a synchrotron source when high-energy electrons are accelerated in a storage ring at relativistic speeds. The ring provides a wide constant-radius circular orbit for the electron beam which is controlled by a series of magnets. Synchrotron installations require a vast amount of space. For example, energy of 100 GeV requires a ring radius of 200 m. The X-ray beam is narrow, extremely intense (of the order of 100 times or more that of a conventional source), and the wavelength can be selected from a very wide range to match experimental requirements. In view of the high intensity it is possible to record complete data sets even from weakly diffracting protein crystals in a matter of minutes. It is also possible to devise time-resolved experiments in order to monitor processes such as modified enzyme–substrate interactions. SR installations are highly specialized research facilities. Experiments to be undertaken with SR radiation require careful planning and time has to be applied for well in advance. SR laboratories are to be found at Cornell, Stanford, Argonne, and Los Alamos in the US, at Daresbury in the UK, Grenoble in France, Hamburg in Germany, and at the Photon Factory in Japan.

5.2 Recording and measuring intensity data

X-ray diffraction patterns are produced in practice by imposing some sort of motion on the crystal whilst retained in the X-ray beam. As a result of this motion a number of reciprocal lattice points are caused to pass through the Ewald sphere and thereby produce X-ray reflections. Several methods are available for recording diffraction patterns.

5.2.1 X-ray cameras

X-ray diffraction patterns can be recorded on X-ray film. This requires the use of one of several types of X-ray camera including:

(a) The Arndt-Wonnacott camera with rotating- or oscillating-crystal.

(b) The Weissenberg camera with a rotating-crystal/translating-film (can also be used to produce rotation or oscillation photographs with no film translation).

(c) The precession camera in which both the crystal and film precess about a common axis.

The precession method is most useful for characterizing the unit cell and symmetry of a protein crystal, monitoring crystal quality, and up-take of heavy atoms for isomorphous replacement phasing. A precession photograph (e.g. *Figure 18*) records a single undistorted layer of the reciprocal lattice. The spots can be indexed on inspection and cell dimensions measured directly (see ref. 21). Although there are methods for measuring I(*hkl*) values from X-ray photographs these are extremely time-consuming and no longer used routinely. However X-ray photography provides a very effective means for teaching the methodology of recording and interpreting diffraction patterns and in many research schools is still considered to be an important part of training in crystallographic techniques.

5.2.2 Diffractometers

Diffractometers employ scintillation counters to detect and measure the X-ray reflections from single crystals.

i. Single counter or serial diffractometers

Traditional diffractometers incorporate a mechanical goniometer to orientate the crystal into the correct reflecting position for each *hkl* and to rotate the counter to receive the scattered X-radiation from this single reflection. The energy is transformed electronically into a form suitable for conversion to I(hkl). Because each reflection is measured individually, with a count time typically of around 60 seconds, the process is very slow, particularly for proteins, which routinely involve the measurement of several thousand reflections. Whilst the accuracy attainable is better than for most of the other methods used for intensity measurement, the limited lifetime of protein crystals in the X-ray beam permits only a fraction of the available data to be recorded from one crystal. The use of several crystals for data collection introduces errors associated with the scale factors required to merge the various collections into a single data set.

Data collection using this type of diffractometer is highly automated and controlled by the manufacturer's computer system. The steps involved include: mounting and centring of the crystal; search for 20–25 reflections to provide initial unit cell data and crystal orientation; refinement of unit cell and orientation; automated data collection and output of I(hkl) values as well as standard correction using manufacturer's software to give F(hkl) values.

The Enraf Nonius CAD4 4-circle diffractometer, in recent years, has been one of the most popular machines available. The crystal is positioned by means of three rotations κ, ϕ, and ω, and the counter is mounted on and rotates about the 2θ circle. Other diffractometers available include models from Siemens, Rigaku, and Mac Science.

5.2.3 Area detectors

The main disadvantages of single counter diffractometry—slow data collection rate and the requirement of several crystals for collection of a complete data set, with the attendant errors associated with scaling and crystal deterioration— have been largely overcome by the use of 'electronic film' area detectors and image plates which have enjoyed rapid development in recent years. For each exposure a series of X-ray reflections is produced by oscillating the crystal through a small angle $\Delta\phi$ and received by the detector, which is an effectively flat surface with a uniform sensitivity to X-rays. The (X,Y) position of each reflection (*Figure 22*) enables its hkl index to be determined, and the signals received are converted electronically into intensity data. Several designs are available:

(a) **Xenon gas/multiwire proportional counter** (MWPC). This type of system occurs in the Siemens or San Diego versions (25).

(b) **FAST area detector** (Enraf Nonius FAST). In this method the X-ray reflections strike a phosphor coated fibre optics screen which converts the signals into light photons which are intensified, integrated, and digitized. The recording device incorporates a television scanning system and the goniometer is essentially that used in the CAD4 (see above).

(c) **Imaging plates**. In this method the image plate consists of a barium halide phosphor doped with europium (Eu^{2+}) which is excited into the metastable Eu^{3+} state by exposure to X-radiation. After completion of each exposure, during which the crystal is oscillated in the X-ray beam for a few minutes, the image plate is scanned with a fine He-Ne laser beam which causes the regions

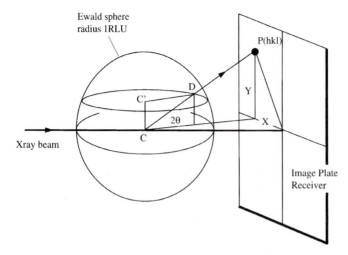

Figure 22 Flat plate recording as used for example with an image plate (IP) diffractometer. A reciprocal lattice point at D when oscillating through the Ewald sphere will give rise to a reflection recorded at P on the IP with co-ordinates P(X,Y). The position of P depends on the following: crystal unit cell; crystal orientation; experimental wavelength λ; crystal to IP distance. For further details see p. 159 of ref. 21 and references in CCP4 documentation.

converted to Eu^{3+} (where the X-ray spots would be located on a photographic film) to emit violet light ($\lambda = 3900$ Å) with an intensity proportional to the absorbed X-ray energy. This light is then detected with a photomultiplier system, integrated and digitized. After reading off the stored data the plate is cleaned by exposure to bright yellow light and the 'film' is then ready to be used to record the next image. These various stages are shown in *Figure 23*.

Advantages of this method include:-

(i) Size. Plates up to 30 cm diameter are available which enable data to be collected to a resolution of about 1.4 Å using Cu radiation and better with synchrotron radiation at smaller wavelengths.

(ii) Intensity range. A very wide range of intensities (approximately 10^5) can be recorded compared to approximately 2×10^2 for X-ray film.

Commercially available image plate systems are listed below.

(d) **Charge-coupled device area detectors** (CCD's).

(i) Phosphor screen-demagnifying fibre optics coupled CCD's. CCD's represent the very latest technology in X-ray diffraction data collection and their use is becoming more and more common. They combine high sensitivity over a wide range of intensities with the capability of very fast readout times (10 s compared with 150 s for laser read phosphor plates), low noise level (through cooling of the CCD chips to -20 to $-50\,°C$), and capacity for storing large amounts of data. In operation the principle is similar to that used with imaging plates. Processing of data, to be discussed in the next section, is also very similar. Amongst the commercially available de-

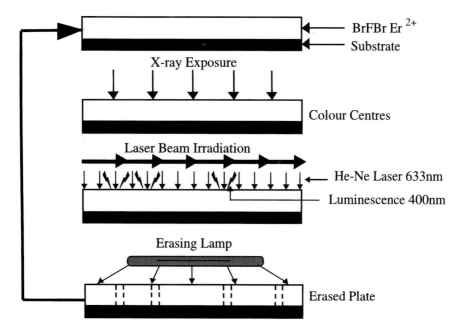

Figure 23 Principle of the MAR image plate recording device.

magnifying fibre optic coupled CCD diffractometers are those produced by Siemens (now Bruker), ADSC, Nonius, and X-ray research.

(ii) Electronically focused CCD's (at ECRF, Synchrotron Facility, Grenoble, France). This device is essentially a CCD incorporating electronic focusing in which the phosphor screen is imaged through an image intensifier and de-magnified using CRT technology. Magnetic shielding is required (26).

i. Area detectors in practice

Data are usually collected using a simple oscillation method, the oscillation or rotation axis being perpendicular to the X-ray beam. Sophisticated software such as the program STRATEGY (27) is available for optimizing the procedure in practice. For a given exposure (frame) the crystal is oscillated through a small angle $\Delta\phi$ (usually set to a value between 0.1° and 1.5°, depending on the unit cell size). The next frame begins where the previous one ended, keeping $\Delta\phi$ at the same setting. The procedure is repeated until the required angular range has been covered. This depends on the Laue group of the crystal (*Table 5*) and the angular disposition of the X-ray beam with respect to the mounted crystal, which may or may not be oscillating about a major crystal axis. Data collection can however proceed even if the orientation of the crystal is unknown. Once the X-ray intensities have been measured, structure analysis software usually assumes that the I(*hkl*)'s cover a single asymmetric unit of the Laue group symmetry (plus Friedel opposites or their equivalents if anomalous dispersion is being utilized).

However, if time permits, it is sound strategy to collect at least two asymmetric units of data and utilize the R_{merge} index (see below) to monitor data quality. For a crystal arbitrarily mounted, 180° of data frames will be more than sufficient for anything but a triclinic crystal. Such an arbitrary strategy can be wasteful of both diffractometer time and crystal lifetime. This can be avoided using the programs DENZO and PREDICT. This software takes autoindexing information from the first frame and then works out the required data collection strategy.

Protocol 5

Setting up a data collection on the Mar image plate

Two models of the Mar IP are available: 18 cm (small-Mar) or 30 cm (big-Mar). A spiral readout is produced, the image plate rotating under a scanning optics device which itself moves inwards radially at constant speed (*Figure 21*). These data are then transformed into Cartesian co-ordinates with a pixel size of $150 \times 150 \ \mu m^2$. Exposure time intervals are 120 s and 240 s respectively for the two models. The minimum crystal to detector distance is 66 mm providing resolution limits of 1.7 Å and 1.4 Å respectively with CuKα radiation.

Method

1 Initially, set the X-ray beam slits (two horizontal and two vertical) to $0.25 \times 0.25 \ mm^2$ (this may require further adjustment once crystal diffraction is obtained).

2 Using 'set distance' on the screen menu move the detector back to 200 mm or more to provide access for goniometer head adjustment, and manually move the backstop away from where the crystal location will be to give access. Set the ϕ rotation position to 0° (use 'set phi' in menu).

3 Screw the crystal on its goniometer head onto the machine's goniometer and centre the crystal using the TV which views the crystal perpendicular to the X-ray beam direction. When centred the crystal may not be central in the crosswires, it must however not change its position relative to the crosswires when rotated. Replace the backstop. By rotating manually ensure that the goniometer is free to rotate over the required rotation range.

4 Set the IP distance. It may be necessary in the first place to use an intermediate value, e.g. 200 mm and to use the PREDICT software to provide a strategy for the data collection once diffraction has been observed.

5 Optimize the X-ray beam flux for the set distance. This should be carried out by a technician or colleague with experience in carrying out this operation.

6 Crystal diffraction may be checked initially by setting up a single oscillation. This requires the following input: ϕ overlap; angular increment between images; complete rotation range required; image type sequence (both still records and oscillation records can be specified); number of oscillations per image (at least one every 300 s to average out beam fluctuations and IP excitation decay) and the exposure time. Initially it is recommended to record two images 90° apart and then work out the best data collection strategy using the software STRATEGY or PREDICT.

7   Further practical points. Use 'end' to stop a series of exposures not 'stop'; do not activate 'stop' when taking a still and ensure that the computer disk will not become full during a long, unattended data collection.

5.3 Image plate data processing

Whilst the procedure for acquiring raw data using an image plate is quite rapid, and this itself is a tremendous advantage bearing in mind the fact that protein crystals have notoriously short lifetimes in the X-ray beam, processing this data to produce the $|F(hkl)|$ values can be very time-consuming. Each frame of data contains information about the (X,Y,Z) co-ordinates of the diffraction maxima relative to the experimental set-up, and digitized information from which a measure of the intensity $I(hkl)$ can be derived. For a given crystal specimen the following procedures are carried out concomitantly:

(a) Determination of an orientation matrix to define the disposition of the crystal axes with respect to the X-ray beam direction.

(b) Definition of the crystal's unit cell, crystal system and space group and measurement and refinement to the highest accuracy possible of the six unit cell parameters, $a, b, c, \alpha, \beta,$ and γ, together with estimates of their standard deviations $\sigma(a), \sigma(b), \sigma(c), \sigma(\alpha), \sigma(\beta), \sigma(\gamma)$.

   Once these procedures have converged, the final stage of data processing can be carried out, that is:

(c) Determination of the indices hkl and intensities $I(hkl)$ for each recorded diffraction spot.

5.3.1 Partially recorded spots and integration of intensity data

Because images are recorded from small oscillations of the crystal and the reciprocal lattice points for a protein crystal are close together, each image will contain a proportion of partially recorded reflections which are completed either on the previous or next record. To obtain the intensity reading the two partials are added together. This effect will inevitably introduce errors in the intensity values. Recording of the diffraction image produces a series of digitized pixels, which can be viewed by computer graphics and processed to provide intensity estimates. Conversion to an intensity reading is achieved by one of two methods:

(a) Summation integration. This method defines a volume containing an individual spot. Summation of the measured counts in all the pixels in the defined volume and subtraction the background, determined by examining surrounding pixels.

(b) Profile fitting of the reflection in three dimensions on a background. In this method an empirically derived model reflection shape is scaled to the data and then integrated. This assumes that the reflection shape is independent of intensity. The observed profiles vary over the detector face, so several model profiles are usually required depending on the location on the detector face. Profile fitting is computationally expensive but produces more reliable results, less susceptible to random errors.

There is a wide selection of software available for data processing. A possible procedure using mainly CCP4 software is indicated in the flow diagram of *Figure 24*.

5.3.2 Problems with data collection

1 **Symptoms**: diffraction is observed but the spots are wide apart and clear to very high resolution. **Diagnosis**: this may indicate that the crystal is not protein but one of the small molecule salts used in crystallization (very often ammonium sulfate). Try the 'click test' with another crystal using a mounted pin or

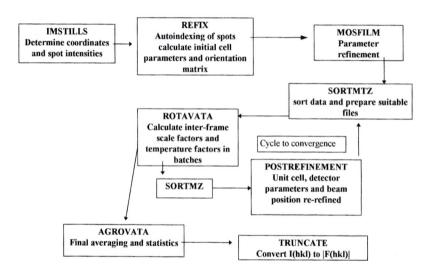

Alternative software which may be used in this scheme is:
DENZO (for IMSTILLS, REFIX and MOSFILM)
SCALEPACK (for ROTOVATA)
SCALEPACK2MTZ and CAD (for AGROVATA)
Further practical points to note are as follows:

DENZO/XDISPLAY/SCALEPACK are part of the HKL suite, Gerwirth, 1995)
DENZO carries out autoindexing, intensity measurement and cell refinement.
XDISPLAY displays the observed and calculated spot positions for comparison and checking. SCALEPACK carries out scaling and merging iteratively (postrefinement)

MOSFLM is similar to DENZO but with the facility to adjust spot size.
REFIX (Kabsch, 1993) allows MOSFILM to autoindex a single image.

XDS or MARXDS includes features of the above programs as required for Mar Research IP's. MARSCALE is an additional scaling programme.

Figure 24 Flow diagram for data processing.

needle to prod the crystal. Salt will be hard and resistant to breakage whereas protein is soft and easily damaged.

2 **Symptoms**: crystals appear to be well formed, optically clear, and with good morphology. The diffraction pattern is typically protein with spots close together, but the individual spots are: diffuse or partially diffuse; 'tailed' i.e. spread out in a tadpole shape; obviously split (twinned); or spread out in a particular direction, e.g. a principal reciprocal lattice direction (statistical or systematic disorder). **Diagnosis**: crystal quality is poor in all of these cases and it is not worth collecting data from such specimens. Review the crystallization conditions, possibly slowing it down by cooling or adjusting the initial concentration. Try crystallization at cold room temperature or even in a colder refrigerator. Look carefully at crystal mounting routines, which may be too rigorous and try to minimize physical contact. Review other conditions: storage time (protein crystals can have a very short shelf-life in crystallization trays); methods of transport (crystals can suffer from rough handling in transit or lengthy journeys).

3 **Symptoms**: crystals are very small and refuse to grow bigger, have good morphology, give good diffraction spots in-house, but the intensities are weak, and θ_{max} is low. **Cure**: use synchrotron radiation (high intensity); try cryo-cooling, or both.

4 **Symptoms**: at least one unit cell parameter is large (> 150 Å) resulting in diffraction spots which are weak and very close together on low intensity in-house diffraction equipment. **Cure**: use high intensity synchrotron radiation which enables a larger crystal to image plate distance to be used, resulting in better spot separation while still maintaining satisfactory image strength.

5 **Symptoms**: crystals diffract well but lifetime in the X-ray beam is too short to allow a full data set to be collected. **Cure**: again use cryo-cooling to $-150\,°C$ (if possible) and/or high intensity synchrotron radiation with a CCD detector, which permits very rapid collection. Cooling not only promotes better resolution in many cases but also improves crystal lifetime almost indefinitely. Crystals are usually loop mounted (*Figure 11*, Section 2.4) and can be stored and transported in a Dewar flask contained in an insulated box.

Note: Cryo-cooling is the method of choice in many current applications and has several advantages in addition to those listed above. These include elimination of crystal slippage and drying out, both of which can occur with capillary tube mounting.

5.3.3 Additional practical tip

Always be meticulous in recording all details of a data collection experiment. These include: crystal; date; station (if synchrotron, SR); operational wavelength (SR); data defining main beam position (wax diffraction ring etc); crystal to detector distance; oscillation range per image; number of frames; temperature at crystal; overall direction of crystal rotation. Always seek help whether at your home laboratory or SR, at the first signs of trouble.

6 Preliminary crystal data

One of the first objectives of an X-ray analysis is to characterize the crystal system and determine the unit cell dimensions, cell volume, and space group symmetry.

6.1 Determination of unit cell and space group

As a by-product of data reduction the unit cell and space group of the crystal will usually be determined, either *ab initio*, or confirmed if known from previous studies (e.g. native protein crystals). Unit cell dimensions will rarely be determined to better than two or three hundredths of an Å unit for most protein crystals unless extreme care is taken and the diffraction pattern extends to < 1 Å resolution. Some image plates allow direct measurement of cell dimensions or other lattice spacings. Subsequent confirmation and least-squares refinement of the cell parameters using the data reduction software (see below) produces the final values together with an estimate of the errors. The space group of the crystal is determined from a consideration of the following:

(a) **Unit cell parameter values**. *Table 1* lists the cell parameter restrictions associated with each of the seven crystal systems. It must be remembered however that '$\neq$' means 'not restricted by symmetry to be equal to'. Consequently it is quite possible for the unit cell parameters of a crystal to appear to have higher symmetry than is actually the case. For example a crystal which is actually monoclinic might have values of α, β, and γ which are all $90°$ within experimental error. The crystal is then apparently orthorhombic.

(b) **Laue symmetry**. The diffraction pattern of a given crystal exhibits the symmetry of one of the 11 Laue Groups (Section 4.3). Provided enough data have been collected, Laue symmetry is usually reliably indicated as a further by-product of data reduction. This serves to pin-point the correct crystal system and possible space groups (*Table 5*) that apply to a given crystal. In the case of an unknown crystal it is best to sample as great a portion of reciprocal space as possible in order to guarantee unambiguous determination of the Laue group.

(c) **Systematically absent X-ray intensities**. As can be seen from *Table 4* there are several possible space groups for a crystal belonging to a given crystal system. For example a monoclinic protein crystal could have one of the three space groups P2, P2$_1$, or C2. Similar considerations apply to the other crystal systems. Further consideration of the X-ray diffraction pattern may enable the exact space group to be indicated. This depends on the fact that translational symmetry elements in the space group give rise to I(hkl) values which are absent in the diffraction pattern (i.e. they have no intensity and are said to be systematically absent). In practice we actually consider the limiting conditions for intensities which are present, i.e. the sets complementary to the systematic absences. For proteins we need to consider limiting conditions associated with (i) the lattice type listed in *Table 2* and (ii) each type of screw axis (*Table 6*).

Table 6 Limiting conditions for screw axes

Screw axis	Orientation parallel to:	Limiting condition[a]
2_1	a	h00: h = 2n
2_1	b	0k0: k = 2n
2_1	c	00l: l = 2n
3_1 or 3_2	c	00l: l = 3n
4_1 or 4_3	c	00l: l = 4n
4_2	c	00l: l = 2n
6_1 or 6_5	c	00l: l = 6n
6_2 or 6_4	c	00l: l = 3n
6_3	c	00l: l = 2n

[a] Systematic absences correspond to those *not* listed for each category, e.g. for h00: h = 2n + 1 etc.

Data reduction software will indicate lattice systematic absences. Screw axis absences may have to be determined from careful consideration of intensity values and may be indicated from graphical plots of the reciprocal lattice {Software: HKLVIEW}.

Protocol 6

Example 1 of space group determination: for mistletoe lectin MLI (8)

1 During data processing the unit cell parameters were found to be: a = b = 110.79, c = 308.53 Å, $\alpha = \beta = 90°$, $\gamma = 120°$, and the Laue symmetry 6/mmm. Together these observations show that the crystals are hexagonal.

2 In addition limiting conditions in the diffraction pattern occur for 00l, l = 6n. This information indicates that there is either a 6_1 or a 6_5 axis in the crystal and the space group is therefore either $P6_1 22$ or $P6_5 22$.

3 Only the X-ray analysis can resolve this remaining ambiguity (the space group was in fact determined to be $P6_5 22$).

4 Note 6_1 and 6_5 screw operations are left-hand right-hand opposites, only one can be correct for a given protein crystal.

Protocol 7

Example 2 of space group determination: ricin agglutinin, RCA (28)

1 During data processing the unit cell parameters were found to be: a = b = 100.05, c = 212.58 Å, $\alpha = \beta = 90°$, $\gamma = 120°$, and the Laue symmetry $\bar{3}$ (*Table 1*). Together these observations show that the crystals are trigonal (referred to hexagonal axes).

2 In addition limiting conditions in the diffraction pattern for 00l, l = 3n indicate the presence of a 3_1 or 3_2 screw axis parallel to **c**. This information indicates that the space group is either $P3_1$ or $P3_2$ (*Table 5*).

3 Only the X-ray analysis can resolve this ambiguity (the space group was in fact determined to be $P3_2$).

4 Note 3_1 and 3_2 screw operations are left-hand right-hand opposites, only one can be correct for a given protein crystal.

6.2 Resolution

In protein crystallography d_{min} is known as the resolution of the X-ray diffraction pattern where, by Bragg's equation, $d_{min} = \lambda/2\sin\theta_{max}$, where θ_{max} is the maximum value of θ for reflections contained within the measured data set. Atomic resolution corresponds to a d_{min} of about 0.8 Å which is usually well out of reach for proteins. Most protein crystals fail to diffract to much better than 1.8 Å resolution. This is sufficient however to define the positions of non-H atoms and many solvent atoms in the structure. At 2.5 Å solvent atoms may be unreliable and at 3.5 Å it may be difficult to refine side chain atoms beyond C_β. Obviously the higher the resolution the better defined the structure will be and it is usually worthwhile to expend the required effort in order to achieve this rather than to cut corners and end up with an inferior analysis. The resolution, as we shall see below, seriously affects the quality of the initial electron density map upon which the ensuing analysis is based.

6.3 Number of reflections in the data set

For a given crystal it is possible to estimate the number of reciprocal lattice points within the range $0-\theta_{max}$. For a crystal with a primitive lattice, ignoring the fact that some reflections will be related by symmetry, this is given as $33.51V_c\sin^3\theta_{max}/\lambda^3 = 4.19V_c/d^3_{min}$. Further details on how to include symmetry information in order to derive the number of unique data are given in Ladd and Palmer, 1993 (21).

6.4 Completeness of a data set at a given resolution

Depending on how the X-ray data set has been measured, the number of reflections present may not be consistent with the number expected from the nominal value of θ_{max} or d_{min}. If the number of reflection data actually measured is significantly lower than the expected value it can be assumed that the resolution of the analysis will be correspondingly lower although it is not easy to estimate by how much. The cause may also be due to an anisotropic intensity distribution, associated with variation of diffracting power with direction. Some data processing programs put out a warning when such an effect is detected. The percentage completeness at nominal resolution will also be indicated. A value less than about 85% complete would indicate that it would be advisable to look into reasons for this.

Protocol 8

Calculation of number of reflections in data set for MLI

1 For mistletoe lectin MLI (8): The space group is P6$_5$22, a = b = 110.79, c = 308.53 Å, α = β = 90°, γ = 120°, therefore V$_c$ = 3.28 × 10^6 Å^3.

2 If d$_{min}$ is 2.9 Å and λ = 0.8 Å (synchrotron radiation, Section 5.1 and see Web Page appended in final section), the expected number of reflections can be estimated as follows:

$$\sin\theta_{max} = \lambda/2 \times 2.9 = 0.1379 \ (\theta_{max} = 7.9°) \tag{14}$$

3 Expected number of reflections = 4.19V$_c$/d$^3_{min}$ = 563 442. This number includes all symmetry-related reflections. Since the Laue symmetry (*Table 5*) is 6/mmm, the number of unique data (i.e. not related by symmetry) is 1/24 times the number in the complete sphere, i.e. 23 477.

4 If only 21 000 reflections are recorded, the data set would be approximately 91% complete at the nominal resolution of 2.9 Å. This corresponds more appropriately to 3.0 Å resolution (working backwards).

5 Note: The above discussion is based on the number of reciprocal lattice points scanned in data collection and processing. Because protein crystals diffract poorly the number of reflections with significant intensities may well be as low as 50%. These weak data do actually contain structural information and will usually be retained in the working data set.

6.5 Internal consistency

As a test of data quality, data processing programs produce an internal consistency index, R$_{merge}$ (also known as R$_{int}$), calculated by comparing I(*hkl*) values which should be equal by virtue of symmetry. To be acceptable, R$_{merge}$ for the complete measured data set should be 9% or better. However it is well known that the accuracy of measurement of diffraction data becomes worse as d(*hkl*) decreases or the Bragg angle θ(*hkl*) increases. This is related to the fact that the intensities are weaker. For outer data (e.g. on the edge of *Figure 18*) R$_{merge}$ might be as high as 20%. Although this may seem to be an inordinately high value, such data may still contain a wealth of structural information and may serve a useful purpose in the analysis if retained, albeit with a low weight.

6.6 Number of molecules per unit cell

Knowing the unit cell volume and space group of the crystal it becomes possible to estimate the number of protein molecules per unit cell. Each of the possible protein space groups is associated with a standard number of molecules per unit cell (given in *Table 5*). If the actual number of molecules n is less than this (usually by a simple factor 2, 3, 4, or 6) deductions can be made about the possible symmetry of the molecule. If n is greater than the standard value, again usually by a simple factor, the structure is oligomeric and contains more than one molecule in the asymmetric unit, not related to each other by crystal symmetry (but frequently by approximate non-crystallographic symmetry known as NCS). In this case the X-ray analysis will involve the determination of the structures of this number of molecules rather than just one. The molecules in this type of situation usually exhibit subtle differences in molecular conformation, particularly in loop regions.

The crystal density may be expressed as:

$$D_c = nM_cm_H/V_c \qquad [15]$$

where n is the number of protein molecules per unit cell, M_c (in daltons) is the molecular weight of one protein molecule and its associated solvent, and m_H is the mass of one hydrogen atom $= 1.66 \times 10^{-24}$ g. If V_c is expressed in Å^3, conversion to cm^{-3} involves multiplying by 10^{-24} and D_c will have the usual units of g cm^{-3}, the two factors of 10^{-24} will then cancel.

Unfortunately there are several unknowns involved here. However, D_c can be assumed, from experience, to be (usually) in the range 1.2–1.4 g cm^{-3} for crystalline compounds composed predominantly of C, N, O, and H (the actual value will depend heavily on the degree of solvation). Let $M_c = M_P + sM_c$, where s is the solvent fraction of the cell mass. Thus $M_c = M_P/(1 - s)$ and $D_c = nM_Pm_H/V_c(1 - s)$, where M_p is the molecular weight of the protein, which we will assume to be known, and M_S is the molecular mass incorporated in the solvent which is not known. We define the fraction $s = M_S/(M_P + M_S) =$ the solvent fraction of the unit cell mass. Finally if the crystal space group is known, as it would normally be after data reduction (Section 5, above), we know that n would normally be a multiple or submultiple, say m of Z, the number of asymmetric units per unit cell (*Table 5*), i.e. $n = mZ$. The density expression then becomes: $D_c = mZM_Pm_H/V_c(1 - s)$. Finally it is known (29, 30) that in crystalline proteins, s is usually in the approximate range 0.27—0.65. Putting all of this information together it is usually possible to determine the appropriate value for m, the number of protein molecules per asymmetric unit.

Protocol 9

Example 1 for analysis of solvent content (MLI)

1 For mistletoe lectin MLI (8): Space group is $P6_522$ (Z = 12, *Table 5*), M_P = 63 kDa, a = b = 110.79, c = 308.53 Å, $\alpha = \beta = 90°$, $\gamma = 120°$, therefore $V_c = 3.28 \times 10^6 \text{ Å}^3$. In terms of the known data:

$$D_c = mZM_Pm_H/V_c(1 - s) = 0.38m/(1 - s) \qquad [16]$$

2 Assuming m = 1 MLI molecule per asymmetric unit and s = 0.68 (i.e. at the top of

the range), $D_c = 1.20$ g cm^{-3}. This is a reasonable result. Note: we could make s = 0.70, slightly higher than normal, and this would give $D_c = 1.28$ g cm^{-3}, which is again quite reasonable. The main result for the structure analysis is that m = 1 or Z = 12.

3   As s is on the high side from the above analysis, let us try m = 2 to see what this implies. So we find that:

$$D_c = 2 \times 0.39m/(1 - s) = 0.78/(1 - s) \text{ thus for } D_c = 1.4 \text{ g cm}^{-3}, s = 0.44.$$

4   This result is, again, reasonable. So there is some ambiguity for this protein. All that can be done is to bear this in mind during the X-ray analysis. The following pointers can be considered important, initially:

(a) MLI crystals are relatively poor X-ray diffractors. This is more consistent with result 1 (step 2 above), i.e. m = 1.

(b) A very reliable analysis was achieved with m = 1 (19). There was no evidence at any stage of the analysis for the presence of a missing protein molecule (i.e. m = 2).

Protocol 10

Example 2 for analysis of solvent content (RCA)

1   For ricin agglutinin, RCA (28): Space group is P3$_2$ (Z = 3, *Table 5*), $M_P = 133$ kDa, a = b = 100.05, c = 212.98 Å, $\alpha = \beta = 90°$, $\gamma = 120°$, therefore $V_c = 1.85 \times 10^6$ Å^3. In terms of the known data: $D_c = mZM_Pm_H/V_c(1 - s) = 0.36m/(1 - s)$.

2   Assuming m = 1 RCA molecule per asymmetric unit and s = 0.65 (i.e. at the top of the range), $D_c = 1.02$ g cm^{-3}. For s = 0.7, $D_c = 1.2$ g cm^{-3}, which is perhaps more realistic. As for MLI the physical wetness of the crystals and relatively poor diffracting power are consistent with this result.

3   Note: RCA is highly homologous with ricin (31) but with twice the molecular weight and forms a BA–AB dimer. The X-ray analysis initially assumes there is non-crystallographic symmetry (NCS) in the form of an approximate 2-fold axis across the molecule.

4   In addition the following alternative should be borne in mind as the X-ray analysis progresses: assuming m = 2, $D_c = 1.19$ g.cm^{-3} for s = 0.40, approximately equal to the average solvation level found in protein crystals, and therefore to be considered reasonable until proved otherwise.

7 Electron density: reconstruction of the molecular structure

7.1 Basic principles

Calculation of electron density maps is a major objective of any single crystal X-ray analysis. The formula for calculating the electron density ρ at the general point with fractional co-ordinates (x, y, z) is given by the Fourier series:

$$\rho\,(x, y, z) = 1/V_c \sum_h \sum_k \sum_l |F(hkl)|\,\cos2\pi\,[(hx + ky + lz) - \phi(hkl)] \qquad [17]$$

where the limits of the summations are: h = hmin to hmax; k = kmin to kmax; l = lmin to lmax (the experimental limits of h, k and l respectively).

This Fourier series has coefficients $|F(hkl)|$ and $\phi(hkl)$, where $|F(hkl)|$ is derived from the measured intensities $I(hkl)$ and the phases $\phi(hkl)$ are obtained indirectly. It is not possible to measure the phases experimentally. For structures involving protein–ligand interactions the phases would usually be calculated from a model derived by molecular replacement (MR), or in more difficult cases using isomorphous replacement (IR). The various cases are discussed below. The Fourier summations are performed over the limits of the hkl values corresponding to the measured intensities. Intensity measurements are made in practice over as wide a range of hkl values as possible (often both positive and negative, and zero). The range of hkl values available is related to the resolution of the electron density map, the more data the better the resolution. V_c is the volume of the unit cell in $Å^3$ and serves merely to act as a scaling factor. {Software for calculation of electron density and other Fourier series: FFT (fast Fourier transform, refs 32–34) in CCP4.}

7.2 Properties of the electron density

(a) When calculated at atomic resolution, $\rho(x, y, z)$ has local maximum values (peaks) at sites corresponding to atom centres, (x_j, y_j, z_j).

(b) Density values at these locations are approximately proportional to Z_j the atomic number of the corresponding atom j.

(c) The locations of all atom positions together define the crystal structure. Atom positions not determined (for whatever reason) lower the quality of the structure model.

(d) For proteins, where the extent of the X-ray data is usually far short of atomic resolution, the electron density tends to be blurred (unfocused). Atom co-ordinates have to be inferred in this case, usually by model building, which makes extensive use of known molecular geometry of bonds and groups involved in the structures.

(e) r.m.s. electron density $\sigma\,(\rho)$. Since clearly defined peaks are not a property of protein electron density the magnitude of given density regions is assessed in terms of the overall r.m.s. value of $\rho(xyz)$. Significant density is taken to have $\rho(x, y, z) \geq 3\sigma$, although a value of 2σ (or even lower) can have useful features but requires very careful consideration.

7.3 Complementarity of the structure factor F(*hkl*) and electron density ρ(*x, y, z*)

We have seen that structure factors can only be calculated if the co-ordinates (*xj, yj, zj*) (for a complete or partial model) are available. This results in calculated model values of $|F_c(hkl)|$ which can be compared to observed $|F_o(hkl)|$ and evaluation of R-factors which gauge the validity of the model. Calculated structure factors also result in calculated values of phases $\phi_c(hkl)$. It then becomes possible to calculate electron density $\rho_c(x, y, z)$ incorporating coefficients $|F_o(hkl)|$ and $\phi_c(hkl)$ which has properties between the true density (from $|F_o(hkl)|$) and the model (from $\phi_c(hkl)$). An outline of the processes involved in the analysis of a protein structure based on these concepts is given in the flow diagram, *Figure 1*.

7.4 Difference electron density functions

Difference electron density is defined as:

$$\Delta\rho\,(x, y, z) = 1/V_c \sum_h \sum_k \sum_l \Delta|F(hkl)|\,\cos 2\pi\,[(hx + ky + lz) - \phi_c(hkl)] \qquad [18]$$

where $\Delta|F(hkl)| = |F_o(hkl)| - |F_c(hkl)|$ and the summation limits are as for [17].
  This function has:

(a) Positive density corresponding to missing atoms not included in the model.

(b) Negative density corresponding to the inclusion of wrong or badly placed atoms in the model.

By adding:

$$\rho(x, y, z) = 1/V_c \sum_h \sum_k \sum_l |F_o(hkl)|\,\cos 2\pi\,[(hx + ky + lz) - \phi_c(hkl)] \qquad [19]$$

to $\Delta\rho(x, y, z)$ we obtain a $[2|F_o(hkl)| - |F_c(hkl)|]$ map which combines the two and shows positive density corresponding to missing atoms and negative density corresponding to incorrect atoms against the background of the original model. This provides a very useful basis for making corrections and additions to the structure model.

7.5 Patterson's Fourier series: the Patterson function P(*uvw*)

Patterson (35) proposed the use of a Fourier series P(*uvw*) for calculation of the positions of the interatomic vectors in crystal. The co-ordinates *u,v,w* are used to represent the positions of one end of an inter-atomic vector, the other end being placed at the origin. The unit cell of the Patterson function is equal in size to the crystal unit cell and the co-ordinate axes are parallel. The formula for calculating the Patterson density P at the general point with fractional co-ordinates (u, v, w) is given by the Fourier series:

$$P(u, v, w) = 1/V_c \sum_h \sum_k \sum_l |F^2(hkl)| \cos 2\pi (hu + kv + lw) \qquad [20]$$

where the summation limits are as for [17].

There are two obvious differences between $P(u, v, w)$ and $\rho(x, y, z)$: the use of $|F^2(hkl)|$; and the absence of phases $\phi(hkl)$ which means that P can be calculated before the structure is known. In fact the use of the Patterson function is central to many techniques employed in structure analysis. *Figure 25* shows the relationship between a three atom crystal structure and its Patterson function. $P(u, v, w)$ can be constructed graphically from the structure by mapping out the ends of interatomic vectors, each time placing one atom at the origin. By definition, all self interatomic vectors AA etc. appear at the origin of the Patterson function, which therefore is always the highest peak.

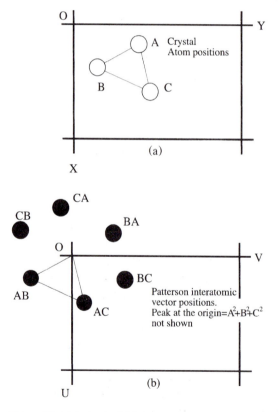

Figure 25 (a) A simple 2D three atom structure and (b) the peak positions in the corresponding Patterson function. Note: (b) can be constructed graphically by placing each of the three atoms in (a) in turn on the origin of the Patterson map while preserving the orientation of the structure.

7.6 Properties of the Patterson function

(a) When calculated at atomic resolution, $P(u, v, w)$ has local maximum values (peaks) at sites corresponding to the ends of interatomic vectors.

(b) Density values at these locations are approximately proportional to Z_1Z_2 the product of the atomic numbers of the two contributing atoms. If both atoms have high atomic numbers (heavy atoms) it is relatively easy to identify their vectors in the Patterson function. This is also facilitated by the use of space group symmetry (see for example ref. 21).

(c) Symmetry related interatomic vectors may also contribute to the same position in the Patterson function, depending on the space group. Data corresponding to this effect are to be found in the *International Tables for Crystallography* (Volume B).

(d) For proteins, where the extent of the X-ray data is usually far short of atomic resolution, the Patterson density tends to be blurred (unfocused).

(e) In protein structure analysis there are two major applications of the Patterson function:

  (i) Determination of the heavy atom co-ordinates in the isomorphous replacement method (IR).

  (ii) In the method of molecular replacement (MR) to determine the location of the molecular envelope by fitting the calculated Patterson of a known structure to the observed Patterson of the target structure, usually a fairly low resolution is satisfactory.

8 Determination of the phases $\phi(hkl)$ for protein crystals

8.1 Introduction

In the previous section we became aware that the first goal in the X-ray analysis of proteins, calculation of the crystal electron density, cannot be achieved unless we have values of both structure factor amplitudes $|F_o(hkl)|$, which can be measured, and the phases $\phi(hkl)$, which cannot. The question is therefore 'how can phase information be derived?'. In this section we look at methods which are available for recovering the necessary phase information.

There are two commonly used methods for phase determination in macromolecular crystallography: (1) isomorphous replacement (IR) and (2) molecular replacement (MR).

In the early years of protein structure analysis (the 1950s, 60s, and 70s) IR was the primary choice for protein structure analysis, and in this period a core of good quality protein structures was established. During this time MR became increasingly popular, as the number of well-defined structures from which to select a search model gradually increased. There are currently about 13 000 protein structures deposited in the Brookhaven Protein Data Bank (PDB). MR is thus the current method of choice for any protein having good structural homology, as judged initially by amino acid sequence, with a known structure. The more experimentally demanding MIR technique is still used for proteins where this condition does not apply and will become again increasingly more popular now that the human genome has been sequenced. The advent of fast, large capacity computers and improved software has also contributed significantly to the ease of application of MR.

8.2 Isomorphous replacement (IR)

By incorporating one or more heavy atoms into a protein crystal, thus forming a heavy atom derivative, changes can be induced in the X-ray intensities I(hkl). For the derivative crystal to be isomorphous the heavy atoms:

(a) Must not disturb the protein structure significantly.

(b) Must not alter the crystallographic space group.

(c) Must not change the unit cell parameters significantly ($> 0.5\%$ in any cell length).

Useful heavy atoms usually have atomic numbers $Z \geq$ iodine. The average fractional change in intensity $\Delta I(hkl)/I(hkl)$ at $\sin\theta/\lambda = 0$ can be estimated as: $\Sigma Z_H^2 / \Sigma Z_P^2$ where the summations are over the number of heavy atoms N_H per unit cell and the number of non-heavy atoms N_P per unit cell respectively. For a protein crystal containing a protein of molecular weight 20 kDa with 1 molecule per unit cell N_P would be equivalent to about 3200 C atoms ($Z_C = 6$). If $N_H = 2$ Hg atoms per unit cell ($Z_{Hg} = 80$) then $\Delta I(hkl)/I(hkl) = 2 \times 80^2 / 3200 \times 6^2 = 11\%$. In practice some I($hkl$) values are increased and some are decreased. Intensity differences of the order illustrated here would be useful for phasing.

8.3 Heavy atoms and compounds for isomorphous replacement

There is a very large repertoire of compounds known to produce heavy atom derivatives of proteins suitable for MIR. With few exceptions (e.g. the homocysteine thiolactone PCMBS derivative of ribonuclease) (13) heavy atoms are non-covalently linked to the native protein molecule, often in surface pockets or other easily accessible regions of the protein structure. The most popular heavy atoms are Pt, Hg, Ag, Au, and U. Reviews of known useful heavy atom compounds can be found in Blundell and Jenkins (10), Petsko (9), and Wood (36).

8.4 Preparation and screening of heavy atom derivatives

Two methods are used for the preparation of heavy atom derivatives for IR: 1) soaking pre-grown native crystals in heavy atom solutions, and 2) co-crystallization of the protein and heavy atoms together from solution.

8.4.1 Soaking method

Soaking can be carried out on crystals in hanging or sitting drops or on crystals mounted for X-ray diffraction in glass capillaries. This method is highly sensitive and dependent on correct pH, concentration, temperature, and time. Heavy atom solution concentrations around 0.5–10 M are frequently used with soaking times from a few minutes to several days or longer. This method can cause (often visible) deterioration of crystal quality, but is not likely to induce changes in unit cell and symmetry, although such effects have been observed.

8.4.2 Co-crystallization in the presence of heavy atoms

Co-crystallization can be undertaken *in situ* in hanging or sitting drops or in test-tubes if (rarely used) batch crystallization is carried out. This method may produce better quality crystals than soaking, but symmetry and unit cell changes are more likely to occur as a consequence of the different crystallization conditions.

8.4.3 Screening

Screening of possible heavy atom derivative crystals can be time-consuming. The easiest method of detecting changes in I(hkl) is from one or more X-ray precession photographs (e.g. see ref. 21). The same photographs provide a sensitive check on symmetry and unit cell dimensions. In the absence of a good precession camera it may be necessary to collect a low resolution (e.g. 5–6 Å) data set. Whichever method is used the technique is, inevitably, experimentally demanding.

8.5 Phase determination in single isomorphous replacement (SIR)

If $\mathbf{F_P}(hkl)$ represents the structure factor for the protein crystal alone, $\mathbf{F_{PH}}(hkl)$ represents the structure factor for the derivative containing protein plus heavy atoms, and $\mathbf{F_H}(hkl)$ represents the contribution to $\mathbf{F_{PH}}(hkl)$ from the heavy atoms alone then: $\mathbf{F_{PH}}(hkl) = \mathbf{F_P}(hkl) + \mathbf{F_H}(hkl)$ or

$$\mathbf{F_P}(hkl) = \mathbf{F_{PH}}(hkl) - \mathbf{F_H}(hkl) \qquad [21]$$

A graphical method (Harker, 1956) for solution of this equation is shown in *Figure 26*. In general there will be two possible solutions for $\mathbf{F_P}(hkl)$ and therefore two possible values for $\phi_P(hkl)$. In practice we assume that one of the two solutions is correct, but it is not possible to determine which one. This is the limitation of the SIR method. It is shown below how the phase ambiguity of SIR is resolved by using not one but a series of heavy atom derivatives.

SIR calculation of $\phi_P(hkl)$ thus requires:

(a) Measurement of $|F_P(hkl)|$ for the native crystal.

(b) Measurement of $|F_{PH}(hkl)|$ for derivative crystal.

(c) Calculation of $\mathbf{F_H}(hkl)$ the structure factor for the heavy atom structure alone.

This requires knowledge of the heavy atom positions which will then enable structure factor components for $A'_H(hkl)$ and $B'_H(hkl)$ to be calculated which in turn will lead to estimation of $\phi_H(hkl)$ from $\tan[\phi_H(hkl)] = [B_H'(hkl)/A_H'(hkl)]$. The

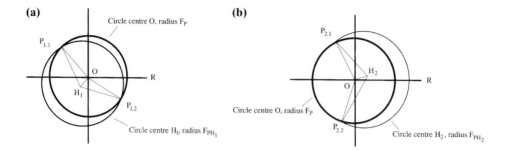

Figure 26 Pedagogical graphical analogue of isomorphous replacement phasing for a general *hkl* reflection using two heavy atom derivatives. The phase angle ϕ is measured from OR (where $\phi = 0°$) in an anticlockwise direction. (a) Phasing for derivative 1. The F_P circle has radius $|F_P|$, centre O; $OH_1 = -F_{H1}$; the F_{PH1} circle has radius $|F_{PH1}|$, centre H_1. Of the two possible phase angles, $P_{1.1}$ and $P_{1.2}$, it can be assumed that one will approximate the required value of $\phi_P(hkl)$. (b) Similar phasing for derivative 2. Here, two possible phase angles correspond to $P_{2.1}$ and $P_{2.2}$. Combining (a) and (b) indicates that the correct phase angle approximates to $OP_{1.1}$ and $OP_{2.1}$ with a value of about 128°. The other two "solutions" are thus eliminated. [In practice, where a large number of *hkl*'s, require to be phased, up to four derivatives may be required in order to provide good quality phasing for the majority of the data.]

question is how do we locate the heavy atom positions to make this calculation possible?

8.6 Location of the heavy atoms

This is usually carried out from a difference Patterson map calculated using, as coefficients, the experimentally derived values:

$$\Delta F^2(hkl) = [|F_{PH}(hkl)| - |F_H(hkl)|]^2 \qquad [22]$$

Interpretation of this Patterson function in terms of the heavy atom co-ordinates is expected to be fairly straightforward provided the number of heavy atoms per protein molecule is no more than about four or five. Software for carrying out these calculations is listed at the end of the section.

8.7 Phase determination in multiple isomorphous replacement (MIR)

In order to overcome the phase ambiguity of SIR a series of heavy atom derivative crystals is required. These should be: (a) isomorphous with the native protein crystals and (b) have different heavy atom positions so that their individual SIR phase ambiguities are different.

Three heavy atom derivatives or more are usually required in order to provide good quality phasing. The heavy atom positions in the different derivatives are determined initially by the difference Patterson method and referred to the same crystallographic origin using difference Fourier techniques (see for example ref. 14), prior to application of the MIR phasing technique.

8.7.1 Phase determination in MIR

The graphical method for determination of $\phi_P(hkl)$ illustrated in *Figure 25a* is extended in *Figure 25b* to illustrate how a second derivative would affect the process. As in the original SIR determination there will be again be an ambiguous result, but this time it is to be expected that out of the four phase solutions (two per derivative) two will be (approximately) equal and the other two, being widely different, will be thus be eliminated. The flow diagram in *Figure 27* indicates the procedure to be adopted for MIR.

8.7.2 Phase probability method for MIR

Whereas the graphical method for MIR discussed previously provides an excellent visualization of the process, in practice it is obviously too cumbersome to use for phasing thousands of different *hkl*'s which may be present in a protein data set. Blow and Crick (38) proposed a probability approach enabling $\phi_P(hkl)$ to

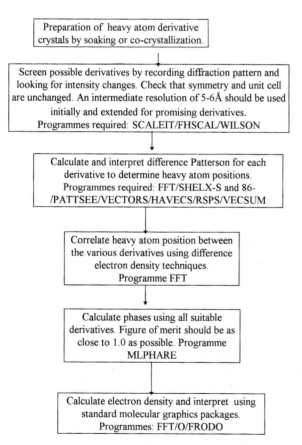

Figure 27 Flow diagram for isomorphous replacement.

be readily computed from the MIR data sets. The Blow and Crick algorithm computes the joint phase probability over all derivatives for each reflection hkl as:

$$P(\phi_T) = \exp[-\Sigma \varepsilon_i^2 (\phi_T) / 2E_i^2]$$ [23]

where ϕ_T is a test phase angle usually varied between 0–360° at 5° intervals; $\varepsilon_i^2 (\phi_T)$ is the calculated lack of closure error for derivative i having a value of 0 at each of the two SIR phase solutions, *Figure 26*, or sometimes a single minimum value; E_i is the overall r.m.s. error in $|F_{PH}(hkl)|$ for derivative i and should be carefully calculated as its value affects the value of m.

8.7.3 Best phase and electron density calculation

The best phase angle $\phi_B(hkl)$ derived from this probability corresponds to the centroid of the probability distribution. Electron density coefficients calculated with $\phi_B(hkl)$ are weighted with a figure of merit (m), calculated from the shape of the probability distribution. The maximum possible value of m is 1 for a given hkl. A mean value of m of about 0.8 at 6 Å resolution fading to 0.5 at 2 Å is acceptable.

Phase probability calculations can be made in SIR analyses but the phases are usually of low quality and subject to pseudo-symmetry effects (39a). Further details of MIR can be found in the article by Dickerson *et al.* (40).

8.8 Molecular replacement (MR)—the Sledgehammer

8.8.1 Introduction

In contrast to MIR, molecular replacement is not an experimentally intense method of phase determination. It requires:

(a) A set of $|F_o(hkl)|$ values for the protein crystal under investigation (the target structure).

(b) The availability of the co-ordinates of a good quality structure (the search structure) expected to be similar to the target structure, usually having different space group and unit cell from the target structure.

(c) A sound understanding of the principles and practices involved.

(d) State of the art software and hardware.

Conceptually the basic principle of MR is quite straightforward. Remember we know that the intensity data I(hkl) as a whole, inherently contain phase information (otherwise it would not be possible to solve crystal structures). The MR method as proposed by Rossmann and Blow (40) involves a critical and quantitative comparison of the Patterson functions of the target and search models. In contemporary software for carrying out MR applications (for example AMoRE and X-PLOR) the method is strengthened through the use of a variety of other crystallographic techniques which are now within the capabilities of modern computing.

As we have seen the Patterson function (Section 7.6) has peaks of high density at locations corresponding to the ends of atom–atom vectors, one atom always being placed at the origin. For complex structures like proteins, interatomic vectors are densely packed in the unit cell and most will not be resolved in the Patterson map; lack of atomic resolution in the X-ray data will also cause a further blurring of the vector distribution density.

The two atoms forming the Patterson vector can be either: (a) in the same molecule (intramolecular) or (b) in symmetry related molecules (intermolecular).

8.8.2 Intramolecular interatomic vectors—molecular orientation

The vectors in this set arise through interactions within one molecule and therefore tend to be shorter than intermolecular vectors, which span structurally related molecules. This self-vector set, as it is also known, is consequently situated around the origin of the Patterson function, the longest vectors arising between atoms at extreme ends of the molecule. Each atom, in theory, images both the structure and its inverse (i.e. vector types AB and BA). Because of the lack of resolution, this will be in the form a blurred molecular envelope of density. There will be one such image of the structure, plus its inverse, per atom, forming a centrosymmetric distribution.

Figure 28 shows two similar (homologous) 2D structures in simplified form and the corresponding Patterson functions as (simulated) vector sheaths are shown in *Figure 29*. *Figure 30* shows the two actual structures in their unit cells. The search structure is based on an inclined unit cell, whilst the target structure has an orthogonal cell. Both incorporate 2-fold symmetry. *Figure 31* shows how intermolecular vectors arise in the target structure and demonstrates the resulting vectors, which arise from an incorrect translation, compared to *Figure 30b* which shows the correct structure and vectors.

8.8.3 Self-rotation function and non-crystallographic symmetry (NCS)

A complication can arise in practice for target crystal structures containing more than one protein molecule in the asymmetric unit. A clue to the presence of NCS may already have been given by the preliminary analysis of unit cell contents (Section 6). This complication is mentioned at this point because in practice it is highly desirable to establish the presence of NCS as early as possible in the MR analysis.

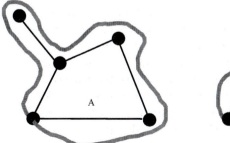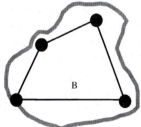

Figure 28 Two similar (homologous) 2D structures A and B in simplified form. A is the unknown target structure and B is the known search structure.

(a)

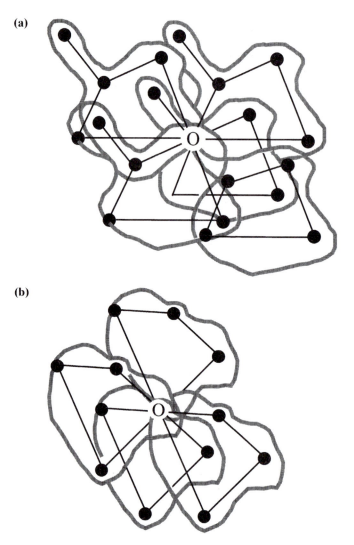

(b)

Figure 29 Rotation stage of MR. The corresponding peak positions in the Patterson functions. (a) Molecule A. (b) Molecule B. The corresponding (simulated) vector sheaths are shown, omitting the Patterson centre of symmetry for clarity. The reader should show that the required rotation to superimpose (a) on (b) is 67°. This can be done using transparent copies of the diagrams.[The extra lobe in Molecule A obviously has no vectors in B].

i. Rotational NCS—self-rotation plots

The non-crystallographically related molecules frequently differ in both position and spatial orientation with respect to each other (often related by a pseudo 2-fold rotation). Consequently around the origin their Patterson functions will contain a differently oriented copy of the molecule's interatomic vectors for every protein molecule. If two exact copies of the Patterson function are superimposed origin to origin, by rotating one copy the replicas can be made to coincide, thus

(a)

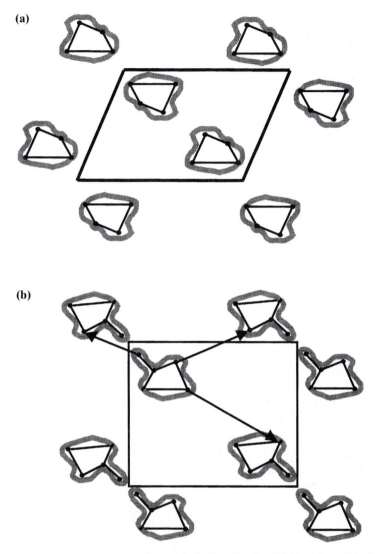

(b)

Figure 30 (a) Known (search) structure in its unit cell and (b) target structure in its unit cell after successful MR (rotation and translation).

determining the relative orientations of the NCS related molecules. Programs are available for plotting the results of this type of self-rotation and calculating the relative molecular orientations in three dimensions. These are listed at the end of the section.

8.8.4 Intermolecular atomic vectors

i. Translational NCS—inspection of the Patterson map

If two NCS molecules differ only by a non-lattice translation, with no rotational component, the Patterson function will contain an outstandingly high peak at a

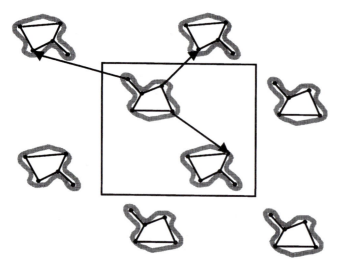

Figure 31 An incorrect translation of the target stricture B gives rise to incorrect (i.e. non-existent) vectors (illustrated). Compare with *Figure 30b*, which shows the correct translation and corresponding vectors.

position corresponding to the NCS intermolecular vector (because all vectors between the two will be lined up). Again it is important to establish this prior to embarking on the full MR analysis, by simply calculating the Patterson function and inspecting the regions of high density. Remember that the origin peak at (0, 0, 0) is always the highest density in the Patterson. Most Patterson density will be a very small fraction of P(000). We are therefore considering here Patterson density which is a numerically significant fraction of P(000), possibly as high as 20–50% (further details of Patterson analysis can be found in ref. 21).

ii. Basis of the translation function

The vectors between atoms in different (symmetry related) molecules contain information about how the molecules are arranged with respect to one another to pack in the crystallographic unit cell. Unit cell translations and space group symmetry operations are involved in this process. *Figure 30b* shows how intermolecular atomic vectors arise for the hypothetical structure used in *Figure 28*.

8.8.5 Further details of MR

MR relies on the presence of both intra- and intermolecular atomic vectors in the Patterson function of the crystal under investigation.

i. Rotation stage of molecular replacement

In the above simulated example the vectors defining the orientation of the unknown structure A can be seen to occur in the Patterson of the known (test) structure B (*Figure 28*). To demonstrate this, a copy of the known Patterson in *Figure 29b* (the search structure Patterson) should be placed over *Figure 29a* (the

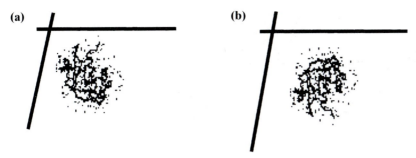

Figure 32 (a) A correctly rotated molecule and (b) an incorrectly rotated molecule (see *Figures 11* and *12*).

Patterson of the unknown structure), with origins in register, and rotated anti-clockwise by 67°. Maximum correspondence occurs for this orientation. This superposition would result in a dominant peak in the rotation function R. Note because the unknown structure, *Figure 28A* contains an additional moiety, vectors for this part of the structure are missing in the search Patterson, *Figure 29b*. In practice this process of matching the orientations of the two Pattersons is carried out, by computation, in the rotation stage of MR, by testing over a series of orientation angles. For 3D structures, it is necessary to perform these rotations about three independent axes. The rotation angles are designated (α, β, γ)—not to be confused with the unit cell angles. The rotation ranges and intervals have to be carefully chosen in order to cover a sufficient number of possibilities to ensure that the correct angular triplet is not overlooked. Examples of correctly and incorrectly oriented solutions are shown in *Figures 32a* and *32b* respectively.

It is extremely important to note that because of the complexity of protein structures, the rotation function $R(\alpha, \beta, \gamma)$ (calculated over the required angular range) generally contains many peaks in addition to the correct solution, which often have values of comparable magnitude. These peaks are simply signifying that there is a degree of correspondence between the two Pattersons in this orientation. The highly popular program AMoRe retains all peaks greater than 50% of the highest peak (in some versions more) for transference to the translation stage of MR.

ii. Translation stage of molecular replacement

Assuming the correct orientation of the unknown molecule has been determined in the rotation stage of MR, the correctly oriented molecule must then be correctly located spatially in position in its unit cell. The origin with respect to this translation process is usually governed by the space group (*International Tables for X-ray Crystallography*, Volume I or Volume A). This translation stage must cover a sufficient number of finely selected 3D translational increments (Δt_1, Δt_2, Δt_3) or (t_x, t_y, t_z) to ensure that the correct location of the molecule is scanned. In essence, the correct structure is recognized initially as corresponding to the highest degree of overlap between the calculated and observed Pattersons when superimposed after applying the given translation vector.

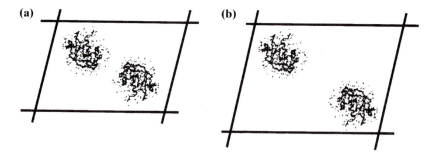

Figure 33 (a) Correctly rotated but incorrectly translated molecules and (b) correctly rotated and translated molecules (see *Figures 11* and *12*).

iii. Packing the rotated–translated molecules in the unit cell

An initial verification of the correctness of the rotation–translation search can be easily achieved by inspecting the packing of the resulting structure using molecular graphics. Several programs are available for this (e.g. MOLPACK) (41). An example of a correctly oriented but incorrectly translated structure solution is shown in *Figure 33a*, and *Figure 33b* shows the correctly orientated and translated solution. Subsequent verification of the choice will depend on successful development and refinement of the model, for example through Fourier and least-squares refinement.

8.8.6 Molecular replacement in practice

i. The search model

One of the following situations may provide a suitable opportunity for using MR:

(a) Determination of a protein derivative structure containing a small ligand, crystallizing in a space group different from that of the known native protein structure. Examples include enzyme–inhibitor and lectin–carbohydrate complexes.

(b) Determination of more complex combinations when one or both components have known structures. It may be possible to carry out successive MR searches in order to provide a starting model for the complete structural combination. If only one component can be located by MR, lengthy Fourier development and refinement may follow. Examples include antibody–protein and protein–nucleic acid complexes.

(c) Functionally similar proteins with high sequence homology, one member of the series having a known structure. Sequence homology as low as 20% may be sufficient especially if many of the sequence changes are conservative, e.g. Ala for Val etc. Sequence homology may also be used in less obvious cases of structural similarity (for example proteins not belonging to the same or related families) but carries a greater risk of failure. It may also be possible to use a partial MR search model which includes only the region of optimum correspondence, again followed by Fourier refinement, or proteins stripped to only main chain atoms (or even less). Examples include bluebell lectin (42).

(d) Recombinant proteins and/or site-directed mutants having known native structures but crystallizing in different space groups.

Database searches. Several protein sequence databases are readily available to accommodate sequence homology searches for suitable MR test structures. A list appears at the end of this article. Depending on the type of protein involved it is often known in advance what likely target structures are available. For example ricin is the parent structure for the ribosome inactivating protein (RIP type II) family and would be the first choice to use in a MR search for determination of the structure of a new member of the family. Sequence matching protocols and algorithms are under active development. Many of these routines are freely available on the Internet and can be accessed in combination with database sequences (e.g. BLAST). Where possible, well defined structures should be chosen as search models. In simple terms this means that the resolution should be 2 Å or better and the R-factor should be 20% or better. It is more risky to use low resolution X-ray structures or NMR derived models which may not be sufficiently refined.

ii. The rotation function

Calculation of the target Patterson function is straightforward in practice as the $|F_o(hkl)|$ are already available from recent diffraction measurements. It has been suggested that different resolution cut-off's should be tried, say between 3–5 Å, i.e. it is best to employ medium to low resolution data in order to blur the differences between the search and target models. An inner low-resolution data cut-off should also be used, e.g. within 15–20 Å as these data are concerned mainly with the solvent structure and therefore do not directly contribute to the protein structure at all. Facilities for easily effecting these cut-offs are available in most software. For the following reasons the Patterson for the search model is not derived from X-ray diffraction intensities:

(a) The $|F_o(hkl)|$ data would usually have to be retrieved, e.g. from a database. This would cause problems, particularly if several search models need to be tried, as is often the case.

(b) The target and search structures may have different space groups and other crystallographic characteristics which make it impractical to compare the actual Patterson maps.

(c) The computations involved would be prohibitive in many cases.

(d) Models derived for example from NMR studies or model building do not correspond to any crystal structure. As this type of model is frequently used in MR studies other approaches to generate interatomic vector sets are required.

One approach is to construct the search Patterson as a vector array from the structure co-ordinates. As this type of approach (used for example in X-PLOR) would generate an extremely large number of vectors it is computationally very costly and slow to apply.

In AMoRe the search model co-ordinates are transformed with respect to an artificial triclinic unit cell with angles of 90° and cell edges of pd_m where d_m is the molecular diameter and the factor p is user selected, frequently as 2.0 or less. $F_c(hkl)|$ values calculated from the triclinic co-ordinates are then used to calculate the search Patterson.

One or both of these procedures by may used in practice. In either case, the need to consider symmetry is eliminated thereby ensuring that only intra-molecular vectors are generated for the search model.

In the case of the target structure it is more difficult to restrict the rotation vectors to intramolecular type alone as the observed Patterson distribution around the origin contains both intra- and interatomic vectors. This problem can be overcome to some extent, in practice, by restricting the volume of the target Patterson used in the search. In most software the radius of summation (or integration) used for this purpose can be selected by the user in an attempt to optimize the signal to noise ratio. Several trials may be required in practice in order to achieve the desired result. It has been recommended (43) to start with a value for this radius of integration approximately equal to $0.5d_m$, where d_m is again the approximate molecular diameter, increasing to about $0.75-0.8d_m$ for further trials if necessary.

Recognition of the correct rotation solution. Several powerful programs are currently available for MR analysis. Of these AMoRe (44), a CCP4 supported program and X-PLOR (24, 45) are commonly used. X-PLOR is a more general program covering most aspects of macromolecular crystal structure analysis and refinement, whereas AMoRe is specific to MR.

Use of these programs to locate the correct rotation function solution involves searching through a large number of trial rotation angles for (α, β, γ). Incre-mental values of 5 or 10° for each of the three angles are typical, resulting in a large number of trials. Programs like AMoRe and X-PLOR optimize the rotation angles around local maximum regions of Patterson overlap. In AMoRe for each promising trial, a correlation coefficient, C_c, indicating the level of agreement between the rotated Pattersons, is calculated by the program. The correct (or even near correct) solution should have the highest C_c values. However at the rotation stage of MR, C_c values do not always differ significantly between potential solutions, and the values tend to be much smaller, typically around 5–15%, than at the later translation and optimization stages. AMoRe also calculates the RMS value (σ) for the Patterson density and gauges the significance of local density regions with respect to this value. A peak $> 3\sigma$ is considered significant. If different search models, resolutions, integration radii, or any of the other para-meters available, have been varied over a series of searches, consistency in the occurrence of solutions (angle triplets α, β, γ) with high C_c is a good indication that this solution is worth carrying over to the translation stage. Because C_c is not an absolutely reliable indicator, all peaks greater than 50% of the maximum value, ranked in order, are normally retained for transfer to the translation stage (some versions of AMoRe retain even more peaks). The density values in terms of σ are also printed out. In this way the probability of overlooking the correct rotation solution should be greatly decreased. AMoRe can accommodate the retained rotation solutions simultaneously into the translation routine.

Subunits and NCS. For structures which contain more than one molecule per asymmetric unit related by non-crystallographic symmetry (NCS), the rotation function would be expected to produce a corresponding number of solutions, equivalent, or nearly so, with respect to both the C_c index and reproducibility under different computational conditions. Genuine solutions are, of course, likely to produce acceptable translated models at the next stage of MR. However there is no guarantee that such translated models will occupy their true relative positions in the unit cell. This problem requires further attention, as explained below.

iii. The translation function

It cannot be over-emphasized that success with the MR technique depends heavily on the precision with which the correct rotation parameters have been established. Programs such as AMoRe and X-PLOR include optimization procedures at the rotation stage to improve the chances of success later on. For each of the potential rotation solutions carried over into this stage of MR a set of atomic co-ordinates is generated. It is necessary to place this oriented model into the correct location in the unit cell. Full attention to the symmetry elements and their relationship with each other and the standard origin (*International Tables for X-ray Crystallography*, Volume I or A) must be maintained. In general only one of the retained rotation solutions will provide a successfully translated model. When NCS is present the user looks for the corresponding number of successfully translated models. Each test model, carried over from the rotation stage is moved by incremental translations called (Δt_1, Δt_2, Δt_3) along three independent axial directions. The small steps used, typically 1 Å or more initially, are further optimized, as with rotation solutions, for the most promising looking models. The translation function routines attempt to correlate, at least conceptually, the observed and calculated Patterson functions. As we have seen optimal correspondence is measured in AMoRe, by the correlation coefficient, C_c which approaches 1.0 for the best solution. In addition it is possible to compare $|F_o(hkl)|$ with $|F_c(hkl)|$ for each translated model and calculate an R-factor (see Section 4.4) R_f which is lowest in value for the best solution. AMoRe employs both the C_c and the R_f to discriminate between potential solutions. As with the earlier rotation stage, consistency under different trial conditions is an important and extremely useful means of generating confidence in the validity of the solution of choice prior to embarking on what can be very lengthy model building and refinement.

Rigid body refinement. Both AMoRe and X-PLOR include refinement routines which fine-tune the rotation and translation parameters by adjusting the position of the search model without making changes to the geometry of the model. This technique is known as rigid body refinement. In this way the validity of the rotated and translated model is given a final check in terms of C_c

(which should approach 1.0) and R_f which should minimize. If successful the model is ready for further development and refinement.

Subunits and NCS. As explained above, individual molecules related by NCS may produce independent acceptable solutions as a result of the rotation and translation stages of MR. Because there is no guarantee that the co-ordinate sets for the individual molecules produced at this stage are properly correlated in the unit cell it is not usually possible to simply combine them to form the final trial structure. The main reason for this is concerned with the existence of more than one equivalent origin in many space groups. For example in the 2D structure of *Figure 12b* it is possible to locate a total of four, 2-fold axes which are different in environment, i.e. at (0, 0), (0, 1/2), (1/2, 0), and (1/2, 1/2). Both AMoRe and X-PLOR have the facility for allowing a rotated–translated model to be included as a rigid group which is subtracted from the unknown structure in order to enhance the signal from the missing NCS related molecule or molecules. If successful this process will result in the NCS related molecules being properly placed relative to one another in the unit cell. This is an extremely powerful facility and will produce a full trial structure to be developed further by refinement procedures.

Protocol 11

Application of the AMoRe algorithms to mistletoe lectin MLI (19)

1 MLI is a type II ribosome inactivating protein (RIP) belonging to the same family as ricin with which it has 52% sequence homology. The molecular weight (M_r) is 63 kDa. The analysis given in *Protocol 9* indicates an ambiguity in the number of molecules per asymmetric unit, either 1 (with 12 per unit cell) or 2 (with 24 per unit cell).

2 The crystals (8) are hexagonal, with unit cell, a = b = 110.79, c = 308.53 Å, $\alpha = \beta = 90°$, $\gamma = 120°$. From 00l reflections, l = 6n present (others absent), the space group is $P6_522$ (or $P6_122$) (*International Tables for X-ray Crystallography*, Vol. I or A). X-ray intensity data were collected on a 30 cm Mar Research Image Plate detector using a synchrotron radiation source, $\lambda = 0.92$ Å. The crystal to detector distance was set to the largest value possible, 450 mm, in order to resolve reflections along the very short c* axis. A total of 130.5° of X-ray data were collected from two crystals to approximately 3.7 Å resolution. These data were indexed using the DENZO software. The X-ray intensity data were scaled and merged using the CCP4 suite of programs. The total number of reflections accepted was 64 381 which reduced to 11 241 unique reflections with an R_{merg} of 0.09 for $d_{min} = 3.7$ Å.

3 The ricin structure previously described by Monfort *et al.* (31), Rutenber *et al.* (46), Katzin *et al.* (47) [Protein Data Bank ID code 2AAI] was used as the search model employing the CCP4 version of AMoRe. Various trials were used to establish the rotation search. The search model Patterson maps were calculated using intensity data restricted to various resolution ranges including 3.8–15 Å by placing the ricin

Protocol 11 continued

structure into an orthogonal cell of P1 symmetry with unit cell parameters 100 Å × 80 Å × 60 Å. Other search unit cells, resolution ranges and search models were used to establish the consistency of the MR solutions produced. The rotation function was stepped over 2.5° and the radii of integration varied between 15–30 Å. In calculating the search model Patterson, an overall temperature factor was set to −20 Å² which has the effect of improving the clarity (sharpening) of the Patterson vectors. In space group P6₅22 a translation function solution was found with a peak height of 5.8σ above the highest noise peak [a value > 3σ is normally considered to be outstanding] with an R of 46.0% and C_c of 54.2% following the rigid body refinement protocol of AMoRe (*Tables* 7 and 8). All symmetry related molecules in the unit cell for this solution were generated and examined graphically using MOLPACK (41). There were no inadmissible intermolecular contacts in the packing of this model, providing further confidence in this result to carry out further refinement.

4   Using the same rotation function results in the enantiomorphic space group P6₁22 produced no translation function peaks comparable with those in space group P6₅22. Subsequent development and refinement of the P6₅22 structure proved the validity of this space group.

Table 7 AMoRe rotation function for MLI in space group P6₅22

	α	β	γ
	48.5	140.6	134.4
	11.4	39.40	314.4
B	**17.2**	**88.04**	**212.6**
A	**42.7**	**91.96**	**32.63**
	10.3	129.0	148.7
	49.6	50.95	328.7
	7.40	98.73	275.8

Data range 15 Å to 3.8 Å. Radius of integration 25 Å. Search cell 100 × 80 × 60 Å. The top seven R-function peaks are shown. Correct solution A and a symmetry equivalent B are highlighted.

Peaks A and B alone persist under other conditions tried including:

Data range 10 Å to 4.0 Å. Radius of integration 35 Å. Search cell 90 × 60 × 50 Å.

Data range 10 Å to 4.0 Å. Radius of integration 22 Å. Search cell 90 × 90 × 90 Å.

8.9 Use of phase information and density modification

Both MIR and MR provide values of ϕ(*hkl*), MIR through phase probability calculations (Section 8.7.2) and MR by calculating structure factors (Section 4.4) from the rotated/translated model placed in the target unit cell (Section 8.8). Errors are inherent in both methods which can affect the quality of the calculated electron density and its interpretation in terms of the protein structure (Section 9.2). Density modification techniques are designed to optimize the electron density

Table 8 Extension of Table 7 to translation stage using space group P6$_5$22

	α	β	γ	tx	ty	tz	C$_c$%	R%	Δσ
	48.58	140.60	134.41	0.0803	0.9280	0.4921	18.0	57.5	
	11.42	39.40	314.41	0.0732	0.5122	0.3640	17.4	57.4	
A	**42.77**	**91.56**	**32.89**	**0.22705**	**0.43659**	**0.01195**	**49.5**	**47.1**	**5.8**
A (refined)	*43.20*	*91.56*	*32.15*	*0.22700*	*0.43660*	*0.01190*	*54.2*	*46.0*	
B	**17.23**	**88.04**	**212.63**	**0.22705**	**0.79405**	**0.32138**	**49.3**	**47.2**	**4.8**
	10.38	129.05	148.70	0.0851	0.6272	0.1617	18.0	56.9	
	49.62	50.95	328.70	0.8746	0.6966	0.3294	17.9	57.7	
	7.40	98.73	275.83	0.0369	0.1143	0.0949	17.5	57.4	

Solution A and its symmetry equivalent B give outstanding C$_c$ (highest) and R (lowest) indicating correctness of the solution. [It was found in space group P6$_5$22, that AMoRe did not recognize the equivalence of solutions A and B. This was shown to be the case by manual calculation. NB AMoRe *does* usually recognize symmetry related solutions and these are then eliminated.]

In italics for solution A the refined values of the R and T parameters are given with the C$_c$ and R indicators which clearly show the improvement in the MR fit.

For the alternative space group P6$_1$22, C$_c$ = 13.7% and R = 56.9%, significantly inferior to the outstanding values for P6$_5$22.

The subsequent use of modified ricin models in the above analysis further confirmed the MR solution. The packing of the molecules from the solution in P6$_5$22 shows no disallowed contacts and also reveals the MLI molecules to be associated into non-covalent dimers (19).

map, particularly in cases where there is reason to believe the initial phasing is of poor quality, to enable a rational interpretation to be carried out. Cases where this might be necessary include: SIR or MIR where the mean figure of merit (Section 8.7.3) is too low ($<$ 0.5 is worrying); MR where C$_c$ is too low, R too high (Section 8.8.6), and/or no outstanding or persistent solution is observed.

8.9.1 Expected properties of ρ(*xyz*) for proteins

Because protein X-ray data fall a long way short of atomic resolution (*Figure 34*), only heavy atoms (with atomic number greater than about 20) are expected to produce distinct peaks in the electron density. Experience has taught us to expect the following general features in protein maps:

(a) Main chain density tends to be continuous and relatively strong. This is because atoms in the polypeptide chain are generally held firmly by inter-actions between neighbouring atoms and are therefore less susceptible to thermal and statistical disorder (Section 4.3). Recognition of strong, continuous density in the map can therefore enable the main polypeptide chain to be traced and the molecular envelope to be outlined.

(b) Side chain atoms tend to be less rigidly held in the structure than main chain atoms, with correspondingly weak electron density.

(c) As we have seen previously, protein crystals can accommodate anything between 35–70% of solvent molecules (Sections 2.1, 6.6). This is the most disordered part of the structure and will consequently be associated with very weak electron density, situated in the intermolecular interstices and solvent

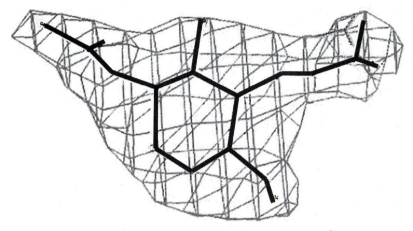

Figure 34 Electron density appearance at a fairly low resolution (2.9 Å) showing the region of a sugar molecule in MLI B-chain (20). The molecule was fitted with reference to known stereochemistry since individual atoms are not resolved.

channels. Consideration of the regions of weak density should therefore enhance the recognition of the protein molecular boundary. Solvent (e.g. water) molecules in the first hydration sphere are expected to be better ordered than bulk solvent atoms, and may be identified once some structure refinement has been effected (Section 9.3.1).

Protocol 12

Application of density modification DM

1 DM is a CCP4 supported program (48) for carrying out density modification. The program applies constraints on the observed electron density, and derives new phases via the Fourier transform. Phases can be calculated for *hkl*'s not involved in the original phasing (phase extension, e.g. from a 6 Å MIR map to 2 Å) which can be used in the subsequent analysis, e.g. calculation of new electron density maps.

2 Routines available include:

(a) SOLV (solvent flattening) (44). Establishes uniformity of density in solvent regions, on the assumption that regions of disordered solvent are essentially without structure.

(b) HIST (histogram matching) (50). Employs the known characteristics of biological structures to predict the histogram of density values in the protein region. The current density map is then systematically modified according to the predicted histogram. This technique is complementary to solvent flattening. Between the two, the image of the protein structure should become much clearer and facilitate model building prior to further refinement (Section 9). The combined process of SOLV/HIST may require 10–20 iterations to converge.

Protocol 12 continued

   (c) SKEL (skeletonization) (51, 52). Provides a sound basis for molecular graphics model building of the structure, by enhancing the connectivity of the electron density in main chain regions.

   (d) SAYR (phase improvement for data at 2 Å resolution or better). Applies phase relationships adapted from a classical small molecule phasing method (53, 50).

   (e) AVER (molecular averaging for structures where either NCS or exact crystallographic symmetry is present) (54, 55). Provides better phasing for initial model building. This constraint should, if possible, be released for the final structure.

3 *Figure 35* is a flow diagram for DM, starting with the phases derived, e.g. from MIR or MR. Output from the DM program provides indicators of how the calculations are progressing.

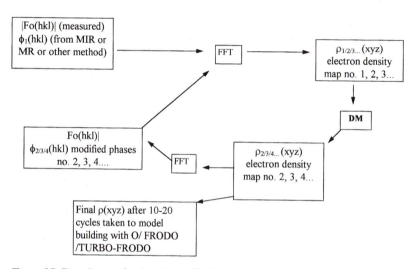

Figure 35 Flow diagram for density modification.

8.10 What to do once phasing is acceptable

Once confidence in the phasing is at a sufficiently high level, whether straight from MIR, MR, or after DM, the analysis proceeds to model building and refinement. The necessary procedures are discussed in Section 9.

9 Structure refinement, ligand fitting, and geometry verification

9.1 Refinement

The process of structure refinement involves optimization of the agreement between the observed and calculated diffraction patterns, represented by $|F(hkl)_o|$

and $|F(hkl)_c|$, and validation of the resulting molecular structure. If IR has been used to derive phase information, the initial structure model will be derived from an electron density map using molecular graphics. For a MR analysis, as mentioned in Section 8, it is possible to carry out limited refinement of the model at an earlier stage. Whichever method has been used for the initial structure analysis, refinement of the model should be undertaken by a combination of the following techniques: 1) Fourier refinement using successive Fourier synthesis, 2) simulated annealing, and 3) least squares analysis.

9.2 Fourier refinement and simulated annealing

9.2.1 Fourier refinement

A successful isomorphous replacement analysis results in the calculation of an electron density map (Section 7). As we have seen, lack of resolution in protein X-ray data results in poor definition in the electron density maps whether from MIR or MR (see *Figure 34*). It will be required to interpret this map by fitting the protein structure to the density. If molecular replacement has been used for the preliminary structure determination, electron density calculations will again play an important role initially, but the crystallographer will have prior knowledge of structural features of the search model, which will help to establish the new structure.

There are three aids to assist the initial process of interpreting the electron density:

(a) The amino acid sequence of the protein. If this is not known, interpretation of the crystal structure will necessarily be difficult and probably not possible unless very good phasing at a resolution of at least 2 Å has been achieved. If MR has been used the target molecule will be subject to sequence changes, and frequently, in addition, insertions and/or deletions may be required. These features have to be built into the density.

(b) Knowledge of the standard geometry of proteins in terms of main chain and side chain bond lengths and angles, and of secondary structural features, particularly α-helix and β-sheet. The principal geometrical features of small peptides have long been established, starting with the studies of the 1940s and 50s which led Pauling and Corey (56) to publish a very reliable summary of the data available at that time. The assumption that features of peptide geometry apply also to proteins has largely stood the test of time. Modern applications rely on structural parameters extracted from the Cambridge Crystallographic Data Base (CCDB) to provide updated values (57). Software such as PROCHECK (58) has been developed as an extremely useful aid to protein structure verification, including bond length, bond angle, and conformational checks.

(c) Molecular graphics. Interpretation of electron density has been revolutionized over the past 20 years due to the development and availability of graphics software. These facilities enable structural features to be built into density maps, following standard geometry protocols, and provide the user with structure files for use in further analysis. Software available for carrying out these procedures include FRODO (59), O (60), and TURBO-FRODO.

The Fourier refinement process is represented in the flow diagram of *Figure 36*. It consists of successive rounds of electron density calculation, graphical interpretation, updating structure files, and calculation of structure factors and R-factors. Improvement in the model and its fit to the electron density should cause significant decrease in the R-factors (Section 4). This process terminates when no further improvement is evident. The structure file at this stage consists of atom names, individual atom fractional co-ordinates (x, y, z), overall or average temperature factor B, for the whole crystal (Section 4), and a scale factor K which converts $|F_o|$ values to the absolute scale. Individual B_j values are not refined at this stage of the analysis.

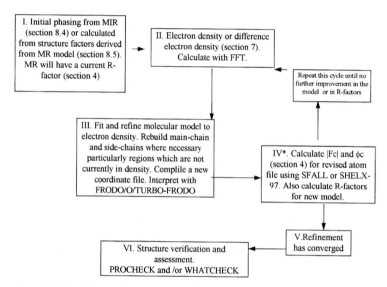

Figure 36 Flow diagram for structure refinement.

9.2.2 Simulated annealing

The program X-PLOR (24) has a facility for carrying out molecular dynamics in the form of simulated annealing. The main objective of this method is to eliminate from the model, regions of structure, which are far away from their true positions. Such regions may have been introduced through the initial search model in MR, or by faulty model building at a later stage. These structural errors are not easily dealt with by Fourier techniques, and least-squares refinement is not able to predict large parameter corrections. It is likely that such regions will be associated with bad interatomic contacts with inadmissibly high potential energy. In the method of simulated annealing the structure is given a large perturbation ('heated to a high temperature') and then allowed to recover ('cool') whilst preserving or re-establishing correct molecular geometry. The X-PLOR protocol involves minimization of an overall total potential energy term, E_{tot}, which includes contributions from empirical energy, E_{emp}, and effective energy, E_{eff} where $E_{eff} = E_{XRAY} + E_p + E_{nb}$. The terms $E_p + E_{nb}$ are included to take into account experimental information about phases and crystal packing, while E_{XRAY} is a 'pseudo energy' term which involves differences in the observed and calculated structure factors. Thus:

$$E_{XRAY} = W_a / N_a \, \Sigma_{hkl} \, W_{hkl} \, (|F_o| - K|F_c|) \qquad [24]$$

where W_a puts the term onto the same basis as 'energy' (established through a dummy dynamics run without E_{XRAY}), N_a is a normalization factor which renders W_a independent of resolution, W_{hkl} provides a weighting scheme, and K is a scale factor.

X-PLOR is sometimes used without further refinement and this can lead to a satisfactory structure. Individual isotropic thermal parameters, B_j, can be included in the refinement if the data collection resolution is sufficiently high, at least 2.5 Å. The conventional R-factor (Section 4) should drop to around 20% or lower unless there are big problems with the data (poor crystal quality or very weak diffraction) in which case it will not be possible to derive a highly significant structure with any refinement protocol. It may be desirable to undertake least-squares refinement of the structural parameters even after simulated annealing has been carried out. In any case routines such as PROCHECK or WHATCHECK (61) should be used to provide rigorous validation of the structure model.

9.3 Least squares refinement

9.3.1 Introduction: constrained and restrained protocols

The method of least-squares is used routinely in small molecule structure analysis, optimizing the fit between $|F_c|$ and $|F_o|$ and providing refined parameters and their standard deviations. Because most small molecule crystals diffract to around 0.8 Å the data set provides about ten $|F_o|$ values per parameter even when an anisotropic thermal displacement model is used. This degree of over-determination is not feasible with poorly diffracting protein crystals for which there may be only one $|F_o|$ (or less) per parameter. Methods have therefore been developed which attempt to:

(a) Economize on the number of parameters defining the structure model.

(b) Provide data in addition to the X-ray data which can contribute towards the refinement of the model.

Software for carrying out such refinements is usually based on constrained or restrained protocols.

i. Constrained refinement and rigid body refinement

This has already been mentioned in Section 8.6.6 in the context of rigid body refinement in MR protocols. In this application the model undergoes repositioning in the unit cell without any further adjustment of individual atom co-ordinates. The method can be useful for refinement of molecules related by NCS by initially averaging the two or more molecules to improve the data/parameter ratio. X-PLOR has facilities for performing this type of refinement as recently applied for example to RCA (62). Variants of the method include facilities for fixing, absolutely, predefined bond length and angle types and planarity of groups such as rings or peptide groups. Whilst possibly effecting some degree of refinement in the structure model constrained refinement will usually require further refinement by restrained least squares if the data/parameter ratio allows.

ii. Restrained refinement

Many programs are available for carrying out restrained least-squares analysis of protein structures. Only two will be mentioned here: RESTRAIN (63); and SHELXL-97 (64b). The method should not be used unless a great deal of effort has been made using Fourier refinement and possibly simulated annealing as well to reduce the errors present in the structure and to include as many atoms as possible in the co-ordinate set. Restrained geometry refinement is more flexible than constrained refinement, each of the standard values of bond lengths, bond angles, and other distances being tagged with a tolerance which effectively specifies an acceptable range for the refined value of the parameter.

As an example of the savings in parameters that can be achieved by applying constraints, consider a flat six-membered ring. In terms of free atom parameters there are $3 \times 6 = 18$ positional parameters. By constraining the six bond lengths, six bond angles, and three cross-ring distances, only 15 parameters are required, a saving of three parameters. When applied repeatedly within a macromolecular structure many times, small savings like this can accumulate to a very worthwhile total, and at the same time ensure that the geometry of the refined structure will be acceptable. The input data set does, however, require a great deal of preparation in order to set up the required constraints.

RESTRAIN: This program employs a least squares algorithm which uses terms involving the differences $(\Delta F)^2 = (|F_o| - |F_c|)^2$, phase data (if used, from MIR), standard geometry values, and planarity of groups. Corrections to structure parameter values are derived by minimizing the function:

$$M = \Sigma W_f (\Delta F)^2 + \Sigma W\phi \, (\phi_o - \phi_c)^2 + W_d \, (d_o - d_c)^2 + \\ \Sigma W_a (a_o - a_c)^2 + \Sigma W_a (b_o - b_c)^2 + \Sigma W_v V \qquad [25]$$

where ϕ_o is the MIR phase (if used); d is a bond length; a is a cross-bond distance, e.g. A-C in the sequence A-B-C and therefore a measure of bond angle; b is a term for non-bonded distances; V is a planarity restraint; the W's are weights for the various types of term and require very careful adjustment during the course of the refinement to ensure proper calculation of standard deviations for the refined parameters. Chirality is preserved by applying restraints to the edges of all chiral tetrahedra. Anisotropic thermal displacement parameters (see ref. 21, p. 418) can be included if the data/parameter ratio allows (anisotropic refinement requires nine parameters for each freely refined atom). RESTRAIN is a CCP4 program. It employs non-FFT calculations for structure factors and partial derivatives and is consequently slow to perform each cycle, but requires fewer cycles to converge than FFT-based algorithms.

SHELXL-97: SHELX programs are well known for their use in small molecule crystal structure analysis. The -97 version has been adapted to accommodate protein refinement via the SHELXPRO interface. The geometry refinement options are similar to those described above. Other facilities include: anisotropic scaling; refinement progress display; thermal displacement analysis; Ramachandran main chain conformation (65); map file for O; input and convert DENZO / SCALEPACK data; generate PDB file for graphics display such as RASMOL (66), SETOR (67), MOLSCRIPT (68), BOBSCRIPT (69); TURBO-FRODO and PDB deposition.

iii. Practical considerations

Initially, small ligands and/or solvent can be excluded but should be added via further electron density maps as the refinement progresses. Least-squares analysis predicts corrections for the structural parameters in the current model (coordinates, temperature factors, and scale factor). As stated previously neither individual B_j values nor anisotropic thermal displacement factors are likely to be refineable in all but high-resolution and very high-resolution protein structures. Selected groups of atoms can be given a single temperature factor, which provides an easy means of saving parameters.

iv. Solvent and small ligands: fitting into the electron density

The assignment of solvent atoms, usually water, into the crystal structure model is one of the last procedures to be carried out. These atoms make only small individual contributions to the X-ray intensities, but taken as a whole, perhaps several hundred water molecules, can considerably influence the values of $|F(hkl)|$. As we have seen (Section 6.6) the level of solvation in protein crystals can range from 35–70% and is usually around 50%. Most of the solvent is unstructured, forming very disordered, fluid regions in the intermolecular channels. However within a layer closest to the protein molecule, it is to be expected that many solvent molecules will form strong interactions with the protein atoms and will consequently acquire an ordered state approaching that of the protein itself. A measure of the ordering may be derived from the refinement if individual atomic thermal displacement parameters B_j are used, but this will only be possible for analyses employing better than about 1.8 Å resolution. Generally speaking solvent atoms will have B_j values 10–20% larger than the protein atom or atoms they are associated with. Hydrogen bonded interactions are responsible for ordering water molecules or other hydrophilic solvent molecules (ethanol for example). The interactions can involve any main chain or side chain atom, which can act as an H-bond donor or acceptor. In order to locate possible water molecules it is necessary to inspect an $(|F_o| - |F_c|)$ or $(2|F_o| - |F_c|)$ density map in detail. Possible water sites may be assigned to significant density regions, say > 2σ (see Section 7.2) located within 2.5–3.3 Å (H-bonding distance) from one or more possible H-bond donor or acceptor atom (mainly OH, NH, O, or N). There should be no other close contacts present. Suitable electron density regions are ideally spherical and small in volume. Since this procedure can be very time-consuming, programs have been developed to expedite the process, including routines in O and SHELPRO. It should be emphasized however that water sites assigned by automatic procedures should be checked manually and, if the analysis is at sufficiently high resolution, their B_j values should be refined. Sites with refined $B_j > 80$ Å^2 should be discarded. Inclusion of significant numbers of solvent atoms, which meet all the requirements, should cause the R-factor to drop by 1–2% or more. Many solvent atoms are functionally important (see the example of Section 10.2) and do not therefore simply provide a means of improving R-factors.

The assignment of small ligands into the structural model follows a similar route to the above procedures for water molecules. Atom co-ordinates should be assigned and added to the data file for as many atoms as possible, paying full attention to the validity of the molecular geometry of the new atoms. As much information as possible should be incorporated into the procedure including:

(a) Known or preconceived stereochemistry of the group.

(b) Known or expected binding region of the protein.

(c) Juxtaposition of protein atoms with ligand atoms and validation of contact distances.

10 Examples of protein–ligand complex studies

10.1 N-acetyl-DL-homocysteinyl ribonuclease derivatives: protein–ligand covalent modification

This example is mentioned because of its potential applications for isomorphous replacement phasing (Section 8.2). The method involves introduction of a highly reactive group into the protein, which will bind easily to heavy metals. Two procedures are possible: 1) modification of the protein followed by crystallization; or 2) chemical modification applied to the actual crystals.

Method 1 is probably the most straightforward to carry out but may result in multiple heavy atom sites and eventually produce crystals which are non-isomorphous with native crystals. Method 2 will discourage the formation of multiple sites, as some will be blocked through crystal contacts, but may also be more difficult to apply to material in the crystalline state.

Bovine pancreatic ribonuclease can be modified in solution by reacting with N-acetyl-DL-homo-cysteinyl thiolactone (70, 71), which targets lysine residues (Figure 37). In the case of RNase it was possible to separate out two distinct products, one of which retained full enzymatic activity while the second was inactive. Avey and Shall (12) undertook X-ray analysis of mercury derivatives of these modified RNases and showed that the Hg atoms occupied distinct single sites. Both derivatives were subject to an increase in the b-axis of the unit cell (approximately 3%) impairing the quality of their IR phasing. One derivative

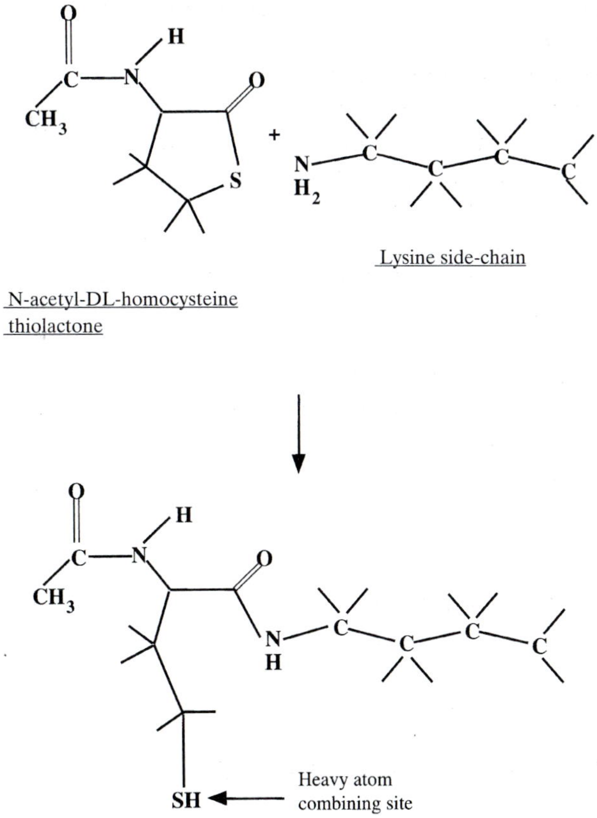

N-acetyl-DL-homocysteine
thiolactone

Lysine side-chain

Heavy atom
combining site

Figure 37 Reaction of *N*-acetyl-DL-homo-cysteinyl-thiolactone with a lysine residue.

(attached to Lys 41) (Plate 1) was found to be useful for the analysis at 2.5 Å
(13, 14).

10.2 Environment of the SO_4^{2-} ion and solvent waters in native bovine RNase A active site at 0.9 Å resolution (18)

Ribonucleases (RNases) are perhaps the most intensely studied of all enzymes.
For a recent review see ref. 72. Bovine RNase A was the first enzyme to be
crystallized for X-ray analysis (73), the crystals being noted for the outstanding
quality of their X-ray diffraction pattern. The unusually high quality of the X-ray
diffraction enabled C. H. ('Harry') Carlisle (see ref. 74) to visually measure over
30 000 I(*hkl*)'s recorded on X-ray film on a modified Weissenberg camera. The
structure, refined to 1.45 Å resolution was published by Borkakoti *et al.* (75, 76).
This analysis confirmed the existence of an SO_4^{2-} ion, located in the active site
close to where the phosphate moiety of inhibitors such as cytidylic acid (2'-CMP)
bind (17). Recently intensity data were measured at room temperature using a
Nonius FAST area detector (Section 5.2.3) with MoKα radiation, λ = 0.71073 Å on
the monoclinic crystals from ethanol with space group P2$_1$. A total of 96 000

reflections was used in the refinement, the model having 1072 atoms and parameters: 972 protein atoms refined anisotropically (Section 9.3.1) = 972×9 parameters, 100 solvent atoms refined isotropically (x, y, z, B) = 100×4 parameters, plus a scale factor = 9149 parameters, and therefore 6.3 reflections per parameter. The final R-factors were 12.94% for the data with $|F_o| > 4\sigma\ |Fo|$ (31000 reflections) and 19.91% for all data. Calculations were carried out on a PC486 computer. The structure was refined using SHELXL-93 (forerunner of SHELXL-97) starting from the 1.45 Å structure of Borkakoti *et al.* (75). A number of ordered solvent molecules both water (Plate 1) and ethanol were located from difference electron density maps, included in the co-ordinate data set and successfully refined by least-squares, with isotropic thermal displacement parameters. Also located in the active site close to His 12 and forming strong H-bonds to two protein residues and one of the water molecules, is the sulfate ion. The hydrogen bonds formed between the sulfate oxygens and Lys 41, His 119, and one of the water molecules are shown in Plate 1, in the active site. Note also that the His 119 ring occupies two positions. The resolution achieved in this analysis is exceptional. We can be confident that reliability of the positions of the atoms involved and associated interactions of the sulfate ion, water molecules and protein residues described is extremely high.

10.3 Enzyme–inhibitor interactions (non-covalent): r-rat RNase A binding with 2′,5′-CpG (85)

Rat ribonuclease, like bovine ribonuclease, is a pyrimidine specific enzyme. The binding of pyrimidine mononucleotide inhibitors to bovine RNase has been studied extensively by X-ray crystallography (see for example: refs. 17 and 77–84). RNases also interact with purines but the process is orders of magnitude slower than for pyrimidines. To date no crystallographic analysis involving a productively bound purine containing nucleotide has been reported. The successful expression of a mutant gene coding for rat pancreatic RNase (84% homology with bovine) has been described by Gupta (85). This work led to the crystallization and successful MR analysis of r-rat RNase using the PDB structure of bovine RNase A (Brookhaven assignment number pdb7rsa) (86) as the search model with AMoRE. The X-ray analysis was carried out at 2.5 Å resolution and refined with X-PLOR to R = 18.6%, R_f = 27.7% (starting values 40.2 and 41.1% respectively). Crystals of r-rat RNase complexed with the purine mononucleotide 2′,5′-CpG were prepared by soaking native crystals for 72 hours in their mother liquor containing 8 mM 2′,5′-CpG. After soaking, a significant change in the c-axis spacing was observed and the complex structure was solved using MR with the native structure as search model. The final R-factors after refinement at 3.3 Å were R = 26.9%, R_f = 28.1%. Electron density for the bound inhibitor was located on $(2|F_o| - |F_c|)$ maps (Section 7.4) and the inhibitor was modelled into the density (Plate 2) using the program TURBO-FRODO. The most significant results of this analysis are:

(a) The electron density clearly shows the flat guanine base and puckered sugar moieties. Density for the phosphate and cytidine moieties is much weaker (not shown in Plate 2) suggesting that there are no specific interactions between the protein and this end of the CpG molecule which protrudes externally into the solvent region.

(b) The $2',5'$-CpG molecule is bound to r-rat RNase in a very similar manner to that of $2',5'$-CpG and $3',5'$-dCpdG (82, 87) to bovine RNase. The binding is non-productive, i.e. not as a substrate would bind during the process of enzymatic cleavage. Furthermore a sulfate (or phosphate) ion is clearly seen to occupy the site where the PO_4 groups of inhibitors bind to bovine RNase (17, 77, 78, 80–84, 87). No rationale for this retrobinding mode (80), where the inhibitor is not bound in the active site has, as yet, emerged. The fit between the guanine base and the protein is however quite remarkable. A recent X-ray study of seminal bovine RNase revealing a similar retrobinding propensity in this member of the RNase superfamily (88; personal communication) may provide a clue to the eventual rationalization of this interesting phenomenon.

(c) H-bond contacts between the guanine base and protein residues are shown in Plate 2. Interestingly the base also forms strong interactions with the phosphate in the active site. The H-bond contact between the active site PO_4 and Lys 41 is the only one common to the SO_4–protein interactions found in bovine RNase (previous example).

This analysis, whilst providing a fascinating new example of the (as yet) unexplained phenomenon of retrobinding in RNases, is of quite low reliability, having been carried out at only 3.3 Å resolution. This is in contrast to example 10.2 at 0.9 Å resolution which provided an extremely detailed picture of the protein–ligand interactions.

10.4 Protein–carbohydrate interactions: the non-covalent sugar binding sites in the lectin MLI (89)

Lectins were first recognized as being present in the seeds of *Ricinus communis* by Herman Stillmark in 1888 at the University of Dorpat. *Ricinus* seed extracts were found to display haemagglutinating activity, hence ricin was called a phytohaemagglutinin. Lectins are proteins of non-immune origin that reversibly bind oligosaccharides or glycoconjugates (glycoproteins or glycolipids) or polysaccharides without modifying their covalent structure (90, 91). The interactions involved are mainly H-bonds with some, at least proposed, hydrophobic contacts being involved. Lectins can often be inhibited by simple sugars and consequently they are often classified on the basis of their sugar-binding properties. Lectins are employed in a number of important applications over a wide variety of immunological and biochemical fields. Moreover they can be regarded as model systems for studying the molecular basis of protein–carbohydrate interactions as they occur in cell–cell recognition events (for reviews see refs. 92, 93). The structures of several plant lectins, including some lectin–carbohydrate complexes, are known at high resolution (see ref. 94 for a review, and ref. 95).

MLI from *Viscum album* (mistletoe) like ricin is a ribosome inactivating protein

(a)

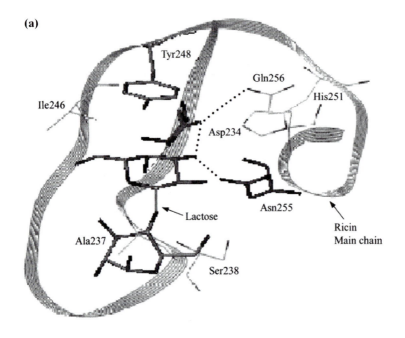

(b)

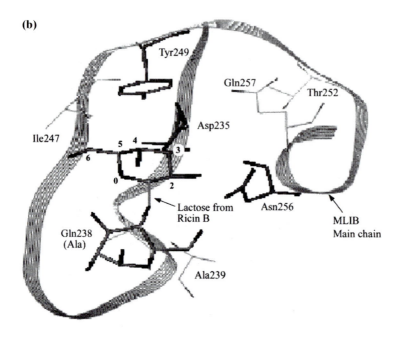

(type II RIP) comprising two protein chains: a toxic A-chain linked to a lectin B-chain by an S-S bridge. The crystallography of MLI has been discussed in Sections 2.3, 2.5, 6.6 (*Protocol 9*) and 8.6.6 (*Protocol 11*). The present X-ray analysis was based on synchrotron data at 2.9 Å, the limit of diffraction for these crystals. The R-factor is currently 22% (R_f 28%). A central point of interest is to establish

(c)

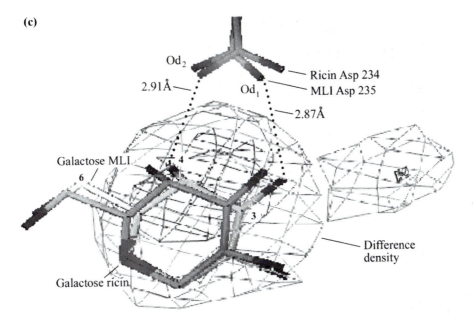

Figure 38 (a) Ricin B-chain sugar binding site 2. The bound lactose molecule and interactions with the protein molecule are indicated. (b) The corresponding site in MLI B-chain showing the ricin lactose modelled in. Note both the similarities and differences in sequence numbering and residue types. (c) Detail of the actual electron density in MLI and H-bonds to Asp 235. The Tyr at 248 in ricin and 249 in MLI may form interactions with the hydrophobic face of the sugar ring. A proposed hydrophobic contact between Ile 246 (ricin) and C6 may also be present in MLI at Ile 247. Similar hydrophobic contacts have been proposed as a general feature of lectin-sugar binding but require detailed quantification (see ref. 116).

the mode of binding of sugars to the B-chain as this is important for understanding how the lectin may attach to sugar residues on the surface of target cells to trigger endocytic uptake of the whole protein. The lectin B-chain of ricin is known to include two sugar-binding sites, in domains 1 (galactose specific) and 2 (both galactose and N-acetylgalactosamine–galNAc specific) respectively (46). Co-crystals of ricin with the disaccharide lactose were employed in this investigation which indicated the interaction points of B-chain atoms with the galactose moiety of lactose. *Figure 38a* shows the sugar binding site of ricin in domain 2, located in a shallow pocket comprising a contiguous single loop (1α) and strand (2γ).

We include this example to illustrate how the PDB database of protein structures can be a useful tool in the study of related proteins. Because of the extreme solvation of MLI crystals (*Protocol 9*) the X-ray diffraction pattern is relatively weak and the analysis is limited to 2.9 Å resolution. Using the ricin sugar binding site 2 as a reference (*Figure 38a*), it can be seen that in the MLI structure (*Figure 38b*), the main chain conformation in this region is very similar but there are differences in both the sequence numbering, and in the types of non-interacting amino acid residues in the vicinity, but not in the important sugar binding residues. These comparisons enabled the corresponding sugar binding site in MLI B-chain to be identified, and close inspection of the difference electron density in the MLI B-chain sugar binding site to be made. This revealed (*Figure 38c*), an isolated density peak with its centre at about 3.6 Å from the important Asp 235. By inserting the ricin galactose ring atoms into the MLI structure (*Figure 38c*) a starting point for refinement of the MLI galactose ring was established. The refined position of the MLI B-chain galactose ring with respect to Asp 235 is shown in *Figure 38c* with the H-bond distances achieved. As in ricin, Tyr 249 (actually Tyr 248 in ricin) is positioned appropriately for making hydrophobic contact with the OH-free side of the sugar ring. A further hydrophobic contact proposed between Ile 246 in ricin (Ile 245 in MLI) and C6 of the sugar also appears to be preserved. (*Figures 38a* and *38b*). Whereas MLI B-chain is both galactose and galNAc specific, the present crystals (Section 2.3) were grown in the presence of β-D-galactose as a stabilizing agent (8). The sugar binding site electron density observed in the current analysis may be weak because not all of the protein sites in the crystal are occupied. It may be possible to achieve better occupancy by adjusting the sugar concentration used for crystallization, but obviously this would be a very time-consuming experiment, as most steps in the analysis would have to be repeated.

10.5 Glycoproteins: covalently bound sugars and the [... Asn, X, Ser/Thr...] binding site; MLI B-chain Asn 136 sugar binding site (89)

In the previous example of MLI, the lectin B-chain in addition to exhibiting sites with specificity for non-covalent binding of sugars also contains sites with covalent sugar attachments and is consequently known as a glycoprotein. In accordance with the general rule that glycosylation sites in glycoproteins occur at triplet sequences of the type [...Asn, X, Ser/Thr...] the MLI B-chain electron density showed a clear peak close to Asn 136 (*Figure 39*). Modelling of a NAG sugar moiety into this density, the link Asn 136Nδ2–NAG C6 and the environment of this glycosylation site in the protein are shown in *Figures 39* and Plate 3. This sugar may be assumed to be part of a complex polysaccharide bound to the protein, regions of which further, from the protein surface are not visible, being disordered due to lack of stabilizing contacts in the crystal structure.

10.6 Protein–steroid interactions: Fab′ fragment of antibody DB3 bound to progesterone and progesterone-like steroids

The crystal structure of the Fab′ fragment of the monoclonal anti-progesterone antibody DB3 and the progesterone–Fab′ complex (16) provides an insight into the important types of interactions which occur between proteins and steroids.

The changes which take place in the antibody on binding to antigens and the

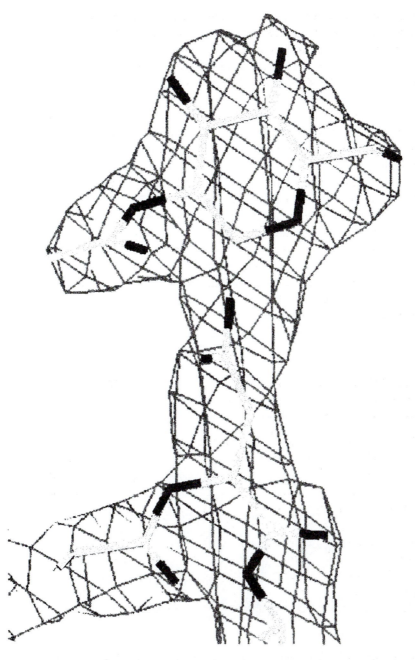

Figure 39 Electron density at the glycosylation site Asn 136 in mistletoe lectin MLI B-chain (Also refer to plate 3).

structural basis of antibody cross-reactivity, i.e. the ability of a binding site to accommodate antigens other than the original immunogen have to be investigated. The site clearly requires some plasticity, particularly in the case of steroids, which are somewhat rigid molecules. It is also of interest to examine in detail the construction of an antibody binding site which has the ability to recognize a relatively hydrophobic and insoluble antigen with high affinity and specificity. In the present case a panel of anti-progesterone mAbs, including DB3, was raised initially with a view to acquiring an understanding of the role of progesterone in pregnancy. The antibodies block pregnancy in mice and other animals through injection shortly after mating. DB3 has a high affinity for progesterone yet cross-reacts with structurally different steroid molecules including 5α-pregnanedione in which the A-ring is co-planar with the B-, C-, and D- rings, instead of being roughly 45° out of the plane, as in progesterone itself (96, 97). Large, stable X-ray quality crystals were grown at 22 °C by vapour diffusion (Section 2.3) from 1.7 M ammonium sulfate at pH 7.4, using 0.002% (w/v) PEG 20 000 and 1% (v/v) ethanol as additives, in 1–5 µl drops containing 5–25 mg/ml protein. Complex crystals were grown by first adding small amounts of each in solution (µg or pg amounts/ml). All crystals including the free Fab' were isomorphous, space group $P6_422$ with a = b = 134.76, c = 124.21 Å, Z = 12, solvent 62%. 17–19 000 reflections to between 2.8–2.7 Å resolution were measured for the various crystals.

The unliganded Fab' structure was solved by MR (Section 8.6). Fab structures are complex, comprising two polypeptide chains, H (heavy) and L (light). Both the H-chain and the L-chain fold into two domains: V_H (variable heavy) and C_{H1} (constant heavy 1); V_L (variable light) and C_L (constant light). The C_{H1}- and C_L-chains are linked by an S-S bridge. The C and V regions are related by an approximate (pseudo) 2-fold rotation or elbow angle, which varies between different Fab' structures. Antigen binding sites occur at the ends of the variable antibody domains. To achieve a MR solution search models of each of the four domains are extracted from the PDB. As search probes for MR, such models can give rise to problems, as they are known to exhibit significant structural differences. Whilst such multi-domain searches are possible the individual signals, even from ideal search models, are weak as each search model constitutes only a small fraction of the total structure. In this form of MR, it may therefore be necessary to employ a series of different search models and carry a large number of rotation solutions through to the translation stage. The use of variations in resolution and Patterson vector cut-off radii apply as usual (Section 8.6). In this case the unliganded Fab' structure was determined and refined to an R-value of 25.4% to 2.7 Å resolution. Studies of progesterone–Fab' and other isomorphous complexes were initiated using the phases calculated from the unliganded structure and then refined. The bound steroid molecules were located using omit maps. This technique is performed by removing atoms in the region of interest (for example protein side chain atoms exhibiting unusually high temperature factors) and then re-refining the structure. Subsequent calculation of $||F_o| - |F_c||$ and/or $[2|F_o| - |F_c|]$ electron density maps then enables re-modelling of the site to be carried out. Application of the above procedures in the enantiomorphous space group $P6_222$ failed to yield a solution to the structure the and space group $P6_422$ may therefore be assumed to be correct.

The Fab' antibody pocket was found to be highly complementary to progesterone, the steroid being almost completely buried in a region formed by the following residues: Asn^{H35}, $\underline{Trp}^{H47}$, $\underline{Trp}^{H50}$, $\underline{Gly}^{H95}$, Asp^{H96}, Tyr^{H97}, $\underline{Trp}^{H100}$, and $\underline{Phe}^{H100}$ (H-chain); His^{L27d}, Ser^{L91}, Ser^{L92}, His^{L93}, $\underline{Val}^{L94}$, and $\underline{Pro}^{L96}$. Of these 14 residues, 7 are hydrophobic (underlined). A total of <u>61</u> atomic interactions between the Fab' and progesterone were identified using cut-offs of 3.4 Å for H-bonds and 4.1 Å for van der Waals contacts, with 15 out of 23 non-H atoms of progesterone (*Figure 40a*) and 49 protein atoms being involved. The progesterone is bound in a narrow slot formed primarily by the indole side chains of $\underline{Trp}^{H50}$ and $\underline{Trp}^{H100}$. The 20-keto group attached to the D ring forms an H-bond with Asn^{H35} and the 3-keto group of the A-ring forms an H-bond with His^{L27d}. The progesterone molecule as a whole is sandwiched between the indole rings of $\underline{Trp}^{H50}$ and $\underline{Trp}^{H100}$. The progesterone binding is accompanied by movements of Asn^{H99} (r.m.s. deviations: 1.4 Å main chain, 2.3 Å side chain), and $\underline{Trp}^{H100}$ (r.m.s. deviations 1.1 Å main chain, 5.6 Å side chain). This large movement of $\underline{Trp}^{H100}$ opens up the binding site to receive the steroid. $\underline{Trp}^{H100}$ is also stabilized by stacking against Tyr^{L32}.

10.6.1 Simulation studies

As an extension to the above Fab'–progesterone studies, Lisgarten and Palmer (98) have carried out a series of molecular dynamics simulations employing synthetic progesterone analogues (99, 100) to investigate their propensity for binding to the antibody. Amongst these are the oxadiazole derivative HS1000 (*Figure 40b*) (101). Related structural studies carried out by Kayser *et al.* (102) involved detailed comparison of proton–proton distances in other oxadiazole steroid derivatives using X-ray, solution NMR, and *in vacuo* modelling studies employing the Tripos Force Field in the SYBYL software (103) Version 6.4. The use of complementary techniques for the study of intermolecular interactions is of great benefit in the rational assessment and design of novel therapeutic compounds. The Fab'–progesterone structures (16) enabled the modelling of HS1000 binding to the antibody as a back up to solution binding studies in which the inhibition of the Fab' protein with various ligands including HS compounds 963, 998, 1000, and 1003 was investigated (104) together with the synthetic steroid, *furazabol* (105). Of these only HS1000 produced significant inhibition, but this was seven orders of magnitude weaker than progesterone. HS1000 was thus selected for further SYBYL dynamics studies in parallel with similar calculations with Fab'–progesterone. The simulation was initiated by inserting the co-ordinates of HS1000 into the progesterone binding pocket, and adjusting the position of the steroid by manual optimization to replace the progesterone molecule. Charges were inserted onto HS1000 atoms through the SYBYL software and dynamics simulations were performed on a localized region of the complex structure by selecting all atoms within a 6 Å extension of the binding pocket. This strategy was necessary in order to reduce the calculations to a manageable time span on

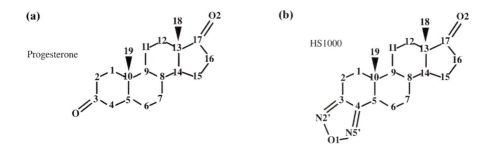

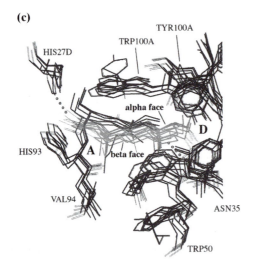

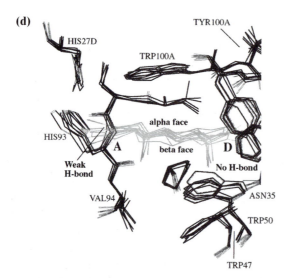

Figure 40 Chemical formulae and atom numbering schemes: (a) Progesterone and (b) HS1000. Overlays for 100 ps trajectories for progesterone (c) and HS1000 (d). The six superposed structures were observed at 20 ps intervals.

the computer (Silicon Graphics Iris Workstation). *Figures 40c* and *40d* show over-lays for 100 ps trajectories for progesterone and HS1000 respectively. The six superposed structures were observed at 20 ps intervals. The results of this study are as follows:

(a) Although some movement of the protein atoms is observed (*Figure 40c*), during the dynamics simulation for progesterone the essential H-bonds between C = O(3)– on the A-ring to His^{L27d}, and C = O(2)– on the steroid D-ring to Asn^{H35} are maintained throughout.

(b) Neither of these H-bonds is formed to HS1000. C = O(3)– on the A-ring is replaced in HS1000 by the oxadiazole ring. Rather than enhancing the bind-ing to the protein this has created a less flexible end to the steroid molecule which fails to bind. C = O(2)– on the steroid D-ring is present in HS1000 but is unable to form the H-bond with Asn^{H35} presumably because the extra bulk on the A-ring has displaced the whole steroid to an inaccessible position.

We conclude that the presence of the oxadiazole ring on ring-A of the synthetic steroids is counterproductive due to its bulk and lack of flexibility. It may be possible to enhance the binding to the Fab′ protein by employing a smaller, more flexible moiety to substitute O(3) such as NH_2, or COOH, both of which could still H-bond to His^{L27d} and this idea is currently under investigation.

10.7 Protein–nucleic acid interactions: procaryotic recognition of helix-turn-helix motifs in DNA (117)

Specialized DNA binding domains occur in proteins that regulate DNA tran-scription. These domains often take the form of a helix-turn-helix motif which binds specific regulatory regions of double helical DNA. Regulation can either be in the form of repression or activation. Detailed structural studies of viral bacteriophage lambda have been undertaken and we employ here the example of 434 Cro repressor N-terminal DNA binding domain protein complexed with a synthetic 20 base pair DNA fragment containing the corresponding operator OR1. [The PDB file 3CRO is supplied with some versions of the RASMOL graphics software. The accompanying figures were produced with RASMOL on a 486 PC.] The 434 Cro molecule employed in this study contains 71 amino acids with 48% sequence identity with the 69 residues of the N-terminal DNA binding domain of 434 repressor. The structure of each of the two 434 Cro fragments employed here comprise five α-helices, helices -2 and -3 in each forming the helix-turn-helix recognition motif, helix-2 (*Figure 41a*) contributing the principal protein–DNA interactions described below. There are two glutamine residues Gln 28 and Gln 29 at the beginning of the recognition helix which provide specific interactions between the 434 repressor fragment and the synthetic DNA fragment. Plate 4 shows details of the H-bond interactions between Gln 28 and the adenine base A1 (paired with thymine base T14'). There are also less specific interactions along two segments of the DNA sugar–phosphate backbone, notably H-bonds between the DNA phosphate groups and peptide backbone NH groups, the protein sub-

(a)

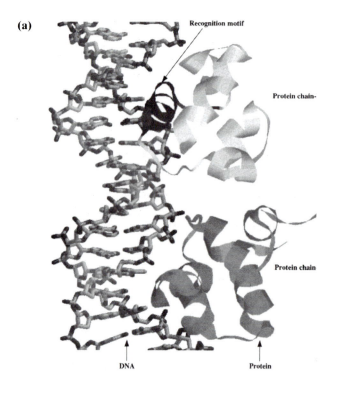

Recognition motif

Protein chain-

Protein chain

DNA   Protein

(b)

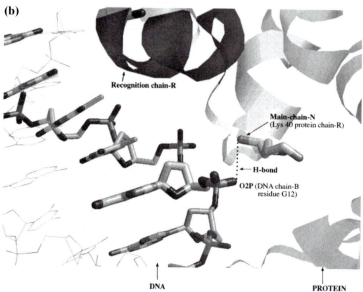

Recognition chain-R

Main-chain-N
(Lys 40 protein chain-R)

H-bond

O2P (DNA chain-B
residue G12)

DNA   PROTEIN

Figure 41 (a) Helix-turn-helix recognition motif of the 434 Cro fragment formed by helices-2 and -3 contributing the principal protein–DNA interactions. (b) H-bonds form between the DNA phosphate groups and peptide backbone NH groups occurring in residues 40–44 of the repressor. The H-bond NH[Lys 40(R)] to O2P (chain B, residue G12) is indicated. The protein subunit is located in the major DNA helix groove.

unit being located in the major DNA helix groove (*Figure 41a*). An example of this occurs in residues 40–44 of the repressor, the NH groups forming H-bonds to P9′ and P10′ (*Figure 41b*). The protein–DNA interactions described here are associated with distortions in the nucleic acid structure with respect to the well known B-DNA conformation.

11 Further recommended reading and software

11.1 Protein–peptide interactions: cyclophilin/cyclosporin complexes

The review by Kallen *et al.* (106) describes the X-ray structures of cyclosporin derivatives with cyclophilin A. These structures provide a structural database for the analysis of protein–ligand H-bonding, van der Waals interactions, and water structure in this biologically important system.

11.2 Protein–protein interactions (e.g. protein–antibody): camel V_H with lysozyme

The *Camelidae* is the only taxonomic family known to possess functional heavy chain antibodies, lacking light chains. The paper by Desmyter *et al.* (107) reports the 2.5 Å resolution crystal structure of a camel V_H complexed with its antigen lysozyme. The results are compared with the human and mouse V_H domains.

11.3 Recommended books

The following are recommended: McPherson (107), Carter (108), McRee (11), and Ladd and Palmer (21).

11.4 International tables

(I) '*International Tables for X-ray Crystallography*' Volumes I, II, III, and IV. Published for the International Union of Crystallography by The Kynoch Press, Birmingham, England, 1952.

(II) '*International Tables for Crystallography*' Volumes A, B, C, and D. Published for the International Union of Crystallography by Kluwer Academic Publishers, Dordrecht/Boston/London, 1989.

Many crystallographers still prefer the old Tables although the new versions are of course updated. The symmetry tables of Volume I are easier to use than Volume A. Volume I also contains useful structure factor expressions, not given in Volume A.

11.5 Synchrotrons Web Page

http://lmb.biop.ox.ac.uk/www/synchr.htt

11.6 Space group diagrams (J. K. Cockcroft)

http://img.cryst.bbk.ac.uk/local/Cockcroft/sgp/mainmenu.htm

11.7 Software for data processing

STRATEGY (27), DENZO (109); The HKL manual, edition 4 (110, 111); PREDICT (112)—see also the web page http://biop.ox.ac.uk/www/distrib/predict.html. The following programs are part of the 'CCP4 suite' which can be found on http://www.dl.ac.uk/CCP/CCP4/main.html: IMSILLS/ REFIX/ MOSFILM/ SORTMTZ/ ROTAVATA/ AGROVATA/POSTREFINEMENT/ TRUNCATE/ SCALEPACK/ SCALEPACK2MTZ/CAD/ XDISPLAYS/ HKLVIEW/ XDS/ MARXDS/ MARSCALE.

11.8 Software for Fourier and structure factor calculations

SFALL (structure factors) and FFT (fast Fourier calculation).

11.9 Software for MIR

CCP4 Programs: SCALEIT/ FHSCAL /WILSON /FFT / VECTORS HAVECS/ RSPS /VECSUM/ MLPHARE/ FFT /SHELX86/-S (64a).

11.10 Software for packing and molecular geometry

MOLPACK (41), PROCHECK (112), WHATCHECK (61). LIGPLOT (113) can be found on http: //www.biochem.ucl.ac.uk/bsm/ligplot/manual/index.html

11.11 Software for graphics and model building

FRODO (114). TURBO-FRODO (Bio-Graphics, Marseilles) can be found on http://almb.curs.mrs.fr.TURBO-FRODO/turbo.html

11.12 Software for molecular graphics and display

RASMOL (66), SETOR, MOLSCRIPT, BOBSCRIPT.

11.13 Software for refinement

X-PLOR (24), RESTRAIN (63), SHELXL-97 (64b) http://linux.uni-ac.gwdg.de/SHELX/ REFMAC.

11.14 Software for molecular dynamics and minimization

SYBYL Version 6.4 can be found on http://www.tripos.com

11.15 Databases

PDB (Protein Data Bank) http://www.rscb.org/pdb/transition_status.html

As of 1 July 1999 there were 9939 co-ordinate entries from 9237 proteins, 684 nucleic acids, and 12 for carbohydrates. BLAST (Basic Local Alignment Search Tool): this searches sequence databases for protein and DNA structures and sequences, and can be found on http://www.ncbi.nlm.nih.gov/BLAST. CCDC (Cambridge Crystallographic Data Centre) can be found on http://www.ccdc. cam.ac.uk. ReLiBase, which finds all ligands for a particular protein family can be found on http://www.pdb.bnl.gov:8081/home.http

References

1. Dounce, A. L. and Allen, P. Z. (1988). *Trends Biochem. Sci.*, **13**, 317.
2. Bernal, J. D. and Crowfoot, D. (1934). *Nature*, **133**, 794.
3. Matthews, B. W. (1968). *J. Mol. Biol.*, **33**, 491.
4. Rosemeyer, M. A. (1961). *PhD Thesis*, University of London. pp. 35–7.
5. King, M. V., Magdoff, B. S., Adelman, M. B., and Harker, D. (1956). *Acta Crystallogr.*, **9**, 460.
6. Vuillard, L., Rabilloud, T., Leberman, R., Berthet-Colominas, C., Madern, D., and Cusack, S. (1994). *FEBS Lett.*, **353**, 294.
7. Fowlis, W. W., Delucas, L. J., Twigg, P. J., Howard, S. B., Meehan, E. J., and Baird, J. K. (1988). *J. Cryst. Growth*, **90, no.1–3**, 117.
8. Sweeney, E. C., Palmer, R. A., and Pfüller, U. (1993). *J. Mol. Biol.*, **234**, 1279.
9. Petsko, G. A. (1985). In *Methods in enzymology*, Vol. 115, p. 147. Wyckoff, H., Hirs, C. H. W. & Timasheff, S. N. (eds) Academic Press.
10. Blundell, T. L. and Jenkins, J. A. (1977). *Chem. Soc. Rev.*, **6**, 139.
11. McRee, D. E. (1993). *Practical protein crystallography*. Academic Press, San Diego and London.
12. Avey, H. P. and Shall, S. (1969). *J. Mol. Biol.*, **43**, 341.
13. Carlisle, C. H., Gorinsky, B. A., Mazumdar, S. K., Palmer, R. A., and Yeates, D. G. R. (1974). *J. Mol. Biol.*, **85**, 1.
14. Carlisle, C. H., Palmer, R. A., and Mazumdar, S. K. (1994). *J. Chem. Crystallogr.*, **24**, 37.
15. Erskine, P. T., Senior, N., Awan, S., Lambert, R., Lewis, G., Tickle, I. J., *et al.* (1997). *Nature Struct. Biol.*, **4**, 1025.
16. Arevalo, J. H., Stura, E. A., Taussig, M. J., and Wilson, I. A. (1993). *J. Mol. Biol.*, **231**, 103.
17. Lisgarten, J. N., Gupta, V., Maes, D., Wyns, L., Zegers, I., Palmer, R. A., *et al.* (1993). *Acta Crystallogr.*, **D49**, 54.
18. Palmer, R. A., Karaulov, A., Mills, A., and Lisgarten, J. N. (2000). *Acta Crystallogr.*, **D** (submitted).
19. Sweeney, E. C., Tonevitsky, A. G., Palmer, R. A., Niwa, H., Pfüller, U., Eck, J., *et al.* (1998). *FEBS Lett.*, **431**, 367.
20. Niwa, H. (1999). Private communication.
21. Ladd, M. F. C. and Palmer, R. A. (1993). *Structure determination by X-ray crystallography* (3rd edn). Plenum Press, NY.
22. King, M. V., Bello, J., Pignataro, E. H., and Harker, D. (1962). *Acta Crystallogr.*, **15**, 144.
23. Cockcroft, J. K. (1999). Space group diagrams. http://img.cryst.bbk.ac.uk/local/Cockcroft/sgp/mainmenu.htm
24. Brünger, A. T. (1992). *Nature*, **355**, 472; Brünger, A. T. (1992). *X-PLOR Version 3.1: A system for X-ray crystallography and NMR*. Yale University Press, New Haven.
25. Garman, E. F. (1996). Chapter 4 in *Methods in molecular biology* (ed. C. Jones, B. Mulloy, and M. R. Sanderson), Vol. 56. Humana Press Inc., New Jersey, pp. 87–126.
26. Moy, J. P. (1994). *Nucl. Instr. Methods*, **A 348**, 641.
27. Ravelli, R. B. G., Sweet, R. M., Skinner, J. M., Duisenberg, J. M., and Kroon, J. (1997). *J. Appl. Cryst.*, **30**, 551.
28. Sweeney, E. C., Tonevitsky, A. G., Temiakov, D. E., Agapov, I. I., Saward, S., and Palmer, R. A. (1997). *Proteins: Struct. Funct. Genet.*, **28**, 586.
29. Matthews, B. W. (1974). In *The proteins* (ed. M. Neurath and R. L. Hill), 3rd edn. Academic Press, New York, London and San Francisco.
30. Matthews, B. W. (1974). *J. Mol. Biol.*, **82**, 513.
31. Monfort, W., Villafranca, J. E., Monzingo, A. F., Ernst, S. R., Katkin, B., Rutenber, E., *et al.* (1987). *J. Biol. Chem.*, **262**, 5398.

32. Cooley, J. W. and Tukey, J. W. (1965). *Math. Comput.*, **19**, 297.

33. Barrett, A. N. and Zwick, M. (1971). *Acta Crystallogr.*, **A27**, 6.

34. Ten Eyck, L. F. (1985). In *Methods in enzymology* Vol. 115, p. 324. Wyckoff, H., Hirs, C. H. W. & Timasheff, S. N. (eds) Academic Press.

35. Patterson, A. L. (1934). *Phys. Rev.*, **46**, 372.

36. Wood, S. P. (1990). In *Protein purification applications: a practical approach* (ed. Harris and Angal), pp. 45–59. IRL Press, Oxford.

37. Harker, D. (1956). *Acta Crystallogr.*, **9**, 1.

38. Blow, D. M. and Crick, F. H. C. (1959). *Acta Crystallogr.*, **12**, 794.

39a. Rossman, M. G. and Blow, D. M. (1961). *Acta Crystallogr.*, **14**, 641.

39b. Dickerson, R. E., Weintzierl, J. E., and Palmer, R. A. (1968). *Acta Crystallogr.*, **B24**, 997.

40. Rossmann, M. G. and Blow, D. M. (1962). *Acta Crystallogr.*, **15**, 24.

41. Wang, D., Driessen, H. P. C., and Tickle, I. J. (1991). *J. Mol. Graph.*, **9**, 50.

42. Wood, S. D., Wright, L. M., Reynolds, C. D., Rizkallah, P. J., Allen, A. K., Peumans, W. J., *et al.* (1999). *Acta Crystallogr.*, **D55**, 1264.

43. Tickle, I. J. and Driessen, P. C. (1996). In *Methods in molecular biology: crystallographic methods and protocols* (ed. C. Jones and M. Sanderson), Vol. 56, pp. 173–203. Humana Press Inc., Totowa, NJ.

44. Navaza, J. (1994). *Acta Crystallogr.*, **A50**, 157.

45. Brünger, A. T. (1990). *Acta Crystallogr.*, **A46**, 46.

46. Rutenber, E. and Robertus, J. (1991). *Proteins*, **10**, 260.

47. Katzin, B. J., Collins, E. J., and Robertus, J. D. (1991). *Proteins: Struct. Funct. Genet.*, **10**, 251.

48. Cowtan, K. (1994). *Joint CCP4 and ESRF-EACBM Newsletter on Protein Crystallography*, **31**, 34.

49. Wang, B. C. (1985). In *Methods in enzymology* Vol. 115, p. 90. Wyckoff, H., Hirs, C. H. W. & Timasheff, S. N. (eds) Academic Press.

50. Zhang, K. Y. J. and Main, P. (1990). *Acta Crystallogr.*, **A46**, 377.

51. Baker, D., Bystroff, C., Fletterick, R., and Agard, D. (1994). *Acta Crystallogr.*, **D49**, 429.

52. Swanson, S. (1994). *Acta Crystallogr.*, **D50**, 695.

53. Sayre, D. (1974). *Acta Crystallogr.*, **A30**, 180.

54. Bricogne, G. (1974). *Acta Crystallogr.*, **A30**, 395.

55. Schuller, D. (1996). *Acta Crystallogr.*, **D52**, 425.

56. Pauling, L. and Corey, R. B. (1951). *Proc. Natl. Acad. Sci. USA*, **37**, 729.

57. Engh, R. A. and Huber, R. (1991). *Acta Crystallogr.*, **A47**, 392.

58. Laskowski, R. A., MacArthur, M. W., Moss, D. S., and Thornton, J. M. (1993). *J. App. Cryst.*, **26**, 283.

59. Jones, T. A. (1985). *Acta Crystallogr.*, **A115**, 157.

60. Jones, T. A., Zou, Y.-J., Cowan, S. W., and Kjegaard, M. (1991). *Acta Crystallogr.*, **A47**, 110.

61. Hooft, R. W. W., Vriend, G., Sander, C., and Abola, E. E. (1996). *Nature*, **381**, 272. (WHATCHECK program).

62. Saward, S. (1999). *Refinement of the RCA structure at 2.9Å resolution* (Personal communication).

63. Driessen, H., Haneef, M. I. J., Harris, G. W., Howlin, B., Khan, G., and Moss, D. S. (1989). *J. Appl. Crystallogr.*, **22**, 510.

64a. Sheldrick, G. M. (1986). 'SHELX-86/-S: *Program for the solution of crystal structures*'. University of Göttingen, Germany.

64b. Sheldrick, G. M. (1997). 'SHELXL-97: *Program for the refinement of crystal structures*'. University of Göttingen, Germany.

65. Ramachandran, G. N., Ramakrishnan, C., and Sasisekheran, V. J. (1963). *J. Mol. Biol.*, **7**, 95.

66. Sayle, R. (1994). *RASMOL, a molecular visualisation program.* Glaxo Research and Development, UK.

67. Evans, S. (1993). *J. Mol. Graph.*, **11**, 134.

68. Kraulis, P. J. (1991). *J. Appl. Cryst.*, **24**, 946.

69. Esnouf, R. M. (1997). *J. Mol. Graph.*, **15**, 133.

70. Shall, S. and Barnard, E. A. (1967). *Nature*, **213**, 562.

71. Shall, S. and Barnard, E. A. (1969). *J. Mol. Biol.*, **41**, 237.

72. Raines, R. T. (1998). *Chem. Rev.*, **98**, 1045.

73. Kunitz, M. (1940). *J. Gen. Physiol.*, **24**, 15.

74. Bernal, J. D., Carlisle, C. H., and Rosemeyer, M. A. (1959). *Acta Crystallogr.*, **12**, 227.

75. Borkakoti, N., Moss, D., and Palmer, R. A. (1982). *Acta Crystallogr.*, **B38**, 2210.

76. Borkakoti, N., Moss, D. S., Stanford, M. J., and Palmer, R. A. (1984). *J. Cryst. Spec. Res.*, **14**, 467.

77. Palmer, R. A., Moss, D. S., Haneef, I., and Borkakoti, N. (1984). *Biochim. Biophys. Acta*, **785**, 81.

78. Borkakoti, N., Palmer, R. A., Haneef, I., and Moss, D. S. (1983). *J. Mol. Biol.*, **169**, 743.

79. Howlin, B., Harris, G., Borkakoti, N., Moss, D. S., and Palmer, R. A. (1987). *J. Mol. Biol.*, **196**, 159.

80. Aguilar, C. F., Thomas, P. J., Moss, D. S., and Palmer, R. A. (1991). *Biochim. Biophys. Acta*, **1118**, 6.

81. Mills, A., Gupta, V., Spink, N., Lisgarten, J. N., Palmer, R. A., and Wyns, L. (1992). *Acta Crystallogr.*, **B48**, 549.

82. Lisgarten, J. N., Maes, D., Wyns, L., Aguilar, C. F., and Palmer, R. A. (1995). *Acta Crystallogr.*, **D51**, 767.

83. Zegers, I., Maes, D., Minh-Hoa, D. T., Wyns, L., Portmans, F., and Palmer, R. A. (1994). *Protein Sci.*, **3**, 2327.

84. Zegers, I., Haikal, A. F., Palmer, R. A., and Wyns, L. (1994). *J. Biol. Chem.*, **269**, 127.

85. Gupta, V. (1999). PhD Thesis, VUB, Brussels, Belgium.

86. Wlodawer, A. and Sjölin, L. (1983). *Biochemistry*, **22**, 2720.

87. Aguilar, C. F., Thomas, P. J., Mills, A., Moss, D. S., and Palmer, R. A. (1992). *J. Mol. Biol.*, **224**, 265.

88. Mazzarella, L., Capasso, S., Demasi, D., Di Lorenzo, G., Mattia, C. A., and Zagari, A. (1993). *Acta Crystallogr.*, **D49**, 389.

89. Palmer and Niwa (2000). To be published

90. Kocourek, J. and Horejsi, V. (1981). *Nature*, **290**, 188.

91. Barondes, S. H. (1988). *Trends Biol. Sci.*, **13**, 480.

92. Lis, H. and Sharon, N. (1986). *Annu. Rev. Biochem.*, **53**, 35.

93. Lis, H. and Sharon, N. (1998). *Chem. Rev.*, **98**, 637.

94. Loris, R., Hamelryck, T., Bouckaert, J., and Wyns, L. (1998). *Biochim. Biophys. Acta*, **1383**, 9.

95. Wood, S. D., Wright, L. M., Reynolds, C. D., Rizkallah, P. J., Allen, A. K., Peumans, W. J., *et al.* (1999). *Acta Crystallogr.*, **D55**,1264.

96. Duax, W. L. and Norton, D. A. (1975). *Atlas of steroid stuctures I.* Plenum Publishing Corp., New York.

97. Griffin, J. F., Duax, W. L., and Weeks, C. M. (1984). *Atlas of steroid structures II.* Plenum Publishing Corp., New York.

98. Lisgarten, D. R. and Palmer, R. A. (2000). in press.

99. Singh, H., Yadav, M. R., and Jindal, D. P. (1987). *Indian J. Chem.*, **B26**, 95.

100. Yadav, M. R., Jindal, D. P., and Singh, H. (1998). *Indian J. Chem.*, **B27**, 205.

101. Lisgarten, D. R. and Palmer, R. A. (1999). *J. Chem. Cryst.*, **28**, 725.

102. Kayser, F., Maes, D., Wyns, L., Lisgarten, J., Palmer, R., Lisgarten, D., *et al.* (1995). *Steroids*, **60**, 713.

103. Clark, M., Cramer, R. D., and Van Opendosch, N. (1989). *J. Comp. Chem.*, **10**, 982.

104. Taussig, M. J. (1996). Personal communication.

105. Lisgarten, D. R. and Palmer, R. A. (1999). *J. Chem. Cryst.*, **28**, 379

106. Kallen, J., Mikol, V., and Walkinshaw, M. D. (1998). *J. Mol. Biol.*, **283**, 435.

107. Desmyter, A., Transue, T. R., Ghahroudi, M. A., Dao Thi, M.-W., Poortmans, F., Hamers, R., *et al.* (1996). *Nature Struct. Biol.*, **3**, 803.

107. McPherson, A. (1989). *Preparation and analysis of protein crystals.* Krieger Publishing.

108. Carter, W. (1990). *Protein and nucleic acid crystallization.* Academic Press, New York.

109. Gerwith, D. (1995). *The HKL Manual* 4th edn. In *Data collection and processing*: Proceedings of the CCP4 Study Weekend, 29–30 January 1993 (ed. L. Sawyer, N. Isaacs, and S. Bailey). SERC Daresbury Laboratory* DL/SC1/R34.

110. Otwinowski, Z. (1993). In *Data collection and processing*: Proceedings of the CCP4 Study Weekend 29–30 January 1993 (ed. L. Sawyer, N. Isaacs, and S. Bailey), pp. 56–62. SERC Daresbury Laboratory* DL/SC1/R34, Warrington, UK.

111. Sawyer, L., Isaacs, N., and Bailey, S. (ed.) (1993). *Data collection and processing*: Proceedings of the CCP4 Study Weekend 29–30 January 1993. SERC Daresbury Laboratory* DL/SC1/R34, Warrington, UK.

112. Laskowski, R. A., MacArthur, M. W., Moss, D. S., and Thornton, J. M. (1993). *J. App. Cryst.*, **26**(2), 283.

113. Laskowski, *et al.* (1995). Personal communication.

114a. Jones, T. A. (1985). *Acta Crystallogr.*, **A115**, 157.

114b. Jones, T. A., Zou, Y.-J., Cowan, S. W., and Kjegaard, M. (1991). *Acta Crystallogr.*, **A47**, 110.

115. Chattopadhyay, T. K., Lisgarten, J. N., Brechtel, R., Rudiger, H. and Palmer, R. A. (1999). *Acta Crystallogr.*, **D55**, 1589.

116. Elgavish, S. and Shaanan, B. (1998). *J. Mol. Biol.*, **277**, 917.

117. Mondragon, A. and Harrison, S. C. (1991). *J. Mol. Biol.*, **219**, 321.

Chapter 2
Molecular modelling

Romano T. Kroemer

Department of Chemistry, Queen Mary and Westfield College, University of London, Mile End Road, London E1 4NS, UK.

1 Introduction

Molecular modelling or computational chemistry have come a long way. From the era when they were restricted to a small number of scientists using very specialized and user-unfriendly software/hardware to the present where we are confronted with a vast amount of well integrated programs and relatively cheap hardware. The recent explosion of the Internet has added significantly to the amount of software available, as it is often accessible directly via the net. Nowadays, many of these programs can be used also by someone without programming expertise, and it is not necessary to have studied theoretical chemistry for years beforehand. However, the program in question should not be treated as a black box either and the user must have an understanding of the underlying theory and concepts.

This chapter is intended to give a flavour of the molecular modelling techniques used in the context of protein–ligand interactions. It is beyond the scope of this chapter to describe all programs/techniques in detail, and in many cases they will be only mentioned briefly. However, the aim is to provide the reader with an overview of some basic techniques in the area of protein–ligand modelling and to give some example applications using selected programs/techniques. The selection of these examples is rather arbitrary and reflects the author's experience and it is not intended to indicate any degree of superiority over other programs with the same or similar functionality. Reference to other programs will be given wherever possible.

In the context of modelling protein–ligand interactions one can think of a variety of issues, which are outlined in *Figure 1*. The first point is that for many applications one needs a protein structure to begin with. This structure can come from experimental determination (usually X-ray or NMR). In cases where there is no structure available, one can resort to the techniques of protein structure prediction. Usually, with the (protein) receptor structure in their hand, one can start to analyse the structure for potential binding sites and/or the characteristics of this site. In cases where the structure(s) of the protein(s) are available, but not the geometry of the complex, elucidating the mechanisms by which proteins or small molecule ligands dock is often required: here one tries to fit two molecules together in energetically favourable conformations, by application of computational procedures.

A further related problem has attracted considerable attention: The design of a new ligand from scratch (*de novo*) with the aim of interfering with a specific biological process. In some cases it is necessary to perform precise calculations or predictions of the binding energies, with the aim to distinguish between several possible ligands. Last but not least there are computational methods for the statistical analysis of binding affinities for a set of ligand molecules. Normally these latter procedures are applied when the receptor structure is not available, but the binding affinities of several different ligands are known.

Before considering how we address these topics in a practical sense it is worth pointing out that the boundaries between many of the topics (see the headings of *Figure 1*) are rather diffuse and that there is sometimes considerable overlap between them.

2 Protein structure prediction

The primary source of experimentally determined protein structures is the Brookhaven Protein Data Bank (PDB, www.rcsb.org) (1). At the time of writing

---

2. Get 3D structure of protein

- From experiment (X-ray, NMR) / database (PDB)
- From modelling

---

3. Analysis of binding sites and/or potential interactions

- Surface properties
- Electrostatic potential
- Molecular probes

---

4. Docking of ligands

- Dock selected ligands
- Scan database

---

5. *De novo* ligand design

- Generate new ligands
- Optimise existing ligands

---

6. Calculation of binding energies

- Free energy calculations

---

7. Statistical analysis of a set of protein ligands

- Pharmacophore analysis
- 3D QSAR

Figure 1 Topics related to the modelling of protein–ligand interactions. The numbers correspond to the sections in this chapter.

the PDB contained approximately 9000 protein structures. This is a considerable amount and the number of experimentally determined structures deposited here per year has been increasing significantly. However, the structure of the protein of interest is not always known and ventures like the human genome project produce many more sequences than solved structures. Therefore, one may need to resort to structure prediction techniques in order to obtain the requisite co-ordinates of the target protein or 'receptor'.

Every protein in its native functional state is folded in a specific way, but the mechanism by which a specific fold is adopted is not understood. The correlation between sequence and fold is low, and many different proteins can share similar folding patterns. Thus it has been suggested that some general features of the sequence determine at least partially the fold and that the number of these folding patterns (or families) is much smaller than the number of proteins (2–4). To date, three distinct categories of protein structure prediction methods exist.

2.1 *Ab initio* predictions

Here one starts from a sequence and tries to predict the secondary or tertiary structure of a protein. Secondary structure prediction methods rely on statistical analysis of known structures. Perhaps the most widely used algorithms are those of Chou and Fasman (5) and of Garnier *et al.* (6). More recent methods (7) are reported to have exceeded 70% three-state prediction (helix, sheet, and other) accuracy (8). Once secondary structural units of a protein have been assigned, one can then attempt to pack them together in three dimensions, using several evaluation criteria for the different packing modes generated. Such an approach was successfully used, for example, in the case of interleukin-4, prior to its experimental structure became available (9). The prediction of tertiary structure directly from the protein sequence is certainly the most ambitious task. Methods include semi-exhaustive searches of the main chain dihedral angles for small proteins and generation of the 3D structures from a predicted set of inter-residue contacts (10).

2.2 Threading, or fold recognition techniques

These techniques try to assess whether a given sequence is compatible with one of the structures in a database of known folds. They account for the possibility that two proteins may share a similar fold despite having no detectable sequence relationship. Since 1991, when the 'three-dimensional profiles' method was developed (11), a number of fold recognition techniques relying on different scoring functions have emerged (12–17).

2.3 Homology (or comparative) modelling

This is usually the method of choice when there is a clear relationship or homology between the sequence of a target protein and at least one known structure. This technique is based on the assumption that the tertiary structures of two proteins will be similar if their sequences are related, and it is the approach most likely to give accurate results.

At first sequences of proteins with known structure have to be identified. This can be achieved by sequence database searching (18, 19). Knowledge of the function or the family of the target protein might prove useful for the identification of homologues. The next task is the alignment of the target sequence with those of the known structure. This is a very important step in the procedure, as serious errors at this stage are very difficult to correct later. If the percentage identities between the compared sequences are high (> 45%) the correct sequence alignment is straightforward. When identity is low (< 25%) alignment becomes difficult. However, knowledge of the family fold (e.g. a four-helix bundle as in helical cytokines, associated with the presence of a hydrophobic core) may allow for a good alignment even in these cases (20).

Two widely used approaches to comparative modelling are discussed below.

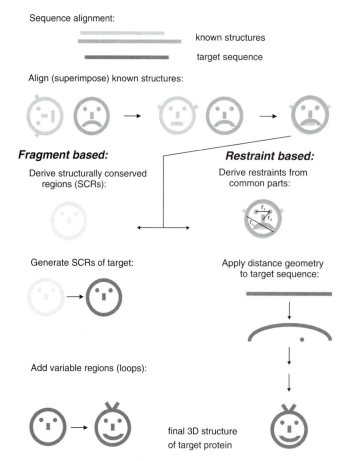

Figure 2 Outline of fragment-based and restraint-based comparative modelling.

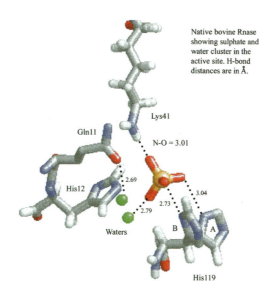

Native bovine Rnase showing sulphate and water cluster in the active site. H-bond distances are in Å.

Plate 1 Contacts to the SO_4^{2-} ion in the active site of native bovine RNase. The water cluster shown forms H-bonds to a sulfate oxygen and to residue Gln 11. His 12 is an established active site residue but does not form contacts to the sulfate. Direct H-bonds form between sulfate oxygens and residues Lys 41 and His 119. A further complication is that the His 119 ring is statistically disordered into two sites A and B respectively (75). On binding to mononucleotides His 119 adopts a single conformation. H-bond distances are shown. (Drawing by RASMOL (66); H atoms inserted by Chem-X.)

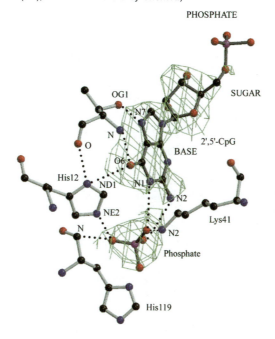

Plate 2 2′,5′-CpG $(2|F_o| - |F_c|)$ difference map for r-rat RNase with the inhibitor modelled into the electron density using the program TURBO-FRODO.

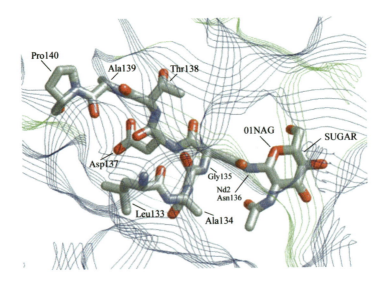

Plate 3 Mistletoe lectin B-chain: General view of the fitted NAG sugar moiety showing the mistletoe lectin MLI B-chain (blue strands) environment and in the background parts of the A-chain (green strands). Note the covalent link between Asn 136Nd2 and C1 of the sugar.

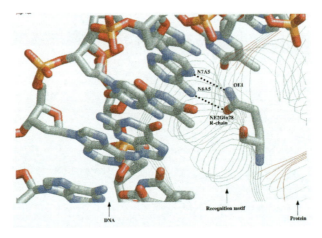

Plate 4 434 Cro repressor fragment: Details of the H-bond interactions between Gln 28 and the adenine base A5 (paired with thymine base T14′).

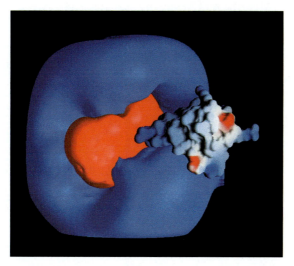

Plate 5 Electrostatic isopotential contours around trypsin. Contour levels: $-1k_BT$ (red) and $+1k_BT$ (blue). BPTI (trypsin inhibitor) is displayed with its molecular electrostatic potential mapped onto its surface representation.

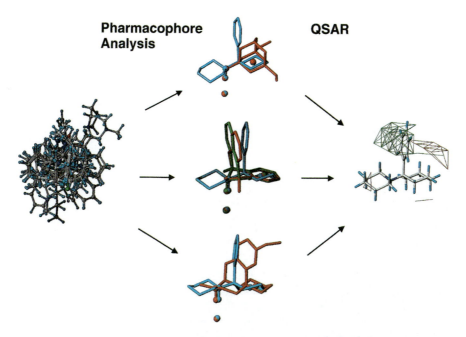

Pharmacophore Analysis

QSAR

Plate 6 Illustration of a 3D pharmacophore analysis combined with *3D QSAR*. At first a pharmacophore model is extracted from the set of randomly superimposed ligands. This model (represented by the spheres) can be used to superimpose the molecules in the data set (middle section). In the second step the superimposed ligands are subjected to a *3D QSAR* analysis. The outcome is represented by a grid map, which is shown in relation to a representative compound. Green contours illustrate a region where substituents increase binding to a putative receptor. Red contours indicate an area where bulky substituents interact unfavourably with the putative receptor.

Quantitative assessment of localised albumin binding to Endothelial Cells

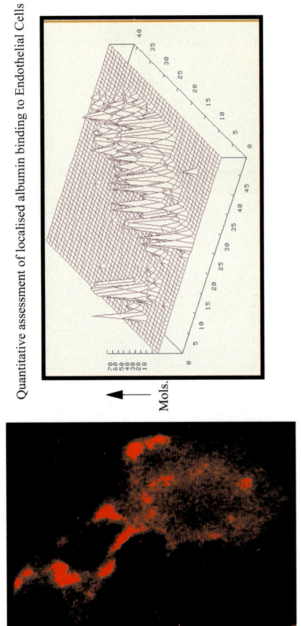

Mols.

Electrostatic difference image of albumin
bound to Endothelial Cells

Plate 7 The spatial localization of the interaction of serum albumin with single endothelial cells. The left-hand side figure indicates a fluorescence difference image of a human umbilical vein endothelial cell labelled with FPE in the absence and presence of 20 μm defatted albumin corrected for the spatial disposition of the dye. The plot shown on the right-hand side indicates that fluorescence changes as related to the surface electrostatic potential (e.g. Equation 12) are related to the surface charge density according to Equation 10 and thus to changes of the latter following the binding of charged molecules such as albumin. Thus as long as the charge on the molecule is known then the number of molecules that become bound at a particular location can be calculated as illustrated on the vertical axis.

2.3.1 Fragment-based comparative modelling

In this approach rigid fragments from other proteins are used to assemble the structure of the target protein (21–24). The main steps are:

(a) A common framework (structurally conserved regions, SCRs) (25) is derived from the superimposed known structures (*Figure 2*).

(b) The side chains of the target protein are 'projected' onto the SCRs, using the sequence alignment generated initially. Side chain conformations can be modelled with respect to the template structures, or—if not applicable—by comparison to rotamer libraries (26, 27).

(c) Variable regions (loops) can be added by employing database searches for suitable loop fragments (21, 28). Other approaches for adding this information include systematic search procedures or methods involving MD/MC simulations.

Frequently used for fragment-based modelling is the software COMPOSER, which was originally developed by T. L. Blundell and co-workers (29, 30). It is also incorporated in the software SYBYL (31). The use of COMPOSER is outlined in *Protocol 1*.

Protocol 1

Fragment-based modelling using SYBYL COMPOSER

Equipment

- SYBYL software including the composer module
- Unix workstation

Method

1 Select the sequence ('Select Sequence').

2 Identify homologous, structurally known proteins and align their sequences with the target sequence ('Find Homologs').

3 Assign seed residues for the start of an iterative 3D-alignment (fitting) procedure of the known structures ('Identify Seeds').[a]

4 Perform the 3D-alignment, i.e. superposition of the known structures, and extract the structurally conserved regions ('Align Structures').

5 Generate the structurally conserved core of the target protein ('Build SCRs').

6 Add the remaining parts of the target protein via database search ('Add Loops').

[a] Alternatively one can perform steps 1 and 2 with other software, i.e. searching in other databases and performing sequence alignments with different methods. In this case the user has to provide three files for step 3: A '.pir' file with the sequence of the target protein, a '.homol' file indicating the homologous sequences, and a '.homlog' file containing the alignment of these sequences with the target sequence. The best strategy in this case would be to initiate first a standard COMPOSER run, to modify the files that have been generated by the program accordingly, and to re-start the procedure with these files.

2.3.2 Restraint-based modelling

Predictions are based on restraints (such as inter-atomic distances) derived from homologous protein structures (32–35). The initial two steps are similar to those for fragment-based modelling, i.e. sequence alignment and structural alignment of the known structures. In the next step distance (and other) restraints are derived from the superimposed structures. A technique referred to as 'distance geometry' (36) is then applied to build an ensemble of models that satisfy the input restraints. An outline of the method is given in *Figure 2*.

An example of this procedure is the program MODELLER, which has been developed by A. Šali and T. L. Blundell (37). Restraints include structural properties at residue positions and relationships between residues. In addition to distance restraints they include solvent accessibility, secondary structure, and hydrogen bonding.

At the end of each structure prediction exercise it is necessary to perform some checks in order to ensure that the structure is free from strain, that there are no clashes between amino acids, and that the chiralities are correct. Energy refinement might be necessary in order to relieve strain and to alleviate slight irregularities in the structure. Verification of the prediction should include information from experiments, such as mutation data, if available.

3 Analysis of protein structures

Once a 3D structure has been obtained for the protein of interest, it can be analysed using different computational techniques. The purpose of such an analysis is many-fold. If the ligand binding site is not known it will be important to determine putative binding sites. Several computational procedures have been developed in order to detect clefts or cavities in proteins (38–41). These cavities can then be explored for binding of selected molecules (i.e. docking) or for the design of ligands. In many cases the binding site is known, and one wants to gain information on the properties of the binding site and how a ligand can bind.

The analysis of known interfaces has contributed significantly to our understanding of protein–protein or protein–ligand interfaces. S. Jones and J. M. Thornton discovered that the properties of the interfaces depend largely on the type of binding partners involved (42): molecules that exist both in complexes and as independent structures tend to form more polar interfaces in the complexes. Small molecules usually bind by docking into a cleft of their binding partner. These and other characteristics then served as a basis for the prediction of putative interfaces (43). Another study has revealed that many protein–protein interfaces are enriched in aromatic and aliphatic amino acids and depleted in certain charged residues (44). From a visual survey of 136 homodimeric proteins T. A. Larsen and co-workers concluded that a significant number of different types of protein–protein interface exist (45). They found that approximately one-third of the interfaces display a defined hydrophobic core, surrounded by a

ring of polar interactions. 61% of the complexes analysed, however, had interfaces composed of a mixture of hydrophobic patches and polar interactions.

The studies mentioned above indicate that it might be of interest to analyse
and display the properties of a binding site in more detail, prior to any docking
or design studies. A commonly used technique is to calculate a surface representation of the protein and to map certain properties such as hydrophobicity
or electrostatic potential onto it (see, e.g. 46–50). Calculation and graphical examination of the electrostatic potential around a molecule can provide other
useful information (51). The location of charged and polar groups in a protein
can have significant influence on the shape of the potential. As an example, this
has been demonstrated convincingly for the enzyme trypsin (Plate 5). In this
case the electrostatic potential around the molecule was calculated using the
finite difference Poisson–Boltzmann method (52). The potential contours reveal
how two proteins having both net positive charges are able to associate in a
complex.

Another procedure for analysing binding sites in proteins is to predict favourable positions for probes or small molecules (53, 54). One of the most popular
tools for this is the GRID program by Peter Goodford (55). Here the binding site
is embedded in a regular grid. A probe (atom, group, or small molecule such as
water) is then placed at the lattice intersections and the interaction energy between the probe and the protein is calculated using an empirical energy function. The resulting energy map can then be analysed for favourable interactions.
The program MCSS combines the analysis of the protein binding site with the
calculation of energetically favourable orientations of small functional groups
(56).

In many cases the analyses we have described so far in this chapter lead on
directly to the following methods: docking and *ab initio* design.

4 Docking of ligands to proteins

In molecular docking one attempts to find the preferred mode of binding of
ligands in a protein–ligand or protein-protein complex. The first approach of this
type was to manually explore this docking by using interactive graphics. This,
however, did not allow for systematic searching of the possible ligand–receptor
orientations. Since then a considerable number of docking methods have been
developed. Nowadays docking is usually performed in an automated fashion
and evaluation of the goodness of fit of the ligand is done by an energy scoring
function, rather than by eye.

In the docking process both the receptor and the ligand can be treated as
rigid structures. This reduces the degrees of freedom in the orientational search
significantly. However, in nature the key-lock principle is only valid for a flexible
lock and a flexible key. Therefore, in order to find a compromise between computational tractability and accuracy, a number of docking algorithms have been
developed where at least the ligand is treated as being flexible. Nevertheless,
rigid docking algorithms have also proven remarkably successful (57).

In principle all docking applications include four steps:

- Identification and preparation of the receptor site
- Preparation of the ligand(s)
- Docking the ligand(s)
- Evaluation of the docked orientations

Criteria for differentiating between the various docking methods include the way the receptor site is described, the type of ligand (small molecule and/or protein), the docking algorithm, and the evaluation procedure (scoring function). A selection of different docking programs/methods is given in *Table 1*.

As described in Section 2, there are three different routes for determining 3D structures of proteins: X-ray crystallography, high-resolution NMR, and homology modelling. The structures coming from these sources have to be sufficiently accurate in order to be of use for the docking experiments. For X-ray structures this implies that their resolution should be in the region of 2 Å or lower and the thermal factors ('R-factors') should be less than 30% (68). NMR structure determination normally results in an ensemble of structures. One can use the average of these structures, in which case the rmsd variation between the structures should be < 1.5 Å. An approach for using the entire ensemble of structures for

Table 1 Examples of docking programs

Program	Ref.	Site description	Search method	Scoring	Type of ligands	Flexibility
AUTODOCK	58	Grid points	MC, GA	FF	Small, proteins	Ligand
DOCK	59, 60	Spheres (site points)	Geometric matching	Shape, FF	Small, proteins	–
DOCKVISION	61	Atoms	MC, GA	FF	Small	Ligand
FLEXX	62	Interaction points	Fragment based	*LUDI* scoring function	Small	Quasiflexible ligand via discrete model
FLOG	63	Site points	Geometric filter	FF terms	Small	Quasiflexible ligand via multiconformer database
FTDOCK	64	Grid	Systematic	Shape, elec. complementarity	Proteins	–
GRAMM	65	Function	Systematic	Energy function	Small, proteins	–
LIGIN	66	Atomic surface	Geometric filter	Steric + Interact + H-bond	Small	Ligand
MULTIDOCK	67	Atoms	Side chain mean field optimization, rigid body minimization	FF	Proteins	Both

docking has been recently reported (69). Structures resulting from homology modelling are likely to have rmsd values higher than 0.7 Å for the C_α atoms (70) and larger errors can be expected for the side chains (71). Nevertheless, protein models structures have been successfully used for docking studies (72, 73).

Once a suitable protein structure has been selected for the docking process, one has to determine the region of interest for the docking procedure. Usually this is the active site of an enzyme, or the binding site of a receptor. In cases where the location of this site is not known, it may be identified by graphical inspection or by using an automated method (41, 59, 74). Many docking programs do not use an atomistic description of the binding site but rely on alternative representations. For example, the molecular surface of a protein binding site can be represented as an array of spheres (*Figure 3*). The centres of these spheres can then be used in the calculation for fitting a ligand atom in the docking process. Another approach would be to represent the binding site by a grid, where the grid points carry information about the interaction energies between probe atoms and the binding site (55).

The ligand(s) have to be considered next. This could involve pre-calculation of different conformers, assignment of rotatable bonds, building of a database of suitable fragments (in order to introduce flexibility), or generation of a grid-representation (discretisation) of the molecule(s). Other tasks may include the evaluation of partial charges for the ligand atoms.

After the preparation has been done, the examination of the docking process can start. Again, depending on the program used, a variety of strategies are possible. The docking step can be performed by matching (fitting) ligand atoms to pre-calculated site points, where the latter represent potentially favourable interactions. By matching the ligand atoms to these points different orientations of the ligand in the protein are generated. Grid or systematic searches fit the ligand into the active site by rotating and translating the ligand in discrete steps. Fragment-based methods can introduce flexibility by docking several ligand fragments in favourable orientations and subsequently reconnecting them. The fragment methods can be used for the evaluation of existing inhibitors or can be applied to so-called *ab initio* ligand design. Last but not least, kinetic docking methods can fit ligands to receptor sites by exploring the potential energy

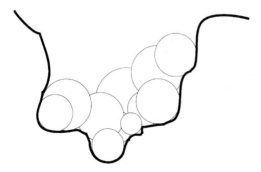

Figure 3 Preparation of the binding site: Representation by spheres.

surface. This is achieved by application of molecular dynamics or Monte Carlo (simulated annealing) procedures. An advantage of these latter methods is that they somehow mimic the actual docking process, i.e. the approach of a ligand molecule to the receptor, followed by binding.

Docking methods are able to generate a large number of possible receptor–ligand orientations. Some of the methods (e.g. the MD or MC procedures) already include the evaluation step because they contain energetic functions. In other methods the different orientations and conformations generated have to be evaluated in a separate step. The 'goodness of fit' can be determined from a variety of scoring functions, based either on geometric or on energetic criteria. Geometry-based scoring functions attempt to evaluate shape or surface complementarity between the receptor and the ligand(s). Examples include evaluation of a correlation function for the grid representations of the two molecules (64) or counting the number of receptor atoms within a specified distance of every ligand atom (75). For energy-based scoring the property of interest is the free energy of binding. However, accurate calculation of this quantity is often not tractable computationally and simplifications have to be introduced. A common approach is to consider only the enthalpy of binding, calculated by a standard force field such as *AMBER* (76) or the *MMFF94* force field (77). In addition to the standard force field terms (such as Lennard–Jones and Coulomb potentials) many of the docking programs include empirical terms for solvation, hydrophobicity, or hydrogen bonding, with the aim to find an approximate value for the free energy of binding.

As an example we illustrate the use of the program *AUTODOCK* (58) in more detail (*Protocol 2*). AutoDock uses a Monte Carlo simulated annealing technique for configurational exploration. The ligand is treated as being flexible by assigning user-specified torsions as rotatable. The degrees of freedom (including internal torsions) of the substrate molecule are then modified randomly by a small amount, starting from the previous configuration. Each of these random orientations is evaluated energetically. If the move is energetically downhill, it is accepted. In case the energy has risen compared to the last configuration, the step is accepted only after comparison with a Boltzmann factor. At high enough temperatures, almost all steps are accepted. At lower temperatures, fewer high energy structures are accepted. This implies that the substrate molecule performs an energy-guided random walk in the space around or in the receptor site. Rapid energy evaluation is achieved by pre-calculating atomic affinity potentials for each atom type in the ligand molecule (55). A probe atom is placed at each point of a rectangular grid around the protein/active site. The interaction energy of the probe atom (typically H, C, O, and N) with the receptor is calculated at each grid point and assigned to it. Steric interactions are calculated using a Lennard–Jones potential (a so-called '12-6' potential), hydrogen bonds are calculated by application of a separate potential (referred to as a '12–10' potential). The electrostatic grid can be evaluated either using a probe charge of +1 and a coulomb potential or by solving the linearized Poisson–Boltzmann equation (51). Solvent screening can be modelled as well (78).

Protocol 2

Docking using *AUTODOCK* version 3.0

Equipment

- UNIX workstation
- *AUTODOCK* software

Method

1   Prepare a file (extension .pdbq) containing the co-ordinates of the receptor (in pdb-format), including polar hydrogens and partial atomic charges.[a]

2   Assign atomic solvation parameters and create a .pdbqs file, which contains co-ordinates, charges, and solvation parameters. (*mol2topdbqs, addsol*)[b]

3   Define ligand torsions and generate the ligand .pdbq file. (*deftors*)

4   Use the receptor .pdbqs and the ligand .pdbq files in order to create the grid parameter file (.gpf) and the docking parameter file (.dpf). (*mkgpf3, mkdpf3*)

5   Calculate the grid maps. (*autogrid3*)

6   Perform the docking. (*autodock3*)

7   Create a pdb formatted file containing all docked conformations. (*get-docked*)

8   View the results using a molecular modelling program.

[a] A .pdbq file has standard PDB format, with the exception that it contains partial atomic charges in columns 71–76 of the file.

[b] Steps 1 and 2 can be performed simultaneously as follows: Use *SYBYL* to add hydrogens ('Biopolymer', 'essential_only') and partial charges to the protein structure of interest. Save the structure as a file with the extension .mol2, use the *AUTODOCK* mol2topdbqs program to create the .pdbqs file.

The way that different ligands are predicted to fit into a receptor provides then a rationale for selecting certain compounds for synthesis. The geometries of docked complexes can also be used as starting points for further studies, such as ligand design (c.f. next section) or *3D QSAR* (79).

The scanning of entire databases in order to identify putative ligands for a known receptor has been reported as well (73, 80) and will become more important with the advent of large compound libraries generated by combinatorial chemistry.

5 *Ab initio* (*de novo*) ligand design

A central problem in current drug design research is to find an inhibitor starting with a known macromolecular binding site. Many procedures have appeared which accomplish this task in an automated fashion. These programs attempt to generate novel ligands from scratch (*ab initio*) to fit the receptor site. A wide variety of methods for generating hypothetical ligand structures exist. The main differences between the programs are related to the main stages of an *ab initio* design procedure as follows:

- Preparation of the active site/derivation of constraints
- Generation of ligand structures
- Evaluation of ligand structures

In the following the reader will find a quick run-down of a number of different procedures available. For further information he is referred to the preceding chapter of this volume (Chapter 1).

5.1 Active site preparation

After identification of an active site (c.f. Section 3 in this chapter) it usually is analysed further in order to extract information necessary for setting up an *ab initio* design procedure. The objectives of such an analysis are to identify particular types of potentially favourable interactions with the receptor (such as hydrogen bonds, hydrophobic or charged interactions) and/or to prepare for efficient algorithms for ligand evaluation. In principle there are two types of approach for this:

1 Potential-based methods use interaction energies calculated by application of a force field in order to derive information about the active site. Representative algorithms of this approach are the programs *GRID* (55), *GREEN* (54), and *GROUP-BUILD* (81). In all these methods a grid is defined within the active site. Different types of probe are placed at the lattice intersections and the interaction energy between the probe and the protein is then computed. The force field used in the computation may include Lennard–Jones, Coulomb, and hydrogen bonding terms. Another method for exploring the interactions with the active site is incorporated in the maximum common substructure (MCSS) procedure (56). At first the active site is filled with many copies of randomly oriented functional groups. Examples of these functional groups include acetic acid, methyl ammonium, and water. A combination of energy minimization and quenched dynamics procedures is then used to determine energetically favourable positions and orientations of the groups within the active site. The minimization is constrained so that the functional groups see only the protein, and not each other. The resulting output is then a functionality map indicating the sites of favourable interactions with the protein. Similar approaches have been incorporated in other programs (82, 83).

2 Rule-based or knowledge-based procedures apply knowledge about geometrical features of ligand–protein interactions in order to explore the active site. These rules are derived from statistical analyses of known structures of protein–ligand or protein–protein complexes. For example information about preferred hydrogen bonding geometries is extracted. The active site is then examined and these rules are applied. In *LUDI* (84), for example, sets of vectors are generated around hydrogen donor or acceptor sites. The orientation of the vectors reflects distances

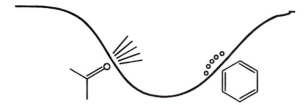

Figure 4 Active site characterization using hydrogen-bond vectors and lipophilic interaction points.

and angles of potential hydrogen bonds at a particular site. Hydrophobic interactions are represented by single points (*Figure 4*). Other programs use similar approaches (85, 86). The vectors and interaction points provide then constraints for the placement of atoms, fragments, or ligands during the next step, the 'design procedure'.

5.2 Generation of structures

For the actual generation of structures within the active site a multitude of different methods exists. One way to differentiate between the approaches would be to ask the question: Growing or linking? Another criterion for classification would be whether the methods follow a deterministic or a stochastic approach.

The linking approach implies that at first several fragments are placed independently into the active site. In the next step linking elements are sought that connect these fragments into a complete molecule. An example for such a strategy is the program *HOOK* (87), which starts with the output generated by *MCSS*. *MCSS* generates a set of functional group sites making favourable interactions with the protein (56). *HOOK* then attempts to link these groups together with molecular skeletons taken from a pre-computed database. The results are new molecules containing multiple functional groups (the original fragments) in potentially favourable positions. Other programs employing similar strategies are *LUDI* (84), *NEWLEAD* (88), and *SPROUT* (89). The advantage of these procedures is that the original fragments can be placed in an optimal fashion into the active sites. A potential problem, however, is to find a suitable linker.

In the growing approach a so-called seed fragment is placed into the binding site. Subsequently, additional fragments are added in a stepwise fashion. The program GROW was originally developed for the design of peptides using amino acid building blocks (90). Here the structures 'grow' from an acetyl group placed as the 'seed corn' into the active site. Subsequent amino acid building blocks are added stepwise via an amide bond linkage. The building blocks are taken from a database of amino acids in various low energy conformations. An advantage of the 'grow' approach is that the resulting structures are more likely to be synthetically accessible. Other variations on this theme are implemented in the programs *LEGEND* (91) and *GROUPBUILD* (81). The program *LUDI* is an example of a method that can build molecules either through linking or through growing (84).

All the algorithms mentions so far contain deterministic features. Earlier constructs predetermine those that follow in a given design sequence, i.e. later constructs are in response not only to the binding site but also to earlier constructs. A continuous metamorphic/non-serial construction of ligands would favour a more exhaustive structural exploration of regiochemical and configurational space. With these ideas in mind a series of stochastic approaches involving molecular dynamics or Monte Carlo procedures and random changes in the structures have been designed (82, 83, 92, 93). Usually all the programs (deterministic or stochastic) use empirical scoring functions or force fields for the evaluation of the structures generated. Recently a method has been reported which combines a stochastic approach with a semi-empirical quantum mechanical Hamiltonian for the assessment of ligand structures (94). In this latter approach ligands are generated on an atomistic basis rather than on a fragment basis. This permits the construction of higher energy transitional structures that bridge between viable ligand structures. Changes are accomplished by random choice to create, destroy, move, or change a randomly selected atom. A semi-empirical Hamiltonian is used to calculate the internal energy of the ligand and its interaction with the receptor. The feasibility of such an approach has been demonstrated using two examples.

5.3 Evaluation of structures

There are several reasons why the evaluation of the structures in *ab initio* ligand design is an essential part of the procedure. First, in the sequential approaches, evaluation provides guidelines for further steps in the design sequence. Secondly, once a number of potential ligands have been generated, it is of interest to produce some kind of ranking of these ligands, in order to decide which of them should be synthesized first. Thirdly, most of the programs are able to generate many thousands of different ligands, due to the number of possible combinations of fragments to form a complete structure. Therefore, a variety of scoring schemes has been devised. Some of these schemes are applied during the design procedure; others are used in a post-processing manner.

Simple scoring schemes rank molecules according to molecular weight, number of atoms, etc. Another way to classify molecules is to consider their synthetic accessibility. In this case the availability of starting materials and the complexity of the molecules can be evaluated in a rule-based fashion (86). Some programs prioritize ligands based on the number of pharmacophoric groups (such as hydrogen bond donors/acceptors, lipophilic sites, etc.) they contain (85). Other methods apply rule-based scoring functions derived from the analysis of protein–ligand complexes (89, 95, 96). A function of this type is given as follows (95):

$$\Delta G_{bind} = 5.4 - 4.7 \sum_{h\text{-}bonds} f(r,0) - 8.3 \sum_{ionic} g(r,0) - 0.17\, A_{lipophilic} + 1.4\, N_{rotatable} \qquad [1]$$

The elements in this function for the estimation of the free energy of binding (ΔG_{bind}) are the number of hydrogen bonds between ligand and receptor (*f* refers to the geometry of the H-bonds), the number of ionic interactions, the lipophilic contact area (A), and the number of rotatable bonds in the ligand. Some programs incorporate mixed functions containing force field like terms and other descriptors, such as rotatable bonds or number of lipophilic contacts (97).

Another example for a *de novo* ligand design program is the *LEAPFROG* module in *SYBYL* (31). It combines a number of ideas incorporated also in the programs considered above. *LEAPFROG* follows a deterministic approach, by repeatedly making some structural changes, and then either keeping or discarding the results, depending on the evolution of the ligand during the run. As a start it can process two different types of input: A receptor structure (from experiment or modelling) or, alternatively, a pharmacophore model. After selection of the input the program can operate in three different modes (c.f. *Protocol 3*): The *OPTIMIZE* mode aims at improving existing ligand structures. In the *DREAM* mode novel ligands are suggested. In the *GUIDE* mode the user can interfere interactively with the design process. New ligand structures are evaluated mainly on *their binding energy relative to their immediate precursor using an approximation of the* GRID procedure (55). Synthetic difficulty can be included in the scoring scheme. The user can decide on the trade-off between variety and quality of the output ligand structures. Therefore, the user may decide to emphasize in a first run the variety. Subsequently, distinctively different structures can be chosen from the output in order to be optimized in a second run. At the end of a run the structures are saved in a *SYBYL* database and are referenced by a molecular spreadsheet containing the *LEAPFROG* binding energies.

Protocol 3

Setting up an *ab initio* ligand design procedure using *LEAPFROG*

Equipment

- *SYBYL* software including the *LEAPFROG* module
- Unix workstation

Method

1 Start LEAPFROG. In the dialog box choose an input structure ('cavity molecule').

2 Select the operating mode ('Guide', 'Optimise', 'Dream', c.f. text).

3 Select whether to run the program interactively or in background.

4 If necessary alter other input data such as the fragment database used ('Data').

5 Choose either to calculate or to read (from a previous run) the sitepoint/box description.[a]

6 Choose starting ligands.[b]

7 Start the run by pressing 'OK'.

[a] In case sitepoints are calculated one can choose to consider only hydrogen bonding and charged atoms in the receptor ('Charged') or to calculate also lipophilic interactions ('All').

[b] In case one wants to build on a previous run (e.g. the 'Optimise' mode) a database with the previously generated structures can be chosen. For the 'Dream' and 'Optimise' modes no starting ligands ('None') are selected by default.

When using any of the design techniques described above the user must bear in mind that they incorporate only crude approximations of the binding energetics and events. They are, however, excellent tools for providing novel ideas in inhibitor design. The most promising candidates have to be synthesized and tested. The test results can then be fed back into the design procedure in order to initiate a new, hopefully improved, design cycle.

6 Calculation of binding energies

Protein–ligand and protein–protein associations are processes which involve flexible molecules and which take place in solution. Therefore, in order to calculate the free energy of binding, the solvent has to be included and conformational changes must be accounted for. The free energy of binding (ΔG_{bind}) is an ensemble average and formally not related to a particular structure. Existing methods to calculate the free energy of binding can be divided into three groups.

Free energy simulations represent a comprehensive approach to the problem. Molecular dynamics or Monte Carlo techniques (98, 99) are used to sample the commonly occurring low-energy structures that give the dominant contribution to the thermodynamic quantities in question. Such simulations typically include an explicit representation of the atoms of both the molecules and the solvent. Due to the necessity for sampling a large number of configurations for a given system they are rather time-consuming, but they can give very accurate results (101–106). In practice the standard free energy of binding can then be calculated with the 'double-annihilation' method (107), applying thermodynamic integration or free energy perturbation methods. A recently proposed modification of this method has been termed 'double-decoupling' (108).

Energy component models are another group of methods for calculating binding affinities which are applicable to biomolecules. These methods avoid the time-consuming problem of conformational sampling almost entirely. Instead, phenomenological scoring functions are applied to rigid or nearly rigid representations of the molecules under investigation (109–111). These scoring functions take into account a number of distinct contributions to the binding free energy. These models are computationally tractable but may be less accurate than the more detailed free energy simulations. The basis of the models is the description of complex formation between two molecules A and B by a thermo-dynamic cycle (*Figure 5*). From this thermodynamic cycle it follows for the free energy of association ΔG:

$$\Delta G = \Delta G^{gas} + \Delta G^{solv.} \qquad [2]$$

with

$$\Delta G^{solv.} = \Delta G_A^{solv.} + \Delta G_B^{solv.} - \Delta G_{AB}^{solv.} \qquad [3]$$

Each of these terms contains different contributions and grouping them together one can write alternatively (109):

$$\Delta G = \Delta G^{elec.} + \Delta G^{cav.} + \Delta G^{conf.} + \Delta G^{vdW} + \Delta G^{rt} \qquad [4]$$

$\Delta G^{elec.}$ contains both the electrostatic solute–solute and the solute–solvent inter-actions. $\Delta G^{cav.}$ refers to the cavitation free energy on binding and can be defined as the free energy required to form a solute-sized cavity in the solvent when all interactions (dispersion and electrostatic) are switched off. $\Delta G^{conf.}$ is the loss of side chain conformational entropy on binding ($T\Delta S$). The van der Waals energy (ΔG^{vdW}) is often neglected, under the assumption that the van der Waals inter-actions at the interface of the associated complex are equal to those with the solvent molecules in dissociated form. The last term (ΔG^{rt}) represents the loss of translational and rotational entropy on complex formation. A number of tech-niques for calculation of all these terms has been reported (52, 112–121).

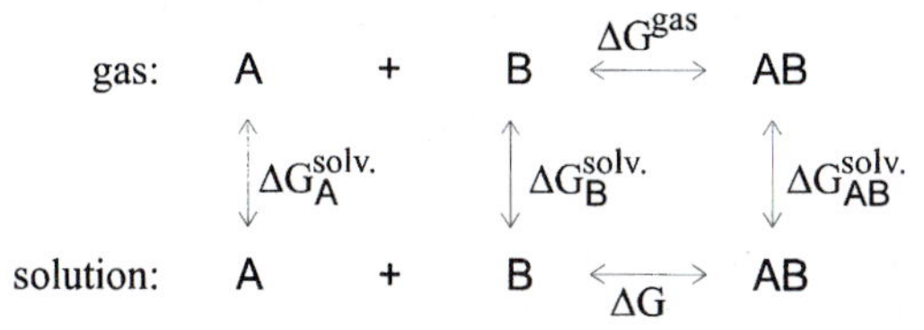

Figure 5 Thermodynamic cycle for complex formation. $\Delta G^{solv.}$ refers to the free energy of solvation. ΔG^{gas} is the free energy of association in the gas phase.

The third method for calculating absolute free energies of binding, referred to as the linear interaction energy (LIE) model, is a hybrid method including elements of the former two. The method was introduced by Åquist *et al.* (122) and employs a linear response approximation to calculate absolute binding free energies as follows:

$$\Delta G_{bind} = \beta \langle \Delta E^{elec.} \rangle + \alpha \langle \Delta E^{vdW} \rangle \qquad [5]$$

$\langle \Delta E \rangle$ is the energy difference between average contributions from ligand–solvent and protein–ligand interactions for the bound an unbound states, as derived from short simulations. α and β represent empirical parameters derived by fitting energy components from a set of ligands to experimental binding affinities. Re-cently, extensions/modifications of this method have been reported (123–125).

7 Statistical analysis of a set of protein ligands

Sometimes affinity data for a set of protein ligands are available, but it is not possible to obtain a reliable receptor structure. However, it would be desirable to perform some kind of rational manipulation of the ligands, with the aim to design improved structures. In this case one has to resort to other methods for modelling potential protein–ligand interactions and to predict novel ligands with higher binding affinities. Most commonly applied under these circumstances are pharmacophore modelling and QSAR (quantitative structure–activity relationship) techniques.

The functionalities of a molecule that are essential for exhibiting its pharmacological activity (i.e. usually binding to a receptor) are referred to as pharmacophores. The idea behind pharmacophore modelling is to analyse a set of molecules and to extract common features for biologically active compounds. The assumption is that these features make favourable interactions with the receptor. In case three-dimensional representations of the molecules are chosen for analysis, the result includes not only the type of functionalities but also their relative orientation or directionality. An excellent overview of pharmacophore modelling and its applications is given in A. K. Ghose's and J. J. Wendoloski's review (126).

QSAR techniques attempt to derive quantitative relationships between the features of a set of ligands and (usually) their binding affinities. Traditional *QSAR* analyses the influence of steric, hydrophobic, and electrostatic effects of substituents on biological activity. It identifies which of these are the dominant features behind the change in biological properties. Statistical evidence for the validity of the proposed relationships is provided. The descriptors used are not directly related to 3D structure and are extrapolated from one reaction to another.

In *3D QSAR* one calculates the descriptive properties directly from the 3D representations of the molecules. Usually these descriptors are calculated in such a way that their 3D distribution is retained in the model itself. The output of a *3D QSAR* analysis is often represented as 3D graphics, which is superimposed on the molecules of the data set. This representation makes it easy to recognize which properties of the molecules investigated are beneficial for binding and which ones are detrimental.

The best known *3D QSAR* method is comparative molecular field analysis, or *CoMFA* (127). In this procedure steric and electrostatic interaction energies between a probe atom and a set of superimposed molecules are calculated at the surrounding points of a predefined grid. The rationale behind this is that the probe atom corresponds to a receptor atom and scans the superimposed molecules for favourable and unfavourable interactions with a putative receptor. The calculated energy values are then correlated with some property of the compounds (usually biological activity). Once a reliable correlation has been found, the *QSAR* model can be used for predicting the activities of novel compounds.

Plate 6 illustrates how a 3D pharmacophore analysis and 3D QSAR can be combined (128, 129). Recent advances in *3D QSAR* and related techniques have been summarized in two volumes of a book series (130) and in a number of papers (131).

8 Concluding remarks

Although many of the computational methods presented in this chapter incorporate various approximations and simplifications, they work remarkably well, as shown by the many examples where experimental results could be reproduced or predicted with good accuracy. The usefulness of these approaches has also been demonstrated by some success stories in the development of new drug molecules (see, e.g. 73, 132, 133). Also, it has become clear that application of these procedures is most useful in combination with experiments, either to obtain confirmation from the latter or to provide guidelines for new ones. In many instances this has proven to be highly synergistic.

The future looks bright for molecular modelling in the area of protein–ligand interactions. New hardware with novel architecture and faster processors and improved generations of software will advance the field significantly. Recent developments such as the advent of combinatorial chemistry will lead to novel computational methods, for example combinatorial docking and virtual screening (134, 135).

Acknowledgements

R. T. K. gratefully acknowledges the help of Martin Parretti for the preparation of Plate 5.

References

1. Bernstein, F. C., Koetzle, T. F., Williams, G. J. B., Meyer, E. F., Brice, M. D., Rodgers, J. R., *et al.* (1977). *J. Mol. Biol.*, **112**, 535.
2. Richardson, J. S. (1977). *Adv. Protein Chem.*, **343**, 167.
3. Ptitsyn, O. B. and Finkelstein, A. V. (1980). *Q. Rev. Biophys.*, **13**, 339.
4. Chothia, C. (1992). *Nature*, **357**, 543.
5. Chou, P. Y. and Fasman, G. D. (1974). *Biochemistry*, **13**, 222.
6. Garnier, J., Osguthorpe, D. J., and Robson, B. (1978). *J. Mol. Biol.*, **120**, 97.
7. Rost, B. and Sander, C. (1993). *J. Mol. Biol.*, **232**, 584.
8. Rost, B. and Sander, C. P. (1995). *Proteins*, **23**, 295.
9. Curtis, B. M., Presnell, S. R., Srinivasan, S., Sassenfeld, H., Klinke, R., Jeffery, E., *et al.* (1991). *Proteins*, **11**, 111.
10. Defay, T. and Cohen, F. E. (1995). *Proteins*, **23**, 431.
11. Bowie, J. U., Luthy, R., and Eisenberg, E. (1991). *Science*, **253**, 164.
12. Godzik, A., Kolinski, A., and Skolnick, J. (1992). *J. Mol. Biol.*, **227**, 227.
13. Jones, D. T., Taylor, W. R., and Thornton, J. M. (1992). *Nature*, **358**, 86.
14. Sippl, M. J. and Weitckus, S. (1992). *Proteins*, **13**, 258.
15. Ouzounis, C., Sander C., Scharf, M., and Schneider, R. (1993). *J. Mol. Biol.*, **232**, 805.
16. Bryant, S. H. and Lawrence, C. E. (1993). *Proteins*, **16**, 92.

17. Krogh, A., Brown, M., Mian, I. S., Solander, K., and Haussler, D. (1994). *J. Mol. Biol.*, **235**, 1501.

18. Altschul, S. F., Gish, W., Miller, W., Myers, E. W., and Lipman, D. J. (1990). *J. Mol. Biol.*, **215**, 403.

19. Pearson, W. R. and Lipman, D. J. (1988). *Proc. Natl. Acad. Sci. USA*, **85**, 2444.

20. Kroemer, R. T., Doughty, S. W., Robinson, A. J., and Richards, W. G. (1996). *Protein Eng.*, **9**, 493.

21. Jones, T. H. and Thirup, S. (1986). *EMBO J.*, **5**, 819.

22. Unger, R., Harel, D., Wherland, S., and Sussman, J. L. (1989). *Proteins*, **5**, 355.

23. Claessens, M., VanCutsem, E., Lasters, I., and Wodak, S. (1989). *Protein Eng.*, **4**, 335.

24. Levitt, M. (1992). *J. Mol. Biol.*, **226**, 507.

25. Sutcliffe, M. J., Haneef, I., Carney, D., and Blundell, T. L. (1987). *Protein Eng.*, **1**, 377.

26. Sutcliffe, M. J., Hayes, F. R., and Blundell, T. L. (1987). *Protein Eng.*, **1**, 385.

27. Ponder, J. W. and Richards, F. M. (1987). *J. Mol. Biol.*, **193**, 775.

28. Moult, J. and James, M. N. (1986). *Proteins*, **1**, 146.

29. Blundell, T. L., Sibanda, B. L., Sternberg, M. J. E., and Thornton, J. M. (1987). *Nature*, **326**, 347.

30. Blundell, T. L., Carney, D., Gardner, S., Hayes, F., Howlin, B., Hubbard, T., *et al.* (1988). *Eur. J. Biochem.*, **172**, 513.

31. SYBYL6.5, Tripos Inc., St. Louis. http://www.tripos.com.

32. Srinivasan, S., March, C. J., and Sudarsanam, S. (1993). *Protein Sci.*, **2**, 277.

33. Brocklehurst, S. M. and Perham, R. N. (1993). *Protein Sci.*, **2**, 626.

34. Fujiyoshi-Yoneda, T., Yoneda, S., Kitamura, K., Amisaki, T., Ikeda, K., Inoue, M., *et al.* (1991). *Protein Eng.*, **4**, 443.

35. Friedrichs, M. S., Goldstein, R. A., and Wolynes, B. G. (1991). *J. Mol. Biol.*, **222**, 1013.

36. Crippen, G. M. and Havel, T. F. (1988). *Distance geometry and molecular conformation.* Chemometrics Research Studies Series 15. New York, Wiley.

37. Šali, A. and Blundell, T. L. (1993). *J. Mol. Biol.*, **234**, 779.

38. Connolly, M. L. (1992). *Biopolymers*, **32**, 1215.

39. Lewis, R. A. (1989). *J. Comput.-Aided Mol. Design*, **3**, 133.

40. Kleywegt, G. J. and Jones, T. A. (1994). *Acta Crystallogr. Sect. D*, **50**, 178.

41. Hendlich, M., Rippmann, F., and Barnickel, G. (1997). *J. Mol. Graphics Model.*, **15**, 359.

42. Jones, S. and Thornton, J. M. (1997). *J. Mol. Biol.*, **272**, 121.

43. Jones, S. and Thornton, J. M. (1997). *J. Mol. Biol.*, **272**, 133.

44. LoConte, L., Chothia, C., and Janin, J. (1999). *J. Mol. Biol.*, **285**, 2177.

45. Larsen, T. A., Olson, A. J., and Goodsell, D. S. (1998). *Structure*, **6**, 421.

46. Wireko, F. C., Kellogg, G. E., and Abraham, D. J. (1991). *J. Med. Chem.*, **34**, 758.

47. Kellog, G. E., Semus, S. F., and Abraham, D. J. (1991). *J. Comput.-Aided Mol. Design*, **5**, 545.

48. Nicholls, A., Sharp, K. A., and Honig, B. H. (1991). *Proteins*, **11**, 281.

49. Sanner, M. F., Olson, A. J., and Spehner, J.-C. (1996). *Biopolymers*, **38**, 305.

50. Duncan, B. S., Macke, T. J., and Olson, A. J. (1995). *J. Mol. Graphics*, **13**, 271.

51. Honig, B. H. and Nicholls, A. (1995). *Science*, **268**, 1144.

52. Warwicker, J. and Watson, H. C. (1982). *J. Mol. Biol.*, **157**, 671.

53. Boobbyer, D. N. A., Goodford, P. J., McWhinnie, P. M., and Wade, R. C. (1989). *J. Med. Chem.*, **32**, 1083.

54. Tomioka, N. and Itai, A. (1994). *J. Comput.-Aided Mol. Design*, **8**, 347.

55. Goodford, P. J. (1985). *J. Med. Chem.*, **28**, 849.

56. Miranker, A. and Karplus, M. (1991). *Proteins*, **11**, 29.

57. Shoichet, B. K., Stroud, R. M., Santi, D. V., Kuntz, I. D., and Perry, K. M. (1993). *Science*, **259**, 1445.

58. Morris, G. M., Goodsell, D. S., Halliday, R. S., Huey, R., Hart, W. E., Belew, R. K., *et al.* (1998). *J. Comp. Chem.*, **19**, 1639. http://www.scripps.edu/pub/olson-web/dock/autodock/

59. Kuntz, I. D., Blaney, J. M., Oatley, S. J., Langridge, R., and Ferrin, T. E. (1982). *J. Mol. Biol.*, **161**, 269. http://www.cmpharm.ucsf.edu/kuntz/dock.html

60. Oshiro, C. M. and Kuntz I. D. (1995). *J. Comput.-Aided Mol. Design*, **9**, 113.

61. Hart, T. N., Ness, S. R., and Read, R. J. (1997). *Proteins*, **S1**, 205. http://www.dockvision.com/

62. Rarey, M., Kramer, B., Lengauer, T., and Klebe, G. (1996). *J. Mol. Biol.*, **261**, 470. http://cartan.gmd.de/FlexX/

63. Miller, M. D., Kearsley, S. K., Underwood, D. J., and Sheridan, R. P. (1994). *J. Comput.-Aided Mol. Design*, **8**, 153.

64. Gabb, H. A., Jackson, R. M., and Sternberg, M. J. E. (1997). *J. Mol. Biol.*, **272**, 106. http://www.bmm.icnet.uk/ftdock/ftdock.html

65. Vakser, I. A. (1997). *Proteins*, **S1**, 226. http://reco3.musc.edu/gramm/index.html

66. Sobolev, V., Wade, R. C., Vriend, G., and Edelman, M. (1996). *Proteins*, **25**, 120. http://swift.embl-heidelberg.de/ligin/

67. Jackson, R. M., Gabb, H. A., and Sternberg, M. J. E. (1998). *J. Mol. Biol.*, **276**, 265. http://www.bmm.icnet.uk/multidock/multidock.html

68. Stroud, R. M. and Fauman, E. B. (1995). *Protein Sci.*, **4**, 2392.

69. Knegtel, R., Oshiro, C., and Kuntz, I. (1997). *J. Mol. Biol.*, **266**, 424.

70. Srinivasan, N. and Blundell, T. L. (1993). *Protein Eng.*, **6**, 501.

71. Šali, A. (1995). *Curr. Opin. Biotechnol.*, **6**, 437.

72. Ring, C. S., Sun, E., McKerrow, J. H., Lee, G. K., Rosenthal, P. J., Kuntz, I. D., *et al.* (1993). *Proc. Natl. Acad. Sci. USA*, **90**, 3583.

73. Li, R., Chen, X., Gong, B., Selzer, P. M., Li, Z., Davidson, E., *et al.* (1996). *Bioorg. Med. Chem.*, **4**, 1421.

74. Liang, J., Edelsbrunner, H., and Woodward, C. (1998). *Protein Sci.*, **7**, 1884.

75. Shoichet, B. K., Bodian, D. L., and Kuntz, I. D. (1992). *J. Comp. Chem.*, **13**, 380.

76. Cornell, W. D., Cieplak, P., Bayly, C. I., Gould, I. R., Merz, K. M., Ferguson, D. M., *et al.* (1995). *J. Am. Chem. Soc.*, **117**, 5179.

77. Halgren, T. A. (1996). *J. Comp. Chem.*, **17**, 490.

78. Mehler, E. L. and Solmajer, T. (1991). *Protein Eng.*, **4**, 903.

79. Gamper, A. M., Winger, R. H., Liedl, K. R., Sotriffer, C. A., Varga, J. M., Kroemer, R. T., *et al.* (1996). *J. Med. Chem.*, **39**, 3882.

80. Böhm, H.-J. (1994). *J. Comput.-Aided Mol. Design*, **8**, 623.

81. Rotstein, S. H. and Murcko, M. A. (1993). *J. Med. Chem.*, **36**, 1700.

82. Pearlman, D. A. and Murcko, M. A. (1993). *J. Comp. Chem.*, **14**, 1184.

83. Bohacek, R. S. and McMartin, C. (1994). *J. Am. Chem. Soc.*, **116**, 5560.

84. Böhm, H.-J. (1992). *J. Comput.-Aided Mol. Design*, **6**, 61.

85. Clark, D. E., Frenkel, D., Levy, S. A., Li, J., Murray, C. W., Robson, B., *et al.* (1995). *J. Comput.-Aided Mol. Design*, **9**, 13.

86. Gillet, V. J., Myatt, G. J., Zsoldos, Z., and Johnson, A. P. (1995). *Perspect. Drug Discov. Design*, **3**, 34.

87. Eisen, M. B., Wiley, D. C., Karplus, M., and Hubbard, R. E. (1994). *Proteins*, **19**, 199.

88. Tshinke, V. and Cohen, N. C. (1993). *J. Med. Chem.*, **36**, 3863.

89. Gillet, V. J., Johnson, A. P., Mata, P., Sike, S., and Williams, P. (1993). *J. Comput.-Aided Mol. Design*, **7**, 127.

90. Moon, J. B. and Howe, W. J. (1991). *Proteins*, **11**, 314.

91. Nishibata, Y. and Itai, A. (1991). *Tetrahedron*, **47**, 8985.

92. Pearlman, D. A. and Murcko, M. A. (1996). *J. Med. Chem.*, **39**, 1651.

93. Miranker, A. and Karplus, M. (1995). *Proteins*, **23**, 472.

94. Rothman, J. H. and Kroemer, R. T. (1997). *J. Mol. Model.*, **3**, 261.

95. Böhm, H.-J. (1994). *J. Comput.-Aided Mol. Design*, **8**, 243.

96. Böhm, H.-J. (1998). *J. Comput.-Aided Mol. Design*, **12**, 309.

97. Head, R. D., Smythe, M. L., Oprea, T. I., Waller, C. L., Green, S. M., and Marshall, G. R. (1996). *J. Am. Chem. Soc.*, **118**, 3959.

98. Leach, A. R. (1996). *Molecular modelling, principles and applications*. Longman, Harlow, UK.

99. Allen, M. P. and Tildesley, D. J. (1987). *Computer simulation of liquids*. Clarendon Press, Oxford, UK.

100. Tembe, B. L. and McCammon, J. A. (1984). *Comput. Chem.*, **8**, 281.

101. Jorgensen, W. L., Buckner, J. K., Boudon, S., and Tirado-Rives, J. (1988). *J. Chem. Phys.*, **89**, 3742.

102. Wong, C. and McCammon, J. A. (1986). *J. Am. Chem. Soc.*, **109**, 3830.

103. Straatsma, T. P. and McCammon, J. A. (1992). *Annu. Rev. Phys. Chem.*, **43**, 407.

104. Kollman, P. (1993). *Chem. Rev.*, **93**, 2395.

105. Mitchell, M. J. and McCammon, J. A. (1991). *J. Comput. Chem.*, **12**, 271.

106. Tidor, B. and Karplus, M. (1994). *J. Mol. Biol.*, **238**, 405.

107. Pranta, J. and Jorgensen, W. L. (1991). *Tetrahedron*, **47**, 2491.

108. Gilson, M. K., Given, J. A., Bush, B. L., and McCammon, J. A. (1997). *Biophys. J.*, **72**, 1047.

109. Chothia, C. and Janin, J. (1975). *Nature*, **256**, 705.

110. Searle, M. S., Williams, D. H., and Gerhard, U. (1992). *J. Am. Chem. Soc.*, **114**, 10697.

111. Weng, Z., Vajda, S., and DeLisi, C. (1996). *Protein Sci.*, **5**, 614.

112. Klapper, I., Hagstrom, R., Fine, R., Sharp, K. A., and Honig, B. (1986). *Proteins*, **1**, 47.

113. Gilson, M. K. and Honig, B. (1988). *Proteins*, **4**, 7.

114. Jackson, R. M. and Sternberg, M. J. E. (1995). *J. Mol. Biol.*, **250**, 258.

115. Gilson, M. K., Sharp, K. A., and Honig, B. (1988). *J. Comp. Chem.*, **9**, 327.

116. Jackson, R. M. and Sternberg, M. J. E. (1994). *Protein Eng.*, **7**, 371.

117. Janin, J. (1995). *Proteins*, **21**, 30.

118. Janin, J. (1995). *Prog. Biophys. Mol. Biol.*, **64**, 145.

119. Creamer, T. P. and Rose, G. D. (1992). *Proc. Natl. Acad. Sci. USA*, **89**, 5937.

120. Sternberg, M. J. E. and Chickos, J. S. (1994). *Protein Eng.*, **7**, 149.

121. Finkelstein, A. V. and Janin, J. (1989). *Protein Eng.*, **3**, 1.

122. Åquist, J., Medina, C., and Samuelsson, J. E. (1994). *Protein Eng.*, **7**, 385.

123. Carlson, H. A. and Jorgensen, W. L. (1995). *J. Phys. Chem.*, **99**, 10667.

124. McDonald, N. A., Carlson, H. A., and Jorgensen, W. L. (1997). *J. Phys. Org. Chem.*, **10**, 563.

125. Hansson, T., Marelius, J., and Åquist, J. (1998). *J. Comput.-Aided Mol. Design*, **12**, 27.

126. Ghose, A. K. and Wendoloski, J. J. (1998). *Perspect. Drug Discov. Design*, **9–11**, 253.

127. Cramer, R. D., III, Patterson, D. E., and Bunce, J. D. (1988). *J. Am. Chem. Soc.*, **110**, 5959.

128. Kroemer, R. T., Hecht, P., Guessregen, S., and Liedl, K. R. (1998). *Perspect. Drug Discov. Design*, **12**, 41.

129. Kroemer, R. T., Koutsilieri, E., Hecht, P., Liedl, K. R., Riederer, P., and Kornhuber, J. (1998). *J. Med. Chem.*, **41**, 393.

130. Kubinyi, H., Folkers, G., and Martin, Y. C. (ed.) (1998). *3D QSAR in drug design*, Vols. 2 and 3. KLUWER/ESCOM, Dordrecht, The Netherlands.

131. (1998). *Perspect. Drug Discov. Design*, Volumes 9–14.

132. Hong, H., Neamati, N., Wang, S., Nicklaus, M. C., Mazumder, A., Zhao, H., *et al.* (1997). *J. Med. Chem.*, **40**, 930.

133. Von Itzstein, M., Wu, W. Y., Kok, G. B., Pegg, M. S., Dyason, J. C., Jin, B., *et al.* (1993). *Nature*, **363**, 418.

134. Sun, Y., Ewing, T. J. A., Skillman, A. G., and Kuntz, I. D. (1998). *J. Comput.-Aided Mol. Design*, **12**, 597.

135. Bures, M. G. and Martin, Y. C. (1998). *Curr. Opin. Chem. Biol.*, **2**, 376.

Chapter 3
Circular dichroism

Alex F. Drake

Franklin-Wilkins Building, Department of Pharmacy, Kings College, Stamford Street, London SE1 8WA, UK.

1 Introduction

Circular dichroism (optical activity) can only be observed for chiral molecules or in the case of molecule–ligand binding only if at least one of the interacting species is chiral. This is usually true in biochemistry particularly if a biological macromolecule such as a protein/peptide is involved. A natural protein/peptide is built from amino acid residues that are preferentially of one hand (normally L-configuration). Monitoring processes by circular dichroism (CD) offers the following main advantages:

(a) Induced circular dichroism occurs only as the result of interaction. In the extreme, a non-chiral molecule does not display any optical activity and CD is only observed on contact with a chiral molecule.

(b) The CD technique is, from a spectroscopic point-of-view, much more select-ive than the related absorption technique.

(c) In certain cases CD can be used for structure determination (absolute stereo-chemistry assignment and protein secondary structure analysis). CD provides an excellent method to assess, at least, the global change in protein folding as the result of ligand interaction.

The concentration range for CD measurements is similar to absorption spectro-photometry although allowance needs to be made for noise, which can be reduced by signal averaging.

Optical activity is often defined as the rotation of the plane of plane polarized light by a chiral molecule. A more complete definition is that optical activity is the result of a differential interaction of a chiral molecule with left and right circularly polarized light. A differential absorption of the two circular polariza-tions by a chiral sample is known as circular dichroism, a differential speed (re-fractive index) gives rise to optical rotation. Plane polarized light (better known as linearly polarized light) is composed of two equal intensity, in-phase, circularly polarized components, one left circular and the other right circular. Differential speeds (refractive indices) of these two components will lead to rotation of the

123

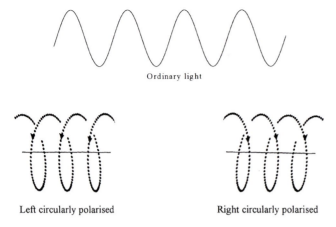

Ordinary light

Left circularly polarised   Right circularly polarised

Figure 1 Linearly and circularly polarized light.

plane of the linearly polarized light. Differential absorption of the two circular components will lead to the linearly polarized light becoming elliptical with the orientation of the major axis of the ellipse defining the optical rotation (*Figure 1*).

Optical rotation offers a means of monitoring molecular interactions involving chiral molecules. The dispersive nature of optical rotation, being dependent on refractive indices, means that at any wavelength all chiral species in the sample contribute to the signal—optical rotation is not very discriminatory. Particularly if a wavelength remote from absorption is chosen, optical activity is not very sensitive—high concentrations and long path lengths (typically 10 cm) are required. On both counts, CD is the preferred technique. The only serious advantage that optical rotation offers is the ability to monitor species with inaccessible absorption for example interacting sugars. The relative merits of CD and optical rotation can be enlikened to the relative merits of absorption and refractive index respectively.

Nowadays, optical activity is best studied as circular dichroism. Circular dichroism is the differential absorption of left and right circularly polarized light (A_L and A_R respectively) as a function of wavelength. It is effectively the difference between two absorption spectra one measured with left circularly polarized light and the other one with right circularly polarized light ($\Delta A = A_L - A_R$). According to Beer's law the differential absorption is given as:

$$\Delta A = (A_L - A_R) = (\varepsilon_L - \varepsilon_R).c.b = \Delta\varepsilon.c.b$$

where $\Delta\varepsilon$ is the differential molar extinction coefficient, c is the concentration in moles per litre, and b is the cell path length in cm.

Typical values of $\Delta\varepsilon$ are between 0.1 and 10 (CD induced as the result of ligand binding can be 10 times smaller). Thus for a solution of 10^{-5} M in a 1 cm cell the measured CD can present ΔA values between 10^{-6} and 10^{-4} (even smaller for induced CD). CD is a very small absorption differential and cannot be measured simply as a difference spectrum using a modified double beam spectrophotometer (see *Figure 2* and later).

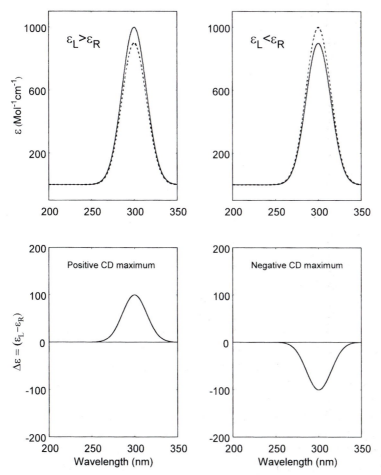

Figure 2 The absorption and CD profile of an isolated electronic excitation.

In the general case, a UV/vis absorption spectrum covers more than one electronic transition. Consider the absorption spectrum for a typical aromatic chromophore with three transitions; each transition can present a positive or negative differential absorption to produce four possible CD patterns (*Figure 3*).

The sign and magnitude of ΔA for the various transitions is sensitive to absolute stereochemistry, i.e. the position of atoms and the orientation of chemical bonds around a chromophore (the molecular centre of light absorption). A UV absorbing ligand sitting in the pocket of a chiral protein is a typical example.

To get the best out of CD measurements it is important to understand the basis of the technique. Measurement conditions need to be set on a study-by-study basis. General measurement conditions cannot be defined although guiding principles can be clearly laid down. In the first instance, the principles of the absorption process need to be understood. For reviews of absorption spectroscopy and optical activity, the reader is referred to the work of S. F. Mason (1).

The four CD patterns associated with three transitions

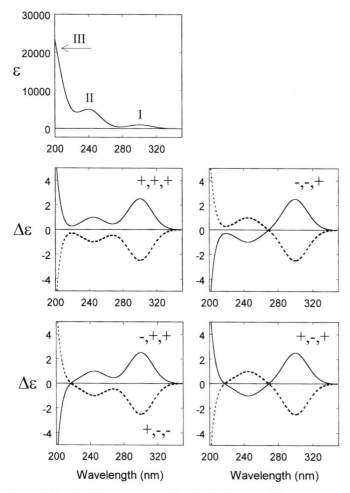

Figure 3 The four CD patterns associated with three transitions.

2 The ordinary absorption of light

The ordinary absorption of light by the transition (excitation) of an electron from a ground state to an excited state is allowed provided:

(a) The energy gap (ground state → excited state) matches the wavelength of light ($E = hc/\lambda$). For an electronic transition the wavelengths fall in the region 1000–350 nm (near IR and visible), 350–250 nm (near UV), and < 250 nm (far UV).

(b) The ground state electron moves in a straight line from the ground state to the excited state as a translation of charge across a chromophore generating a transition electric dipole moment that dictates the potential extent of absorption quantified as the extinction coefficient, ε. In the simple example of ethylene, the oscillating electric field of the incident light beam, provided it has the appropriate energy ($E = hc/\lambda$), will induce an electron in the π-bonding orbital to translate along the double bond and become excited in the π^*-orbital (*Figure 4*).

The absorption of light by ethylene

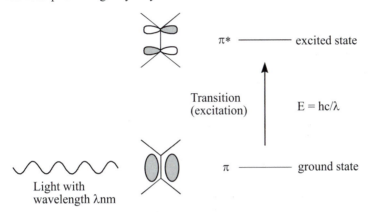

The direction of the oscillating electric field of the light correlates with the electron movement during excitation

Ethylene transition electric dipole moment :

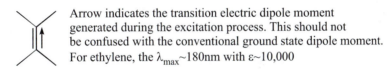

Arrow indicates the transition electric dipole moment generated during the excitation process. This should not be confused with the conventional ground state dipole moment. For ethylene, the $\lambda_{max}\sim180nm$ with $\varepsilon\sim10,000$

Figure 4 The absorption of light by the ethylene chromophore.

The absorption process in this case is said to be electric dipole allowed. The oscillating electric field of the incoming light must correlate with the electron movement for resonant absorption to take place. Linearly polarized light matching the orientation of the transition in an oriented molecule (chromophore) gives rise to linear dichroism.

The absorption process follows Beer's law:

$$A = \log (I_o / I_t) = \varepsilon.c.b$$

where ε, the extinction coefficient, is a measure of the strength of the interaction between the light and the transition. The extinction coefficient is also a measure of the transition electric dipole moment, and dictates the molar concentration, c, required to give a particular absorbance A in a cuvette of path length b cm. The larger the value of ε, the greater is the amount of light absorbed by a given concentration in a given path length cell (see *Figure 5*).

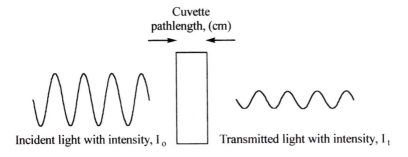

Figure 5 The absorption of light and Beer's law.

Electronic transitions are usually localized within a group of atoms collectively known as a chromophore where the absorption process is centred, such as the phenyl group, the amide group (proteins and peptides), and the heterocyclic bases of nucleic acids. An electronic transition with an extinction coefficient of 10 000 or greater is spectroscopically allowed. Transitions with $\varepsilon < 1000$ are said to be forbidden without an inherent ability to absorb light. That absorption does occur is due to the perturbation of pure electronic energy states by the chromophore environment.

3 The spectroscopic basis of circular dichroism

Circularly polarized light is helical either left- or right-handed. For preferential absorption to take place the excited electron must trace out a helical path on going from the ground to the excited state. The helical electronic excitation can be either left- or right-handed and is produced by a combination of a translational motion and a rotational motion (clockwise or anticlockwise, *Figure 6*).

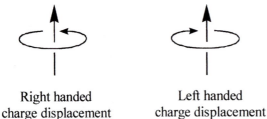

Right handed
charge displacement

Left handed
charge displacement

Figure 6 Spectroscopic chirality.

How a left and right light circularly polarized light is differentially absorbed by a helical charge displacement is illustrated in *Figure 7*.

An example of a transition possessing the character of a rotation is the n → π* transition of a ketone at about 300 nm (*Figure 8a*). The transition does not behave as a translation and the extinction coefficient is low ($\varepsilon \sim 100$).

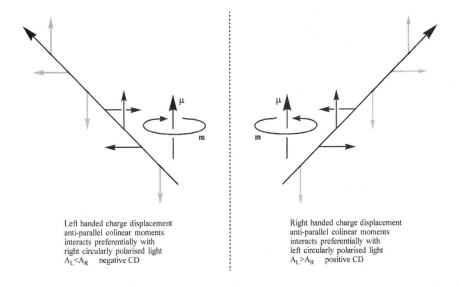

Left handed charge displacement
anti-parallel colinear moments
interacts preferentially with
right circularly polarised light
$A_L < A_R$ negative CD

Right handed charge displacement
anti-parallel colinear moments
interacts preferentially with
left circularly polarised light
$A_L > A_R$ positive CD

The interaction of circularly polarised light with a helical charge displacement

The emphasised electric vectors of the light beam correlate with the appropriate helical motion.
The propagation direction of the light beam with respect to the chromophore is the same as the ordinary absorption

Figure 7 The interaction of light with a helical charge displacement.

Only chiral chromophores present helical transitions inherently. The disulfide group -S-S- is the most frequent example found in biological systems; the sign of the CD originating from the disulfide chromophore is therefore influenced by the -S-S- dihedral angle (2).

More generally, a transition possesses only an electric or a magnetic dipole moment or neither. The origin of optical activity lies in the ability of the molecular framework (environment) of the chromophore to induce the missing character. Whether, the resultant transition is a left- or right-handed helical charge displacement and the CD is negative or positive, respectively, depends on the absolute positions of the atoms in the chromophore environment. This is illustrated in *Figure 8a*. The location of the methyl group in 3-methyl cyclohexanone electrostatically dictates the relative directions of the inherent magnetic dipole and the induced electric dipole moments. The two enantiomers will induce oppositely signed CD with one enantiomer producing a left-handed and the other a right-handed charge displacement (spectroscopic chirality).

A correlation between molecular environment and the CD sign of a particular transition is the crucial factor. This is achieved by quantum mechanical calculations or experimental comparison of similar structures. Both topics are outside the scope of this article. In practice, the empirical approach is the more usual. The CD spectra of two very similar compounds can in suitable cases be used to demonstrate that the two compounds have the same or opposite absolute configuration (both R-, both S-, or one is R- and the other is S-). One of

The absorption of light by the carbonyl group

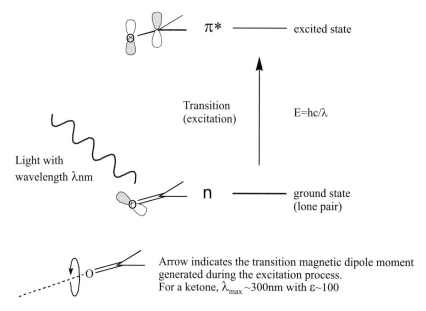

π* ——— excited state

Transition (excitation)

$E = hc/\lambda$

Light with wavelength λnm

n ——— ground state (lone pair)

Arrow indicates the transition magnetic dipole moment generated during the excitation process. For a ketone, $\lambda_{max} \sim 300$nm with $\varepsilon \sim 100$

Figure 8 (a) The nature of the $n \rightarrow \pi^*$ transition in a ketone.

the two compounds will need absolute configuration determination by another technique such as X-ray crystallography.

The overall analysis scheme for interpreting a CD spectrum is given in *Protocol 1*.

Protocol 1

Overall strategy for analysis of a CD spectrum

1 Spectroscopic assignment.

2 Spectroscopic chirality (handedness of excitation and sign of CD).

3 Relationship between spectroscopic moments and molecular environments.

4 Stereochemistry (absolute configuration, protein/peptide conformation).

5 Nomenclature (R-, S- configuration, biopolymer secondary structure composition).

CD spectroscopy provides the most convenient method to monitor protein conformation in solution. Knowing the base spectra of the α-helix, β-sheet, β-turns, the P_{II} helix, and the disordered (dynamic or random) conformation, the secondary structure content of a protein can be estimated by a mathematical components analysis of the CD spectrum. The secondary structure elements cannot be located directly nor described with the atomic precision of X-ray crystallography or NMR. However, changes can be detected easily and routinely as the result of changes in pH, buffer salt, protein/peptide concentration (aggregation), and solvent. CD generally needs about 1/10th the concentration used in NMR or vibrational spectroscopy.

In binding studies, the spectroscopic chirality is controlled by the pocket geometry and the structure about the ligand chromophore. Consider the ketone in *Figure 8b* with the keto group sitting in a binding pocket. The keto $n \rightarrow \pi^*$

The two enantiomers of 3-methyl-cyclohexanone

S-3-methyl-cyclohexanone   R-3-methyl-cyclohexanone

The two possible charge displacements during the absorption of light are :

Right handed charge displacement   Left handed charge displacement
positive CD   negative CD

It is not trivial to predict which enantiomer will give which charge displacement

The two possible induced CD cases resulting from S-3-methyl-cyclohexanone sitting in a binding pocket :

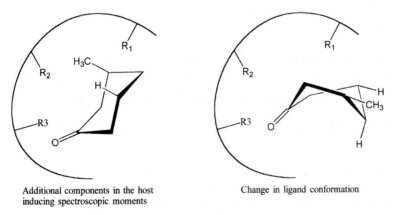

Additional components in the host   Change in ligand conformation
inducing spectroscopic moments

Figure 8 (b) The CD sign associated with the ketone $n \rightarrow \pi^*$ transition.

spectroscopic chirality will be dependent on the disposition of *host* atoms around the pocket wall as well as the methyl group. CD cannot readily determine the precise geometry of this interaction without recourse to quantum mechanical methods. Nevertheless, the CD of the ketone will be sensitive to the binding geometry. Two classes of induced CD can be realized (see later). One is simply the influence of the environment and the second is the effect of a change in the ligand conformation as the result of binding.

The excitation mode (translational, rotational, or helical displacement of charge) is the character of a spectroscopic transition. The selection rules for ordinary absorption spectra and CD spectra are clearly different. A transition that shows a strong UV absorption may show a weak CD feature. On the other hand a transition with an inherent rotational character may be clearly observed in the CD spectrum but not apparent in the absorption spectrum. A symmetric molecule lacking chirality and optical activity will display CD on entering a chiral binding pocket.

However, the measurement of the CD spectrum is controlled by the ordinary absorption spectrum. A small absorbance presents a very small absorbance differential. A larger absorbance gives a larger absorbance differential but the transmitted light intensity will now be very low and the detected signal will be weak and noisy. A compromise is reached by ensuring that the CD spectrum is measured with an $A_{max} \sim 0.8$.

4 The circular dichroism spectrometer

Intense light from a xenon lamp is first rendered monochromatic and linearly polarized. The linearly polarized, monochromatic light is polarization modulated by an electro-optic device from left circular through elliptical and unchanged linear to elliptical and right circular polarization. The λ-drive changes the wavelength of light incident on the sample and ensures the polarization modulator operates at the condition appropriate for the selected wavelength. During

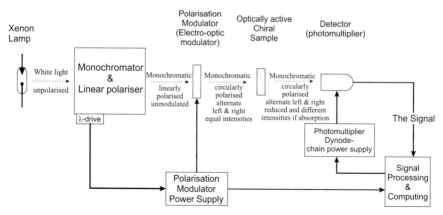

Figure 9 The CD spectrometer.

absorption, provided the sample is optically active (chiral), the detector presents a signal with two components: an AC signal at the polarization modulation frequency associated with the differential intensities of left and right circularly polarized light (circular dichroism) and a DC component related to the light level throughput (instrument characteristics and light absorption by the sample). Phase sensitive detection, linked to the modulator power supply and locked to the polarization modulation frequency, allows the AC signal to be rectified as the CD signal. The AC signal component needs to be measured with respect to a constant light level (DC) signal. This is effectively achieved with a feed-back that varies the detector amplification (dynode chain voltage) that amplifies the AC (and DC) signal so that the magnitude of the AC signal is with respect to a constant DC signal component and, accordingly, directly proportional to the circular dichroism. The instrument component responsible for maintaining the constant DC level (e.g. photomultiplier dynode chain voltage) can provide a measure of light absorption. Accordingly, the spectrometer is capable of in principle measuring the solution absorption and optical activity simultaneously.

Although spatially a CD spectrometer is a single beam instrument, it behaves as a double beam spectrometer in time with temporal modulation between left and right circular polarization ensuring the stability of the differential absorption measurement. The magnitude and sign of the circular dichroism is set by calibration with a solution of known CD characteristics (e.g. ammonium camphor sulfonate in water with a 1 cm path length cell). For more detailed information on the design of a CD spectrometer the reader is referred to two reviews by A. F. Drake (3).

A CD spectrometer has the appearance and footprint of a large spectrophotometer although costing more at between 100–200 k$. The main suppliers are JASCO, AVIV, and APL Ltd. The APL instrument is specifically designed to obtain stopped-flow solution kinetic CD measurements. Images of the JASCO J810 and the APL π^*-CDF180 are provided in *Figures 10* and 11.

The CD spectrometer operates in a similar manner to a conventional spectrophotometer. The operator must control a number of features.

Figure 10 The JASCO J810 Spectropolarimeter.

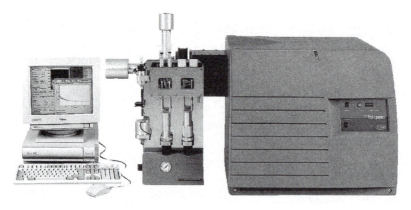

Figure 11 The APL π*-180 CDF Spectrometer.

Protocol 2

What the operator must control in a CD experiment

1 **Sample concentration and cell path length**. The combination should be chosen to ideally give an $A_{max} \sim 0.8$. A_{max} includes absorption from all species in the measured solution including solvent, buffer salts, counterions, and other solutes. For interaction studies, concentration may be set by the experiment in which case only the path length is a variable. Path lengths between 0.1 mm and 100 mm can be chosen. Narrower path lengths (< 0.1 mm) may be practically difficult (but not impossible).

2 **Baseline correction**. Every sample spectrum should be corrected by subtracting the corresponding spectrum obtained for the solvent (including buffer salts and counterions) in the same cell and measured during the same work session with the same instrument settings as the sample measurement. In the ideal world, the spectrum of a non-chiral or non-absorbing solvent (including buffer salts and counterions) in a perfect cell should be flat with a ΔA value of zero. In practice the background, baseline spectrum is neither flat nor non-zero and correction is necessary particularly at high sensitivity. For this reason, strain-free spectrophotometer cells are used, which are expensive. These are provided and certified strain-free by Hellma.

3 **Spectral range of measurement**. The spectrum scan should run from about 20 nm before the onset of absorption to ensure that where the sample does not absorb, the absorbance and the circular dichroism is zero to confirm baseline position.

4 **Spectrometer slit width (spectral bandwidth, SBW)**. Typically, the entrance and exit slits of the monochromator are set to give a λ spread of about ± 0.5 nm passing through the sample. The larger the slit the more light is available and the better is the signal-to-noise. An SBW = 1 nm is preferred for spectroscopy, although the low noise advantage of larger slits offer benefits for monitoring processes. Polarization can be very restrictive and the SBW may have to be constrained to 1 nm

for optical reasons. The APL CD instrument can offer the noise benefits of wider slits.

5   **Data resolution**. Data is stored in digital form on a computer. The operator must decide how often a data point needs to be recorded. A data resolution throughout the spectrum of a data point every 0.2 nm is a good starting point.

6   **Measurement time**. In analogue terms, this refers to the scan speed, which in the main is dictated by the measurement time constant. The longer the time constant (the greater the damping) the lower is the noise but the slower must be the scan. A time constant of 4 sec with a scan rate of 10 nm per minute is typical enabling a 100 nm CD spectrum to be measured in about 10 minutes. CD spectra are noisier than ordinary absorbance measurements and the spectrum measurement time is longer. Noise can be further reduced at the expense of time by averaging a number of scans with the noise reduction being proportional to $\sqrt{}$ (number of scans). A more modern instrument may be entirely digital. In this case the operator must decide how long to spend at each data interval (typically 0.2 nm) accumulating data until acceptable noise characteristics are achieved. Larger slits with greater light throughput can help reduce measurement time.

7   **Internal instrument environment**. O_2 absorbs below ~205 nm; intense Xe lamps convert O_2 to O_3. Ozone formation is deleterious to health, corrodes mirror optics, and absorbs in the 260 nm region. Accordingly, CD spectrometers are kept continuously flushed with pure nitrogen at a flow rate of about 3 litre/min.

8   **Instrument sensitivity**. The weaker the signal, the greater is the sensitivity required of the spectrometer. Modern instruments are capable of measuring differential absorbances of $\Delta A \sim 1.10^{-5}$. An appropriate sensitivity range may need to be selected.

Since the early 1960s, CD spectra have been measured directly in terms of differential absorbances. Before the patented use of polarization modulation, CD spectra were measured as the change in ellipticity of a linearly polarized light beam passing through an optically active, absorbing sample. Unfortunately, many scientific publications (and instruments) have retained the older units for data presentation. To better appreciate CD, it is necessary to convert the ellipticity data and think in terms of absorption. The appropriate parameters and units are:

(a) Molar differential extinction coefficient, $\Delta\varepsilon$, defined as:

$$\Delta\varepsilon = (\varepsilon_L - \varepsilon_R) = (A_L - A_R)/(c.b) = \Delta A/(c.b) \; l.mol^{-1}cm^{-1} \qquad [1]$$

where c = molar concentration (M or mol. l^{-1}), b = cell path length (cm).

(b) Molar ellipticity, $[\Theta]$, defined as:

$$[\Theta] = \theta.c.b \qquad [2]$$

where θ = observed ellipticity reported as millidegrees (mdeg).

(c) Conversion between the two units is given by:

$$\Delta\varepsilon = [\Theta]/3300 \text{ and } \Delta A = \theta/33000 \qquad [3]$$

(d) g-factor. To judge the magnitude of a differential absorbance the CD signal per absorbance can be quoted at the g-factor:

$$g_\lambda = \frac{\Delta A_\lambda}{A_\lambda} = \frac{\Delta\varepsilon_\lambda}{\varepsilon_\lambda} \qquad [4]$$

5 The merits and uses of CD spectroscopy

CD is a spectroscopic technique bringing many of the benefits of spectroscopic techniques to ligand binding studies. In many respects CD can be treated as a sensor of conformation or better chromophore environment. As the result of binding to a protein, CD monitors the amount of species bound and unbound as the result of relatively small changes in the electronic structure of the chromophores involved but relatively large changes in the 3D environment without fully determining the structures of the environments involved. A list of the merits of CD is given below:

(a) **Measurement**. CD measurement is akin to the measurement of ordinary UV/vis electronic absorption spectra. Measurements are relatively straightforward, are routine, and can be easily automated. Beer's law is important.

(b) **Chirality**. Uniquely amongst spectroscopic techniques, CD is directly sensitive to chirality. This allows the ready monitoring of chiral molecules and in many cases the assignment of absolute configuration. CD spectroscopy is sensitive to the molecular co-ordinates of molecules. However, complete structure determination is too complex to be of general use. At present the absolute stereochemistry of only relatively small molecules is realistic with in the main relative assignments based upon qualitative comparison of the CD spectra of very similar compounds. Typically, a technique like NMR would be used to determine molecule structure and CD would be used to determine which mirror image is involved.

(c) **Biopolymer conformation**. The CD of proteins/peptides and nucleic acids present a special case of optical activity with CD arising from the stereospecific spectroscopic interaction of neighbouring chromophores in individual residues. In proteins/peptides this implies the amide bond; in nucleic acids the heterocyclic bases are involved. Changes in chromophore/chromophore orientation produce specific CD features. The sensitivity of CD to 3D structure and the ease of measurement in solution are key factors. For example, although the precise location of α-helices in a protein is not realistic, the estimation of the percentage of α-helix in the whole protein is straightforward. The ability to follow the gain or loss of helical character as the result of a change is routine.

(d) **Chromophore specificity**. CD will be detected only for transitions that have relatively strong optical activity. The selection rules for ordinary absorption and CD are different and different transitions may dominate in the two spectra. For example, the strong aromatic absorption of human serum albumin around 280 nm can mask the absorbance of bound ligands. On the other hand, the aromatic CD of HSA is relatively weak allowing the ready detection of ligand based CD. The CO_2H and the -S-S- are examples of chromophores that may be detected in CD but not readily in ordinary absorption.

(e) **Concentration**. The CD spectrum can be normally obtained at a relatively low concentration (sub-milligram per ml concentrations are normally sufficient). Taking the cell path length into consideration, a wide range of concentrations can be covered. This is important for the faithful evaluation of binding constants.

(f) **Overlapping CD components are often of opposite sign**. Overlap cancellation will in practice lead to peaks whose wavelength separation is artificially exaggerated. Overlapping bands of opposite sign are easier to distinguish in the CD technique (*Figure 12*).

(g) **Bandwidth of the CD peak**. The selection rules for CD are relatively restrictive with the excitation requiring the character of a collinear rotation and translation. Ordinary absorption requires the excitation to proceed simply as a translation. Accordingly fewer vibronic components are often observed in the CD measurement and the CD peak is therefore narrower than the corresponding ordinary absorption peak.

(h) **Conformation**. The CD associated with a chromophore is more sensitive to conformation and chromophore environment than is the ordinary absorption.

(i) **Induced CD**. When a non-chiral species comes into contact with a chiral molecule the disymmetric environment of the host is sensed and optical activity is induced into the absorption bands of the ligand (or vice versa). Whilst CD is blind to the unbound non-chiral species, it can detect the species bound to a chiral host (or vice versa). There are two forms of induced CD, one is associated with a ligand that remains conformationally unchanged on binding and senses the host chirality—this generally leads to small CD (g-factors $\leq 1.10^{-4}$). This small optical activity generally benefits from the improved sensitivity of newer instruments and very careful choice of optimum measurement conditions. The second form of induced optical activity is the result of the ability of the host to change the shape of the ligand on binding to an extent that the ligand is trapped as a chiral conformation even though it is not formally chiral in isolation. This second origin of induced CD is expected to be much stronger, more typical of natural optical activity and includes the long known Pfeiffer effect associated with the Δ and Λ form of metal complexes. Examples in organic molecules include the changes associated with the binding of bilirubin or diazepam to HSA with the preferential trapping of a chiral conformation.

(j) **Light scattering is not generally polarization selective**. Ordinary light scattering is less likely to distort a CD spectrum than an absorption spectrum. This is important for the higher protein host concentrations.

(k) **Fast time scale**. The time scale of an electronic absorption/relaxation process is typically $\leq 10^{-8}$ sec, sufficiently fast that each species in an equilibrium is observed in the spectrum separately albeit spectral responses can overlap making discrimination difficult. Techniques like NMR spectroscopy can present time averaged results. Time resolved measurements are readily achieved with great potential for kinetic measurements in the msec time scale.

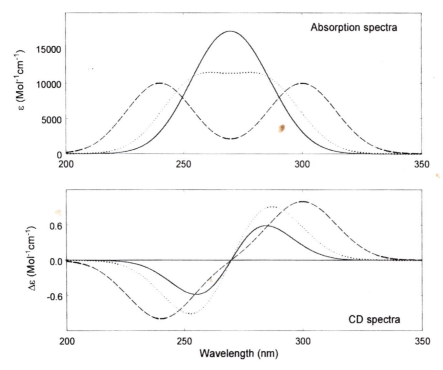

Figure 12 The effect of the separation of two transitions on the absorption and the CD spectrum. Transition separation 60 nm —— ; 30 nm − − − ; 15 nm ....

These factors lead to the range of applications listed in *Table 1*. For further reading the reader is referred to the work of Mason (1) and the book edited by Fasman (6) and Nakanishi *et al.* (4) and the articles by Drake (5).

6 Protein/peptide conformation as studied by CD spectroscopy

CD spectroscopy provides the most convenient method to monitor protein conformation in solution and the changes that can occur in response to environ-

Table 1 Applications of CD spectroscopy

1	**Chirality**
	Detection and characterization of chirality (registration, patent, identification)
	Determination of absolute stereochemistry (configuration and conformation)
	Enantiomeric purity
2	**Biological macromolecule conformation**
	Protein/peptide and nucleic acid secondary structure
3	**Binding/interaction studies**
	Drug/biopolymer
	Metal/ligand
4	**Time resolved studies and solution dynamics**
5	**Other**
	Liquid crystal characterization
	Molecular aspects of drug formulation

ment (solvent, solvent mixtures, pH, buffer salt, counterion, and protein/peptide concentration) and interaction.

The CD spectrum of a protein can be subdivided into three regions (see *Figure 13*). Therefore, a complete CD spectrum will be a composite derived from measurements in cells of different path lengths. CD associated with side chain aromatic residues will provide local information about the aromatic residue such as local changes in environment/shape of a binding pocket. Changes in the CD associated with aromatic residues on a protein surface can infer changes in tertiary structure. Complementary fluorescence studies can be revealing. It is stricter to say that aromatic CD changes monitor local events that may reflect changes in tertiary structure.

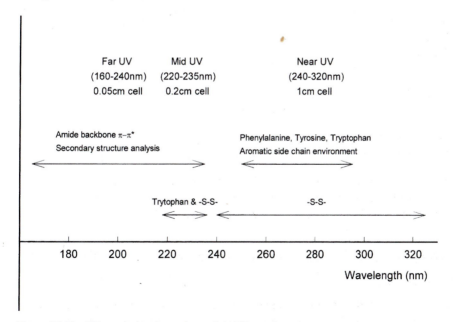

Figure 13 The UV spectral regions of a protein CD spectrum.

The CD in the far UV associated with the amide backbone provides information about backbone conformation (see ref. 6, and references therein). From a CD point-of-view, there are five classes of protein/peptide conformation with the typical CD profiles presented in *Figures 14* and *15*. All other conformations are best treated as subdivisions in this classification.

Protein Conformation and CD Spectra

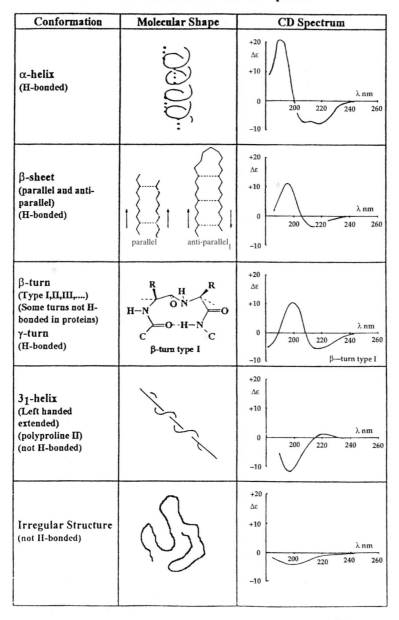

Figure 14 The major secondary structure classes and their associated CD spectra.

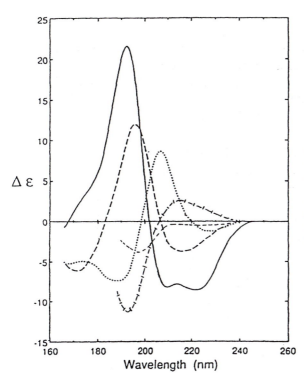

Figure 15 CD spectra associated with various secondary structures: α-helix (——), antiparallel β-sheet (– – –), turn type 1 (······), and left-handed P$_{\parallel}$ helix (–ı–ı–). Redrawn from reference 5.

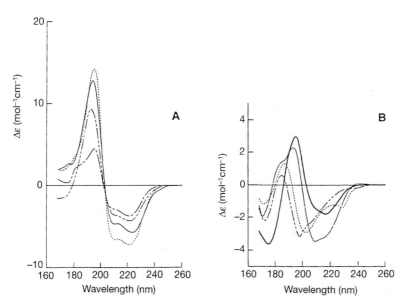

Figure 16 Representative CD spectra of proteins with secondary structure content defined by X-ray crystallography (work of W. C. Johnson Jr.). **(A)** α-helix rich: (——) T4 lysozyme (67% H, 10% S, 6% T, 17% O), (- - - -) hemoglobin (75% H, 0% S, 14% T, 11% O), (– - –) cytochrome c (38% H, 0% S, 17% T, 45% O), (– - - –) lactate dehydrogenase (chicken heart) (41% H, 17% S, 11% T, 31% O). **(B)** β-sheet rich: (══) prealbumin (7% H, 45% S, 14% T, 34% O), (- - - -) α-chymotrypsin (10% H, 34% S, 20% T, 36% O), (– - - –) elastase (10% H, 37% S, 22% T, 31% O), (——) ribonuclease A (24% H, 33% S, 14% T, 29% O). Conformation contents are given by H = α-helix, S = β-sheet, T = β-turn and O = other.

The CD of a typical protein (*Figure 16*) is a combination of CD contributions from each conformation class. In their original paper, Greenfield and Fasman (6) considered that the observed CD of a protein at a given wavelength (λ nm) is given by the expression:

$$\Delta A^{\lambda}_{obs} = \Delta A^{\lambda}_{\alpha\text{-}helix} + \Delta A^{\lambda}_{\beta\text{-}sheet} + \Delta A^{\lambda}_{irregular} \tag{5}$$

where $\Delta A^{\lambda}_{element}$ is the CD contribution of the indicated conformational component. Applying Beer's law to each element ($\Delta A = \varepsilon.c.b$) where c is the molar concentration of the secondary structure element and b is the measurement cell path length:

$$\frac{\Delta A^{\lambda}_{obs}}{b} = (\Delta \varepsilon^{\lambda}_{\alpha\text{-}helix} \cdot c_{\alpha\text{-}helix}) + (\Delta \varepsilon^{\lambda}_{\beta\text{-}sheet} \cdot c_{\beta\text{-}sheet}) + (\Delta \varepsilon^{\lambda}_{irregular} \cdot c_{irregular}) \tag{6}$$

To enable the comparison of proteins/peptides of varying molecular weight/length, concentration here refers to amino acid residue (see Section 7). With a total protein molar concentration (per residue) $c_{protein}$, the percentage of each secondary structure element is given by the respective $[(c_{element})/(c_{protein})].100\%$ based upon $\Delta\varepsilon$'s from a reference data set. To solve *Equation 6* data from at least three wavelengths is required. In practice, values were taken at several wavelengths and fed into a computer program that effectively solved simultaneous equations to provide the best fit $[(c_{element})/(c_{protein})].100\%$ values. However, it soon became apparent that this approach was too simplistic. Several factors made the analysis less precise than required (*Table 2*).

Table 2 Factors affecting the analysis of protein/peptide CD data

1	**Instrumental (measurement)**
	Instrument wavelength calibration
	CD magnitude calibration
	Wavelength range of measurement
	Instrument spectral bandwidth
	Stray light
	Speed of measurement/instrument response time
	Quality of experimental data (noise)
	Protein concentration and purity
	Mean residue molecular weight
2	**Spectroscopic**
	Non-amide contributions to the CD spectrum (tyrosine, tryptophan, disulfide, and prosthetic groups)
	Spectroscopic solvent effects (e.g. threitol and erythritol)
	Light scattering
	Conformational equilibria (is a dynamic α-helix $\leftrightarrow$ β-sheet equilibrium an irregular structure?)
3	**Data analysis**
	Definition of secondary structure components
	Transferability of CD data (e.g. length of α-helix, α-helix in isolation or buried in a structure)
	Wavelength data resolution (CD data point every 1 nm or 2 nm …etc.)
	Assumptions made by the mathematical model
	Mathematical random noise associated with the number of mathematical factors used in the mathematical model

6.1 Instrumental measurement factors

As in all experimental methods, there are practical aspects to CD that have to be taken into account:

(a) Wavelength calibration is important, a blue or red shift will lead to errors.

(b) The instrument must be calibrated with respect to CD magnitude. This is particularly true because intensity data is so crucial to the analysis.

(c) The wavelength range of the instrument is extremely important in that it is the 'eye' of the instrument. The greater the wavelength range the more desirable the instrument. Measurements below 170 nm are effectively impossible because of solvent absorption. Secondary structure analysis measurements are generally made between 190 nm to 260 nm, though measurements started at 250 nm make little difference to the estimation of secondary structure content. The measurement cell path length may need to be reduced to 0.5, 0.2, or 0.1 mm (with an appropriate increase in protein concentration if necessary). Solvent absorption in the very far UV (< 195 nm) must be kept to a minimum. The A_{max} of the whole solution (solute, solvent, and additives) must remain ideally around 0.8. As measurements are extended to lower wavelengths measurements become progressively more difficult.

(d) The spectrometer spectral bandwidth is normally set to 1 nm.

(e) The amount of stray light must be minimized because transmitted 'white' light effectively has no CD signal. This becomes increasingly important at shorter wavelengths as all components in the measurement sample begin to absorb strongly and the instrument light throughput is inherently weak.

(f) CD spectrum measured too quickly will be 'distorted'. A balance between noise, measurement time constant, and scan speed is required.

(g) Noise must be kept to a minimum so that the quality of the data is high. This can be done by ensuring the gratings and mirrors are not damaged, and the lamp is in good condition, the correct spectral bandwidth is chosen, and the appropriate time constant to scan speed ratio taken.

(h) The concentration of the sample being measured must be known with accuracy, otherwise the analysis results obtained will inevitably be in error. This is a very important, often underestimated aspect of secondary structure analysis.

(i) Mean residue weight must be known precisely since CD data are expressed as per mean residue.

6.2 Spectroscopic factors

For secondary structure component analyses, the CD associated with the amide backbone in the far UV (240–170 nm) is analysed:

(a) Contributions from non-aromatic electronic absorptions will result in analysis errors (6). These are due to the aromatic groups of the tryptophan, phenylalanine, histidines, tyrosines, as well as prosthetic groups.

(b) Solvents are known to produce red and blue shifts due to stabilization of either the ground or excited states on going from polar to non-polar solvents. Erythritol is often used as a S-S bond reducing agent, however, over a long period of time it breaks down, to chiral products that can effect the CD spectrum. Threitol is preferred. The pH of a solution can also affect the form of the CD spectrum, for example, at pH 12 the tyrosine group is ionized to the phenoxide, which has a quite different absorption.

(c) Light scattering reduces light levels and increases noise. Polarized light scattering as in chromosomes can lead to macroscopic effects that swamps the molecular signal.

(d) NMR and X-ray structure determination tend to focus on a single conformation of a protein. CD spectroscopy on the other hand provides 'global' secondary structure contents. If a protein is only partially folded the CD analysis will give a reduced secondary structure content. Thus a 40% α-helix result may correspond to a protein with 40% of its residues in a fully folded α-helix or for example a 66 residue per cent α-helix that is partially unfolded (36.7%).

6.3 Data analysis methods

The analysis of protein/peptide secondary structure also depends on the following factors:

(a) The original CD classification of protein/peptide secondary structure elements as three classes is inadequate.

   (i) At least two subclasses of β-sheet are recognized (parallel and antiparallel).

   (ii) There are various β-turns with different CD profiles; some writers erroneously group them together as a single component.

   (iii) The designated CD spectrum of the *random coil* has been reinterpreted as a mixture of contributions from the polyproline II type left-handed conformation and what is now better termed as the unordered or irregular conformation (15, 16). The modern CD classification of protein/peptide secondary structures is presented in *Figure 14*.

(b) The definition of the amount of a secondary structure element is also open to debate. There are two ways of defining the α-helix length. In the first approach (9), the secondary structure recognition algorithm is based on hydrogen bonding patterns. There are two types of patterns, 'n-turns' with an hydrogen bond between the CO of residue i and the NH of residue $i + n$, where $n = 3, 4, 5$, and 'bridges' with hydrogen bonds between residues not near each other in sequence. Repeating four turns then define α-helices, and repeating bridges define β-sheet. The second approach (10) makes use of a single torsion angle α_i defined by the path of the Cα atoms belonging to residues $i - 2, i - 1, i$, and $i + 1$. The two methods give different answers.

(c) Currently, the most widely used reference data set is the one reported by Hennessey and Johnson (11). However, the precise form of the CD of any secondary structure element remains an open question. Factors such as helix length, α- or 3_{10}-helix, environment effects, and β-sheet shape are not clearly defined in CD spectroscopy even though differences there must be.

(d) Data resolution is not a problem for secondary structure analysis, so long as the number of data points is greater than 25 (12).

(e) Measurement noise can effect the quality of the analysis. Theoretically it can be shown that in the absence of noise, the number of factors needed to analyse data is identical to the number of secondary structure components. However, introduction of noise increases the factors needed to analyse data. Interfering components have the effect of simulating data that are not present in the calibration data set (13).

Because of the difficulties listed above the simple Greenfield and Fasman method of analysis (7) is generally considered to be inadequate and many more sophisticated mathematical models have been produced. These have been well reviewed and discussed at length elsewhere (6, 14). Most of the methods can be applied using public domain software available from the instrument suppliers or directly from the authors.

All of the methods have relative strengths and weaknesses.

In practice, the precision of CD secondary structure analysis is less dependent on mathematical model than on the other factors listed above. The choice of method is largely a personal one.

However, CD spectroscopy cannot locate directly protein/peptide secondary structure elements nor describe it with the atomic precision of X-ray crystallography or NMR. However, changes can be detected easily and routinely as the result of changes in pH, buffer salt, protein/peptide concentration (aggregation), solvent and ligand interactions. CD generally needs about one-tenth the concentration used in NMR or vibrational spectroscopy.

In terms of percentages, only an α-helix content of $> 30\%$ can be quoted with a precision much greater than 10%. In peptides, CD provides the only spectroscopic method capable of directly detecting the polyproline PII conformation in solution. The P_{II} conformation is an important representation of the binding conformation of a linear peptide in many biological systems (16).

7 Monitoring ligand binding by CD spectroscopy

CD is a spectroscopic technique. Therefore, to generate the optimum experimental conditions and to interpret the results informed decisions have to be made. The change in CD as the result of interaction is normally small and good spectra can take up to 10 minutes to accumulate. A 10-point titration requires the order of 100 minutes measurement time which coupled to solution handling means that a single titration can take at least two hours. With data analysis and presentation only two binding constants can be realistically determined in a working day. An experimental flow chart is given in *Figure 17*.

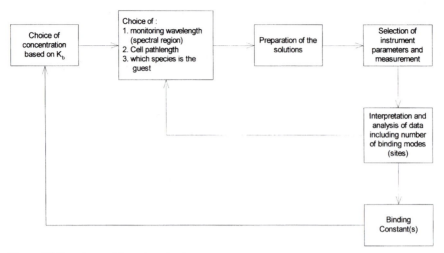

Figure 17 Experimental flow chart for CD monitored binding studies.

7.1 Measurement conditions and instrumental parameters

When two species bind to each other; one is known as the *host* the other is the *guest*.

In practice, the larger of the two is normally treated as the *host*, for example a protein acts as the *host* for smaller molecules (*guests*). The *guest* is often referred to as the ligand. However, this can be misleading. In inorganic chemistry, metal complexes are treated as metal centres with ligands (e.g. amino acids) bound to them. The ligand in this case can be large, e.g. peptides and in the extreme can be very large for example haemoglobin. Does haemoglobin act as a ligand and bind to iron or is it better to treat haemoglobin as the *host* with the iron atom as the binding ligand.

Mathematically, the choice is arbitrary provided the decision is maintained consistently throughout the binding study. The choice of which species is to be treated as the *guest* and which is the *host* is set largely by spectroscopic arguments. Ideally a monitoring wavelength should be chosen that provides a *guest* CD signal clear of the *host* and any other interfering signals (e.g. strongly absorbing species presenting no CD but inducing noise due to low consequent light levels).

In CD spectroscopy, like other forms of spectroscopy, the species being monitored is best treated as the *guest*. Ideally, the *guest* concentration should be maintained constant throughout a titration and the concentration of the *host* varied up to a molar ratio excess of ~ 5 to ensure that all the *guest* has been bound. This marks the end of the titration. It is important to have spectra of both the bound and unbound *guest* in order to facilitate the mathematical analysis of the results. Holding the monitored species concentration constant ensures that observed spectral changes are due to binding interactions and measurement precision is improved.

The choice of CD measurement conditions is not trivial. Three criteria apply in the case of simple 1:1 binding and are give in *Protocol 3*.

Protocol 3

The three criteria for analysis of simple 1:1 binding

1 For a titration the *guest* (monitored species) should have a concentration given by:

$$[G_T] \bullet K_B = 0.5 \text{ to } 5.0$$

where G_T is the total *guest* concentration and K_B is the binding constant.

To drive the titration to completion the *host* concentration must run from 0 to $5 \bullet [G_T]$ i.e. $[H_T]/[G_T] = 5$.

2 To achieve a 1:1 complex with all binding sites occupied with an equimolar total concentration of both species:

$$[G_T] \bullet K_B = 50 \text{ to } 100.$$

3 The CD measurement must be made such that the following is satisfied:

$$\begin{bmatrix} extinction \\ coefficient \cdot (Mol^{-1} \ cm^{-1}) \end{bmatrix} \times \begin{bmatrix} concentration \\ (Mol^{-1}) \end{bmatrix} \times \begin{bmatrix} cellpathlength \\ cm \end{bmatrix} \approx 0.8$$

Obviously absorbance values between 0.1 and 1.5 are acceptable provided the CD is strong enough. This can be set by the cell path length (10 cm to 0.02 cm) or the choice of monitoring wavelength with associated extinction coefficient. It needs to be stressed that in practice the absorbance refers to that of the whole solution. Considerations may have to be given to overlapping absorption from different sources (*guest*, *host*, and solvent).

Inevitably conflicts will arise and informed decisions and compromises may need to be made.

7.2 Further aspects of the concentration used for absorption and CD measurements

The definition of concentration in the Beer's law expression depends on circumstances. For a small isolated molecule, concentration is based on the conventional molecular weight. However polymers have molecular weights that dependent on chain lengths and comparison of optical data will reflect this. Accordingly, in optical spectroscopy concentrations of biopolymers are quoted in terms of residue. This is not the case for other techniques such as NMR spectroscopy. It is certainly not appropriate for interaction studies.

Consider a peptide with 10 residues and a whole molecule molecular weight of 1130 giving a per residue mean molecular weight of 113:

(a) A solution of 5 mg/ml corresponds to a whole molecule concentration of 4.43 $\times 10^{-3}$ M; this is a typical NMR condition.

(b) A solution of 0.5 mg/ml corresponds to a per residue molar concentration of 4.43×10^{-3} M; this is a typical condition for secondary structure conformation analysis by CD. In practice, CD requires a peptide concentration at least one-tenth that of the NMR experiment although this may not be explicitly apparent in literature reports. To convert literature CD values for proteins/peptides, the reported values ($\Delta\varepsilon_{res}$) should be multiplied by the number of residues n_{res}.

$$\Delta\varepsilon_{whole} = \Delta\varepsilon_{res}{}^{*}n_{res} \qquad [7]$$

(c) For interaction studies, concentrations must be quoted in terms of whole molecule molecular weight.

Consider a CD study of human serum albumin:

Human serum albumin (HSA) has a whole molecular weight of 66 438 with 585 residues giving a mean residue molecular weight of 113.6. For secondary structure analysis based on the amide chromophore in the far UV (250–170 nm region), the amide backbone absorption ($\varepsilon_{amide} \sim 10\,000$) of a 0.23 mg/ml solution gives an $A \sim 1.0$ in a 0.05 cm (0.5 mm) cell. This corresponds to an HSA concentration of 2.02×10^{-3} M on a per residue basis and 3.46×10^{-6} M on a whole molecule basis. On a per residue basis the CD for HSA is $\Delta\varepsilon_{192} = +10.7$, $\Delta\varepsilon_{208} = -6.4$, $\Delta\varepsilon_{220} = -5.9$ which analyses for an α-helix content of 60% as expected from the X-ray. The extinction coefficient based on a whole molecule molecular weigh (66 438) would read $\sim 5.85 \times 10^{6}$ and the CD would read as $\Delta\varepsilon_{192} = +6260$, $\Delta\varepsilon_{208} = -3740$, $\Delta\varepsilon_{220} = -3450$. These are unnaturally high values for the optical spectroscopist.

In the near UV, the concept of a molar extinction coefficient is not readily defined as differing numbers of aromatic amino acid residues will all be contributing to the absorption spectrum. Thus the absorbance is either quoted as a specific absorbance (A_1^1 (λ nm) being the absorbance of a 1% (1 g per 100 ml) solution in a 1 cm cell) or as the absorbance of a 1 g.l^{-1} solution. For HSA the absorbance of a 1 g.l^{-1} solution at 279 nm is 0.53 in a 1 cm cell. This means that a concentration of the order 1 mg/ml (8.8×10^{-3} M in terms of residues or 1.5×10^{-5} M in terms of whole molecule) in a 1 cm cell is required to have an optimal CD measurement. The ideal measuement conditions for the near and far UV are different.

This question of clarifying concentration in terms of whole molecule molecular weight is important. For proteins this is relatively straightforward provided good attention is paid to units. However, this is not necessarily so for nucleic acids were absorption data can be with respect to nucleic acid base or base pair. Thus some nucleic acid studies may be misleading as 1:1 drug : base pair is not the same as 1:1 drug : whole DNA molecule.

Consider a study of the interaction of diazepam with HSA; a 1:1 bound mole ratio will need to be achieved:

The ideal concentration for studying the backbone conformation of HSA is 2.02×10^{-3} M on a per residue basis, 3.46×10^{-6} M on a whole molecule basis. For a 1:1 molar ratio this corresponds to a diazepam concentration of 3.46×10^{-6} M

(0.1 mg/ml). The binding constant for diazepam binding to HSA is 2.5×10^5 M. Application of the criterion for 100% binding, namely (3.46×10^{-6}). $(2.5 \times 10^5) = 0.87$, implies that most of the HSA binding sites will be unfilled and studies of the effect of binding on protein conformation will not be appropriate. To guarantee 100% occupation of the binding sites at the set HSA concentration requires a diazepam concentration of 3.46×10^{-4} M (0.1 mg/ml). The ideal concentration for titration is $\sim 2.5/K_B$ which translates as an HSA concentration of 1×10^{-5} M. The backbone far UV CD measurement at this HSA concentration will require a path length of 0.1 or 0.2 cm. At this concentration a five molar excess of diazepam will ensure at least most of the binding sites are filled. A 5×10^{-5} M solution of diazepam corresponds to 0.0142 mg/ml. Now the stability and nature of the HSA peptide backbone can be studied by CD albeit the path length may inhibit CD measurement to about 195 nm due to solvent and buffer salt absorption. At this concentration the spectroscopic contribution of diazepam can be ignored.

In practice, there is negligible change in the global secondary structure of HSA as the result of binding diazepam. Therefore, the induced CD associated with diazepam needs to be monitored. Following the arguments outlined above for good titration, the concentration of diazepam needs to be 10^{-5} M. Using an ε_{max} for diazepam of 5000 presents a diazepam absorbance of 1 with a cell path length of 20 cm. This path length is unreasonable. The compromise we have employed involves a diazepam concentration of 2×10^{-5} M with a path length of 2 cm. The diazepam absorbance is now 0.25. A five-fold molar excess of HSA (10^{-4} M, 1.66 mg/ml) now ensures that all the diazepam is bound. This is the condition used to monitor the kinetics of drug binding.

Several factors arise from this analysis:

(a) Much forethought must go into the design of the experiment.

(b) The components of the titration may not be sufficiently soluble in the ideal concentration range. This is particularly true if the monitored *guest* is a small molecule and the *host* is a large polymer.

(c) Precious amounts of material may not be available to achieve the optimum conditions. At the required concentration levels large protein solutions may not be very clear, scattering may need to be reduced by filtration/centrifugation. The protein and any impurities are likely to produce considerable background absorption of light to interfere with the spectroscopic measurements.

(d) The spectra of the *guest* and the *host* may overlap. A discriminating wavelength for analysis may be difficult to find. This is particularly true if the *host* is a large molecule experiencing local changes that are small on a global scale that mask the relatively more pronounced changes with very small change the large change associated with the *guest* molecule.

Consider the case of two peptides/proteins interacting. A small peptide is likely to undergo a relatively large change in conformation (and CD) on binding.

This could be monitored in the far UV 200–240 nm region. However, if the *host* is a large protein giving only local changes in conformation at the binding site. The CD associated with the bulk of the *host* protein will mask any changes in CD derived form local changes. Hopefully, either the *host* or the *guest* has a prominent aromatic based CD in the near UV around 275 nm, particularly tryptophan, associated with the binding centre. The aromatic based CD now provides a good monitor feature. The protein carrying the monitored aromatic residue should now be treated as the *guest*.

Consider the case of copper binding to a cyclic peptide. Complexed Cu^{2+} ions present potentially selective CD at about 600 nm. However, the extinction coefficient is low $\varepsilon \sim 100$ and to maintain $A \sim 1.0$ means a *guest* concentration $\sim 10^{-2}$ M with and *host* concentration of $\sim 10^{-2}$ M. This is obviously unreasonable. On the other hand, the peptide and the copper complex show strong absorption and CD about 225 nm. The free Cu^{2+} charge transfer at this wavelength has no associated CD. Therefore, the binding process can be monitored by the CD at 225 nm with the peptide treated as the *guest* (fixed concentration) and Cu^{2+} as the *host*. The extreme points of the titration are the free peptide CD and the CD of the complex. See ref. 17 for an example of Cu–peptide binding.

The optical properties of the *guest*, *host*, and the complex may mean that ideal concentrations are outside the path length range to produce an absorbance of ~ 1.0. The choice of measurement concentrations/path lengths is fundamental to good CD measurements. This may not be trivial and consideration is required on an individual basis. Judicial compromises may be required in the real case.

Ideally, a CD titration experiment can be undertaken in one of three ways:

(a) Take a single solution with the appropriate *guest* concentration and add very small aliquots of a concentrated *host* solution. Make measurements after each addition covering the range zero *host* to about five molar excess of the *host*. The concentration of the *guest* should remain effectively unchanged (minor corrections for larger *host* solution additions may be appropriate). A plot of measured CD against *host* concentration will now yield the titration curve.

(b) Two solutions are prepared one of the *guest* and one of the *host*. A series of solutions is now prepared containing a fixed concentration of *guest* and a variable concentration of *host* (up to about five molar excess). Separate measurements are made of these solutions.

(c) A third method more suitable for automation and less consuming of reagents involves preparing two solutions: one containing the *guest* at the required measurement concentration (solution A) and the other containing both the *guest* at the required measurement concentration and the *host* at a five molar ratio excess (solution B). A known volume of solution B (full complexation) is taken for measurement and the titration proceeds by judicial additions of solution A or by the removal of a known volume of the measurement solution followed by the addition of a known volume of solution A. The procedure can also be undertaken starting with solution A

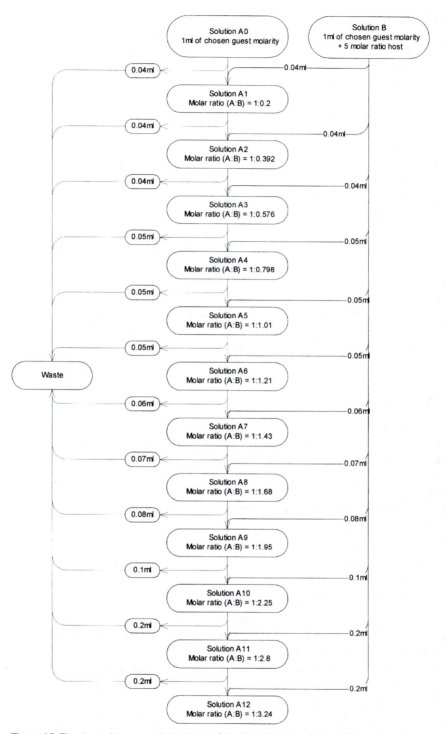

Figure 18 Titration scheme employing 1 ml of the *host* (solution A) by withdrawals and additions of appropriate solutions of B (1 ml) containing a fixed concentration of the *guest*.

increasing the proportion of *host* by the appropriate addition or removal and addition of solution B. This latter procedure is presented in flow diagram form in *Figure 18*.

Note: all measurements must be baseline corrected with a measurement of the appropriate solution (solvent, buffer salts, ...etc). This usually involves a single measurement at the outset of the experiment. It is important to ensure that the total absorbance of the measured solutions should remain as close to 0.8 as possible. This should not be too difficult as the concentration of the monitored chromophore should remain constant.

8 Analysis of CD ligand binding titration studies

By tradition, biochemists have been interested in the effect of a ligand on a protein or receptor. Accordingly, many traditional titrations of the Scatchard type have been done by keeping the *host* (protein or receptor) concentration constant and varying the ligand concentration. The methodology employed in this paper prefers the *guest* (ligand/monitored species) concentration to be constant and titrating with *host* until an excess of the *host* witnesses all the *guest*/ligand having been bound to give a limiting CD associated with the *guest*/ligand absorptions. The mathematical analysis of the CD ligand binding data is based upon the Law of Mass Action, Beer's law, and the non-linear fitting of the binding equation expressed as the observed CD as a function of *guest* and *host* concentrations.

The application of CD to ligand binding studies, as a method in its own right has not been extensively reported. The preliminary results from our laboratory reveal that with proteins like HSA the simple 1:1 binding case is frequently too simplistic and multi-site binding algorithms need to be developed. This is the result of the specificity that CD spectroscopy offers. Accordingly, only the simple case of 1:1 binding will be explored in this article.

Three situations can be examined based on the presence of chirality:

(a) Class 1A: Non-chiral *guest* / chiral *host*. Single binding site.

(b) Class 1B: Chiral *guest* / non-chiral *host*. Single binding site.

(c) Class 1C: Chiral *guest* / chiral *host*. Single binding site.

In order to monitor the CD of the *guest*, the *guest* must have an accessible ordinary electronic absorption. The free *guest* may or may not be chiral. That is the free *guest* may or may not have an associated natural CD in free solution. At

Figure 19 The 1:1 binding equilibrium of *guest* and *host*.

the same time the *host* may or may not have a CD at the chosen wavelength. Notwithstanding, at least one of the *host/guest* pair must be naturally chiral. Therefore, in terms of analysing titration data, four cases can be distinguished for completeness.

(a) Case IA:

Free *guest* shows no CD signal at *guest* based detection wavelength.

Free *host* shows no CD signal at detection wavelength.

The bound *guest* (the complex) shows a CD signal at *guest* based detection wavelength.

(b) Case IA′:

Free *guest* shows no CD signal at *guest* based detection wavelength.

Free *host* shows a CD signal at detection wavelength.

The bound *guest* (the complex) shows a CD signal at *guest* based detection wavelength.

(c) Case IB:

Free *guest* shows a CD signal at *guest* based detection wavelength.

Free *host* shows no CD signal at detection wavelength.

The bound *guest* (the complex) shows a CD signal at *guest* based detection wavelength.

(d) Case IB′:

Free *guest* shows a CD signal at *guest* based detection wavelength.

Free *host* shows a CD signal at detection wavelength.

The bound *guest* (the complex) shows a CD signal at *guest* based detection wavelength.

8.1 Case IA

The binding of a single *guest* (ligand) to a single *host* site is adequately described by the standard formalism:

$$host + guest \leftrightarrow (host/guest) \text{ complex}$$

Initially, H_T = total *host* concentration; G_T = total *guest* concentration. At equilibrium: H_f = free *host* (protein) concentration; G_f = free *guest* (ligand) concentration; HG = *(host/guest)* complex concentration.

$$H_f + G_f \leftrightarrow HG$$

$$H_T = H_f + HG \tag{8}$$

$$G_T = G_f + HG \tag{9}$$

The association constant (often referred to as the binding constant or affinity constant) is defined in terms of free *guest* (ligand) concentration G_f, free *host* concentration H_f, and *host/guest* complex concentration HG as:

$$K = \frac{HG}{H_f \cdot G_f} \tag{10}$$

Strictly K should be written as K_a. Note: A dissociation constant is often defined $K_d = K_a^{-1}$.

The calculated CD (theoretical) induced signal of the bound *guest*, based upon measurements in a 1 cm path length cell, is:

$$\Delta A_{calc} = \Delta A_{HG} = \varepsilon_{HG} \cdot HG \qquad [11]$$

It is important to note that ε_{HG} is the molar CD of the complex at the *guest* based detection wavelength.

Note: For a b cm path length, the associated CD will be:

$$\Delta A'_{calc} = \Delta A'_{HG} = b \cdot \varepsilon_{HG} \cdot HG \qquad [11a]$$

Dividing both members of *Equation 11a* by b, the expression *Equation 11* will be obtained.

The methodology employed prefers the *guest* concentration to be constant and titrating with *host* until an excess of the *host* witnesses all the *guest* (ligand) having been bound to give a limiting CD associated with the *guest* (ligand) absorptions.

There is a system of four equations (*Equations 8–11*) with four unknowns: H_f, G_f, HG, and ε_{HG}. The total *host* concentration H_T and the total *guest* concentration G_T are pre-set quantities, therefore known. The independent variable is the total *host* concentration. Rewriting *Equations 8–10* will give the expressions:

$$H_f = H_T - HG \qquad [12]$$

$$G_f = G_T - HG \qquad [13]$$

$$HG = K \cdot H_f . G_f \qquad [14]$$

Bringing *Equations 12–14* will provide the following solutions:

$$HG_1(H_T) = \frac{K \cdot H_T + K \cdot G_T + 1 - \sqrt{(K \cdot H_T + K \cdot G_T + 1)^2 - 4 \cdot K^2 \cdot H_T \cdot G_T}}{2 \cdot K} \qquad [15]$$

$$HG_2(H_T) = \frac{K \cdot H_T + K \cdot G_T + 1 + \sqrt{(K \cdot H_T + K \cdot G_T + 1)^2 - 4 \cdot K^2 \cdot H_T \cdot G_T}}{2 \cdot K} \qquad [16]$$

Introducing the expressions (*Equations 15 and 16*) into *Equation 11* leads to two solutions for ΔA_{calc}:

$$\Delta A_{calc1} = \varepsilon_{HG} \cdot HG_1 \qquad [17]$$

$$\Delta A_{calc2} = \varepsilon_{HG} \cdot HG_2 \qquad [18]$$

Using the expressions (*Equations 15 and 16*) in the formulae above, the two solutions obtained are:

$$\Delta A_{calc1} = \varepsilon_{HG} \cdot \frac{K \cdot H_T + K \cdot G_T + 1 - \sqrt{(K \cdot H_T + K \cdot G_T + 1)^2 - 4 \cdot K^2 \cdot H_T \cdot G_T}}{2 \cdot K} \qquad [19]$$

$$\Delta A_{calc2} = \varepsilon_{HG} \cdot \frac{K \cdot H_T + K \cdot G_T + 1 + \sqrt{(K \cdot H_T + K \cdot G_T + 1)^2 - 4 \cdot K^2 \cdot H_T \cdot G_T}}{2 \cdot K} \qquad [20]$$

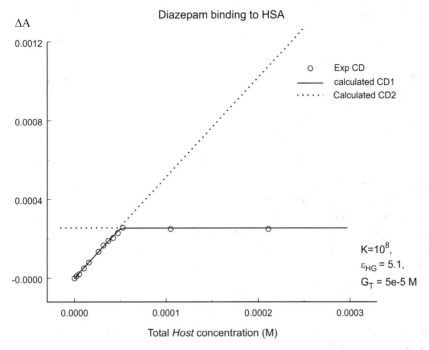

Figure 20 The diazepam CD / human serum albumin binding curve (Case 1A).

A graphical representation of the calculated CD signal against the total *host* concentration is shown in *Figure 20*. An experimental ΔA_{obs} (diazepam binding to HSA) is also represented graphically against the total *host* concentration H_T in *Figure 20*. The experimental data was taken from the work conducted by Mrs Beauleah Banfield (PhD project with Dr A. F. Drake, King's College, London). A repeat of this work with more appropriate concentration conditions is presented in *Figure 26*.

Clearly ΔA_{calc1} (H_T) is the mathematical function which provides the realistic solution. Typical examples of this case include Cu^{2+} binding to a protein, a non-chiral drug binding to a single site on a protein or a nucleic acid, a non-chiral drug binding to cyclodextrin.

8.2 Case IA′

$$Host + Guest \leftrightarrow (Host/Guest) \text{ complex}$$

Initially, H_T = total *host* concentration; G_T = total *guest* concentration. At equilibrium: H_f = free *host* (polymer) concentration; G_f = free *guest* (drug) concentration; HG = (*host/guest*) complex concentration; $H_f + G_f \leftrightarrow HG$.

At equilibrium:

$$H_T = H_f + HG \qquad [21]$$

$$G_T = G_f + HG \qquad [22]$$

155

The association constant is:

$$K = \frac{HG}{H_f \cdot G_f}$$
[23]

Following the same logic as in Case IA the calculated (theoretical) induced CD signal of the bound *guest* is:

$$\Delta A_{calc} = \Delta A_{HG} + \Delta A_{Hf}$$

$$\Delta A_{calc} = \varepsilon_{HG}. HG + \varepsilon_H \cdot H_f$$

$$\Delta A_{calc} = (\varepsilon_{HG} - \varepsilon_H) \cdot HG + \varepsilon_H \cdot H_f$$
[24]

It is important to note that ε_{HG} is the molar CD of the complex at the *guest* based detection wavelength and has to be determined, ε_H is the molecular CD of the *host* alone (free) and has a known value.

The solutions for the complex concentration are given, as previously stated by *Equations 15* and *16* which when introduced in *Equation 24* will give two solutions for ΔA_{calc}:

$$\Delta A_{calc1} = (\varepsilon_{HG} + \varepsilon_H) \cdot HG_1 + \varepsilon_H \cdot H_T$$
[25]

$$\Delta A_{calc2} = (\varepsilon_{HG} - \varepsilon_H) \cdot HG_2 + \varepsilon_H \cdot H_T$$
[26]

The graphical representation of the calculated CD signal against the total *host* concentration is shown in *Figure 21*.

$\Delta A_{calc1}(H_T)$ is the mathematical function, which will more realistically represent the experimental results. An example of this case would be a *host* showing

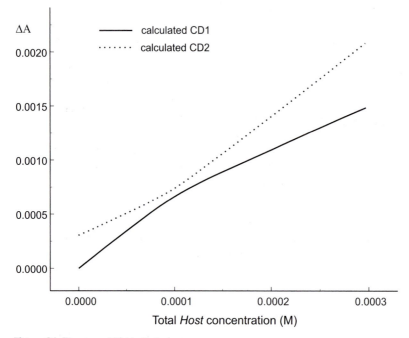

Figure 21 The case 1A′ binding curve.

a signal at the same wavelength as the monitored one characterizing the bound complex. However, it is, experimentally, a bad case because it will be difficult to achieve a good estimate of ε_{HG}.

8.3 Case IB

The free *guest* (ligand) is itself chiral and now displays optical activity which is modified on binding to the *host*. Providing there is no *host* CD being monitored, following the same principles as in Case IA and keeping the same notations, the calculated CD of the bound complex will be given by the *Equation 27*:

$$\Delta A_{calc} = \Delta A_{HG} + \Delta A_{Gf}$$

$$\Delta A_{calc} = \varepsilon_{HG}\, HG + \varepsilon_G \cdot G_f$$

$$\Delta A_{calc} = (\varepsilon_{HG} - \varepsilon_G) \cdot HG + \varepsilon_G \cdot G_f \qquad [27]$$

It should be noted that ε_{HG} is the molar CD of the complex at the *guest* based detection wavelength and has to be determined, ε_G is the molecular CD of the *guest* alone and has a known value. The solutions for the complex concentration are given by the same *Equations 15* and *16*.

Introducing *Equations 15* and *16* into *Equation 27*, the theoretical CD will have the two solutions:

$$\Delta A_{calc1} = (\varepsilon_{HG} - \varepsilon_G) \cdot \frac{K \cdot H_T + K \cdot G_T + 1 - \sqrt{(K \cdot H_T + K \cdot G_T + 1)^2 - 4 \cdot K^2 \cdot H_T \cdot G_T}}{2 \cdot K}$$

$$+ \varepsilon_G \cdot G_T \qquad [28]$$

$$\Delta A_{calc2} = (\varepsilon_{HG} - \varepsilon_G) \cdot \frac{K \cdot H_T + K \cdot G_T + 1 + \sqrt{(K \cdot H_T + K \cdot G_T + 1)^2 - 4 \cdot K^2 \cdot H_T \cdot G_T}}{2 \cdot K}$$

$$+ \varepsilon_G \cdot G_T \qquad [29]$$

The experimental ΔA_{obs} (observed CD) is represented graphically against the *host* concentration H_o in *Figure 22*. The graphical representation of the calculated CD signal against the *host* concentration is also shown in *Figure 22* for the binding of ibuprofen to cyclodextrin (the experimental data is taken from an MSc project by Mrs Jarna Khan working with Dr A. F. Drake).

Clearly $\Delta A_{calc1}(H_T)$ is the mathematical function which realistically represents the experimental results. Typical examples of this case include a chiral ligand such a peptide binding to Cu^{2+} (a non-chiral *host*).

8.4 Case IB′

Using the assumptions adopted in the previous cases and keeping the same notations, the theoretical differential absorption of the bound complex will be given by the following *Equation 30*:

$$\Delta A_{calc} = \Delta A_{HG} + \Delta A_{Gf} + \Delta A_{Hf}$$

$$\Delta A_{calc} = \varepsilon_{HG}\, HG + \varepsilon_G \cdot G_f + \varepsilon_H \cdot H_f$$

$$\Delta A_{calc} = (\varepsilon_{HG} - \varepsilon_G - \varepsilon_H) \cdot HG + \varepsilon_G \cdot G_f + \varepsilon_H \cdot H_f \qquad [30]$$

157

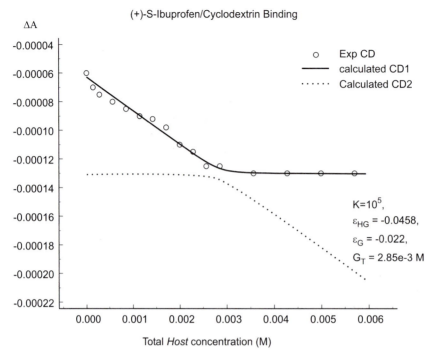

Figure 22 The (+)-S-ibuprofen / cyclodextrin binding curve (Case 1B).

It is important again to note that ε_{HG} is the molar CD of the complex at the *guest* based detection wavelength and has to be determined, ε_G is the molecular CD of the *guest* alone and has a known value, ε_H is the molecular CD of the *host* alone and has a known value.

The solutions for the concentration of the complex are given by the same expressions *Equations 15* and *16*. After introducing *Equations 15* and *16* into *Equation 30* the theoretical CD will have the two solutions:

$$\Delta A_{calc1} = (\varepsilon_{HG} - \varepsilon_G - \varepsilon_H) \cdot HG_1 + \varepsilon_G \cdot G_T + \varepsilon_H \cdot H_T \qquad [31]$$

$$\Delta A_{calc2} = (\varepsilon_{HG} - \varepsilon_G - \varepsilon_H) \cdot HG_2 + \varepsilon_G \cdot G_T + \varepsilon_H \cdot H_T \qquad [32]$$

The graphical representation of the calculated CD signal against the *host* concentration is shown in *Figure 23*.

$\Delta A_{calc1}(H_T)$ is the mathematical function, which will more realistically represent the experimental results. An example of this case is the dimerization process (with the role of *guest* and *host* being played by the same monomers) of two chiral molecules.

The experimental results in the examples above were fitted with the non-linear equations describing the theoretical CD signal using Levenberg–Marquardt fitting procedure in the Mathsoft MathCad computer program. The method has become the standard of nonlinear least-squares routines.

The treatment presented here is explicitly with respect to CD. However, the

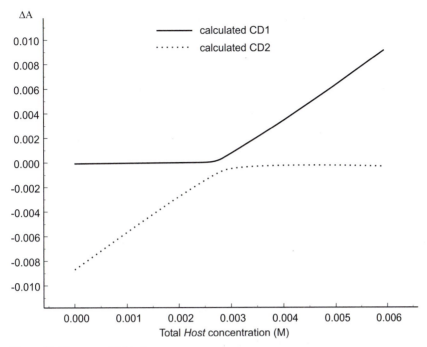

Figure 23 The case 1B′ binding curve.

treatment is general and can be applied to other techniques such as ordinary absorption or fluorescence. In terms of ordinary absorption the relevant equations can be derived simply by replacing $\Delta\varepsilon$ with ε and ΔA with A. (Note: The molecular differential absorbance $\Delta\varepsilon$ was named ε for simplification of this Chapter.) There will remain three cases respectively replacing the terms *chiral* and *non-chiral* with *absorbing* and *non-absorbing*.

Provided it is well defined in the mathematical treatment, from the physico-chemical point of view it does not matter which of the two partners is defined as the *host* and which is the *guest* (ligand). If the experiment were done keeping the *host* (protein) concentration fixed with an increasing concentration of the *guest* (ligand), the expressions for ΔA_{calc} in the four cases would remain the same. The variable in this context will not be the *host* concentration, but the *guest* concentration. The mathematical functions representing better the experimental results will be $\Delta A_{calc1}(G_T)$ (all with index 1).

To fit *Equations 19* and *20* there are two unknown parameters at the beginning of a titration: the molar CD, ε_{HG} associated with the bound complex and the binding constant K. The value of ε_{HG} is obtained from the limiting value of ΔA_{obs} (ΔA_{lim}) reached as the result of titrating a fixed concentration of the *guest* (ligand) with increasing amounts of the *host*.

To the detriment of the CD spectroscopist, it may be difficult to distinguish *guest* (ligand) and *host* signals. Thus for example with protein studies the *host* (protein) may contain aromatic residues that spectroscopically overlaps the absorption of the single chromophore in the *guest* (ligand). Therefore analytical wavelengths < 290 nm can be restrictive.

Traditionally, CD is considered to be an instrumentally 'noisy' technique. Modern instruments, to a large extent, rectify this. Nevertheless the choice of monitoring conditions (concentration, path length, and λ) can prove to be crucial.

Consider the case of a Cu^{2+}/ peptide study. The experimenter could choose to monitor the Cu^{2+} absorption between 600 and 800 nm. The Cu^{2+} would be viewed as the ligand with a constant concentration and the peptide *host* concentration would be increased until the induced CD levelled. However, Cu absorbs very weakly at this wavelength and high concentrations (large sample quantities) are required. Cu^{2+}, the peptide, and the Cu/peptide complex absorb more strongly in the UV and smaller quantities will be required. Monitoring at $\sim$ 230 nm means treating the peptide as the ligand and Cu^{2+} as the *host*. Now the peptide concentration needs to be fixed as the experimenter follows the effects of increasing the Cu^{2+} concentration. It is interesting to note that the ordinary absorption changes at this wavelength are relatively complex, including contribution from free Cu^{2+}. However, only the *guest* (ligand or peptide) and the *guest/host* complex (or peptide/Cu) have CD at this wavelength and Case IB applies (see ref. 17).

8.5 The effect of concentration on binding constant determination

For a given *guest:host* concentration ratio, the shape of the binding curve is dependent on various parameters of the binding equation, e.g. different extinction coefficients, different total *guest* concentration, different binding constants. The effect of the magnitude of the binding constant on the Case IA binding curve will be explicitly discussed here. Fitting various K values in *Equation 19* for a *guest* concentration of 5.10^{-5} M gives the graphical outputs shown in *Figure 24a*.

The 'sharpness' of the binding curve increases with an increase in the binding constant value. The higher the binding constant the more readily the free *guest* molecules are co-ordinated in the formation of the bound complex. The binding curve becomes sharper at a 1:1 molar ratio (total *host*: total *guest* concentration) for higher values of the binding constant. In the illustrated example K values $> 10^6$ cannot be experimentally distinguished.

The need to design a procedure in which it is possible to readily assign a significant K_a value whilst ensuring that ΔA_{lim} can also be determined with precision leads to the questions what is a significant K_a value and when is ΔA_{lim} sufficiently precise. Reference to *Figure 24* indicates that it is necessary to distinguish curves between different K_a values whilst not requiring an unrealistic huge excess of *Host* in order to determine the ΔA_{lim} value. Thus according to *Figure 24b*, a G_T concentration of $\sim 5 \cdot 10^{-5}$ M is suitable for determining K_a values of the order of 10^4. A 95–98% detection of ΔA_{lim} would seem to be a reasonable target. K_a determination should be undertaken with the condition that $G_T \cdot K_a$

(a)

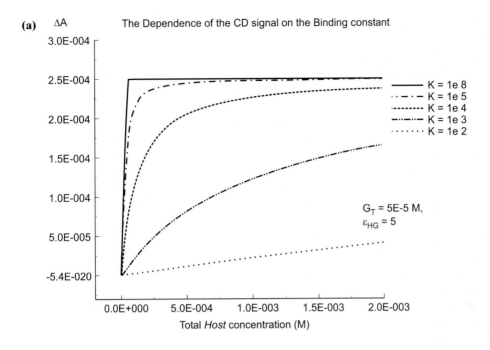

(b)

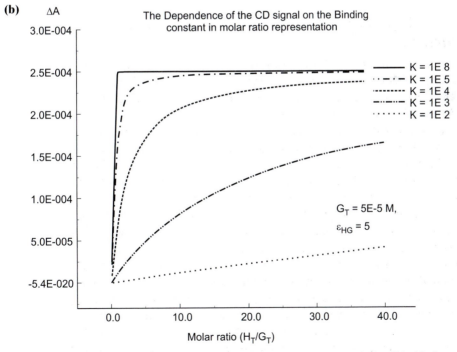

Figure 24 (a) The dependence of the Case 1A binding curve on the magnitude of the binding constant presented in terms of *host* concentration for a fixed *guest* concentration. (b) The dependence of the Case 1A binding curve on the magnitude of the binding constant presented in terms of *host* : *guest* mole ratio.

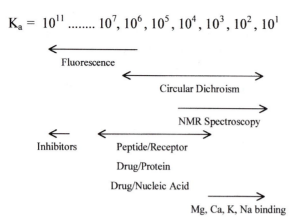

$$K_a = 10^{11} \dots \dots 10^7, 10^6, 10^5, 10^4, 10^3, 10^2, 10^1$$

Figure 25 The binding constant regimes of various spectroscopic techniques.

takes values between 0.5 and 5. Optimum values should be chosen between 0.5 and 2.5 ensuring a good discrimination between plots especially in the 0.5 to 4 *host/guest* molar ratio range (see *Figure 24b*).

Taking the value of 2 for a $K_a = 10^8$ M^{-1} leads to the conclusion that a concentration of $G_T = 2 \cdot 10^{-8}$ M is required for a good titration. This is an unrealistically low concentration for optical spectroscopy. Applying Beer's law for ordinary absorption $\varepsilon = A/c \cdot l$ in this condition gives for a small $A = 0.1$ measured in a 10 cm cell an $\varepsilon = 5 \cdot 10^5$. The largest K_a that can be reasonably reached is likely to be $\sim 10^7$ M^{-1}. Higher values of K_a will need to be determined by techniques such as fluorescence. The lowest conceivable limit for K_a based on the same arguments is ~ 1.0 M^{-1}. A comparison of different spectroscopic techniques is revealing. Fluorescence is only proportional to concentration if $A < 0.02$ this means that for an $\varepsilon = 10\,000$ the concentration is $2 \cdot 10^{-6}$ M which corresponds to the determination of a $K_a = 10^6$ M^{-1} as a typical lowest K_a value that can be determined. CD spectroscopy is an excellent technique for deter-

Table 3 Summary of the criteria required for determining binding constants by CD spectroscopy

Titration	The guest (monitored species) concentration needs to be in the range: $[G_T] = (0.5/K_b)$ to $(5.0/K_b)$
1:1 complex	A minimum concentration range for 1:1 complexation is given by: $[G_T] = (50/K_b)$ to $(100/K_b)$
CD measurement Condition	The following condition needs to be satisfied: $$\left[\begin{array}{l}Molar\ extinction \\ coefficient \cdot (Mol^{-1}\ cm^{-1})\end{array}\right] \times \left[\begin{array}{l}Molar \\ concentration\ (mol/litre)\end{array}\right] \times \left[\begin{array}{l}cell \cdot pathlength \\ (cm)\end{array}\right] \approx 0.8$$ The ability to vary the measurement path length from 0.02 to 10 cm in optical absorption spectroscopy (CD and absorption) increases the concentration range of K_b determination.

mining K_a values in the range 10^3 to 10^6. NMR spectroscopy can, at a limit, monitor concentrations of the order $2 \cdot 10^{-4}$ M which correlates with the strongest $K_a \sim 10^4$ M^{-1}. This provides the scale shown in *Figure 25*.

The arguments presented above lead to the guidelines presented in Section 7 as being the criteria for determining binding constants by CD spectroscopy. These are reiterated here in the form of *Table 3*.

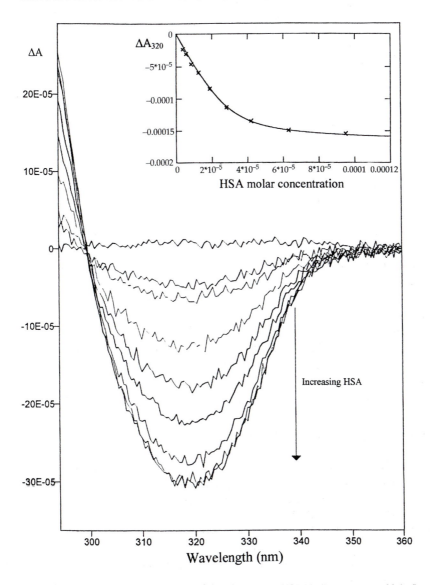

Figure 26 An improved determination of the diazepam / HSA binding constant. Main figure: CD spectra associated with a diazepam concentration of $2.87.10^{-5}$ M as a function of HSA concentration (see insert). Insert: CD at 320 nm as a function of HSA concentration. −x−x−x− experimental data—curve produced for a Case 1A model with $\varepsilon_{HG} = -5.8$ and K = $2.4.10^{-5}$.

Following these arguments the determination of the diazepam/HSA binding constant presented in *Figure 20* is better presented by the set of conditions shown in *Figure 26*. Further examples specifically concerned with the monitoring of peptide/protein interactions can be found in the work of Siligardi (18).

9 CD kinetics—drug binding to proteins

Human serum albumin (HSA) is the major drug-transporting agent in the blood stream enabling the targeting of drugs to active sites. The interaction of drugs with HSA is therefore important in understanding the factors affecting their efficacy. CD binding studies have been initiated to determine the kinetics of drug binding to proteins by observing the induced CD signal that results during the binding process.

The example described here concerns the binding of diazepam to HSA. The CD spectrum of HSA in the near UV is presented in *Figure 26*. A weak featureless signal is observed at around 265 nm—most likely associated with the -S-S- chromophores of HSA.

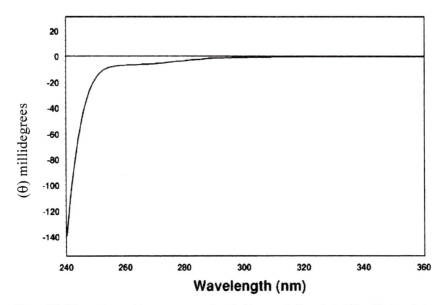

Figure 27 CD spectrum of human serum albumin (1 cm cell, 5 mg/ml, 100 mM phosphate buffer pH 7).

Diazepam

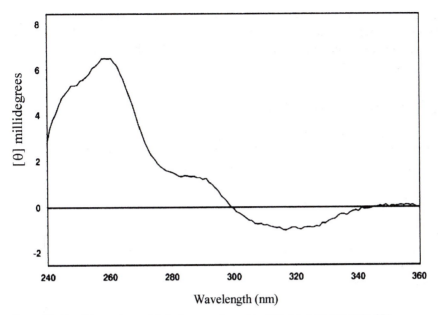

Figure 28 The CD spectrum of 1:1 diazepam bound to HSA (corrected for HSA CD).

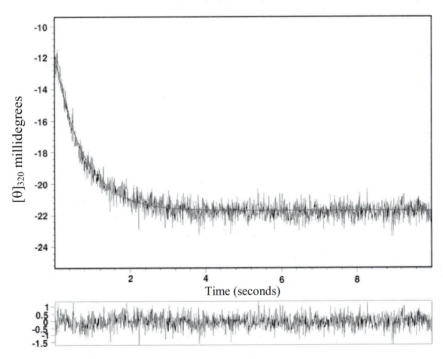

Figure 29 The kinetics (upper curve) describing the binding of diazepam to HSA as followed using the CD signal at 320 nm. The lower curve presents the residual associated with a single exponential kinetic curve fit.

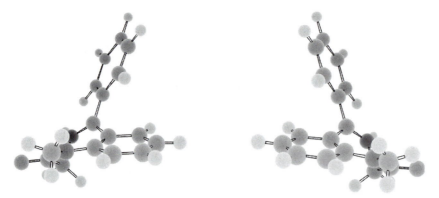

Figure 30 The two mirror image conformations of diazepam.

The molecular structure of diazepam is on page 164. Diazepam is not chiral and will not display optical activity in isolation. However, on binding to HSA prominent CD associated with the UV absorptions of the aromatic groups is observed (*Figure 28*).

Figure 29 shows the kinetic record obtained when diazepam (0.03 mg/ml in 100 mM phosphate buffer) was mixed (1:1 ratio) with HSA (10 mg/ml in 100 mM phosphate buffer pH 7) using the APL stopped-flow unit attached to the APL π^*-CDF180 Spectrometer.

Diazepam exists in a conformational equilibrium of mirror image conformations (*Figure 30*).

In free solution, there is a 50/50 conformer mixture with no preference and no CD. In our experience, the normal binding of a drug to HSA is complete in under 100 msec (A. F. Drake, unpublished data). In the case of diazepam, only one conformer binds to HSA and after about 100 msec there is a 50/50 mixture of bound and unbound species. There is still no major CD. As the free solution conformation mixture requilibrates so more diazepam is available for binding. There is now an excess of one diazepam conformation and inherent CD (as imposed to induced CD) is developed. The CD is in effect monitoring the free solution kinetics and the time taken for diazepam to convert to a form that can be bound to HSA.

References

1. Mason, S. F. (1961). *Q. Rev.*, **15**, 63; Mason, S. F. (1963). *Q. Rev.*, **17**, 20; Mason, S. F. (1982). *Molecular optical activity and the chiral discriminations*. Cambridge University Press, Cambridge, UK.

2. Siligardi, G., Campbell, M. M., Gibbons, W. A., and Drake, A. F. (1992). *Eur. J. Biochem.*, **206**, 23.

3. Drake, A. F. (1986). *J. Phys. E*, **19**, 170; Drake, A. F. (1988). *Physical methods of chemistry, Vol 3B: Determination of chemical composition and molecular structure* (ed. B. W. Rossiter and J. F. Hamilton), pp. 1–41. Wiley, New York.

4. Nakanishi, K., Berova, N., and Woody, R. W. (ed.) (1994). *Circular dichroism, principles and applications*. VCH Publishers, New York.

5. Drake, A. F. (1994). In *Methods in molecular biology, Vol 22: Microscopy, optical spectroscopy and macroscopic techniques* (ed. C. Jones, B. Mulloy, and A. H. Thomas). Humana Press, New Jersey; Drake, A. F. (1993). *Spectroscopic assignments in the CD spectra of proteins and peptides*, pp. 21–46. International CD Conference Colorado State University, Pingree Park, Colorado, USA.

6. Fasman, G. D. (ed.) (1996). *Circular dichroism and the conformational analysis of biomolecules*, Plenum Press, New York.

7. Greenfield, N. and Fasman, G. D. (1969). *Biochemistry*, **8**, 4108.

8. Johnson, W. C. Jr. (1990). *Proteins: Struct. Funct. Genet.*, **7**, 205.

9. Kabsch, W. and Sander, C. (1983). *Biopolymers*, **22**, 2577.

10. Levitt, M. and Greer, J. (1977). *J. Mol. Biol.*, **114**, 181.

11. Toumadje, A., Alcorn, S. W., and Johnson, W. C. Jr. (1992). *Anal. Biochem.*, **200**, 321.

12. Ledermann, W. (1981). In *Numerical methods, Vol. 3* (ed. R. F. Churchhouse). John Wiley & Sons Ltd., Chichester, UK.

13. Geladi, P. and Kowalski, B. R. (1986). *Anal. Chim. Acta*, **185**, 1; Geladi, P. and Kowalski, B. R. (1986). *Anal. Chim. Acta*, **185**, 19.

14. Greenfield, N. (1996). *Anal. Biochem.*, **235**, 1.

15. Drake, A. F., Siligardi, G., and Gibbons, W. A. (1988). *Biophys. Chem.*, **31**, 143.

16. Siligardi, G. and Drake, A. F. (1995). *Biopolymers*, **4**, 281.

17. Freeman, D. J., Pattenden, G., Siligardi, G., and Drake, A. F. (1998). *J. Chem. Soc. Perkin Trans II*, 129.

18. Siligardi, G. and Hussain, R. (1998). *Enantiomer*, **3**, 77.

Chapter 4

Generic techniques for fluorescence measurements of protein–ligand interactions; real-time kinetics and spatial imaging

Josep Cladera and Paul O'Shea

School of Biomedical Sciences, Medical School, University of Nottingham, Nottingham NG7 2UH, UK.

1 Introduction

The non-covalent interactions of small molecules (i.e. the '*ligand*') with a larger molecule (the '*receptor*'—usually a protein or nucleic acid) underlies much of the 'process' of cell biology and physiology as illustrated by the broad range of examples covered in this volume. Whilst there are techniques available that offer very accurate quantitative and unequivocal measurements of ligand–receptor interactions (see, e.g. ref. 1, and other chapters in the present volumes), many such techniques are highly specialized and/or dedicated to a particular biological phenomenon. A number of more simple and routine techniques have also evolved to measure or indicate this behaviour and have tended to be grouped within one or more of the following general areas:

(a) The ligand–receptor interaction leads to a biological response that is assayed; examples include signalling responses (e.g. intracellular Ca^{2+} mobilization), i.e. the ligand–receptor association is inferred from the response. Strictly, this is then more of a dose–response reaction rather than a ligand–receptor binding reaction. With localized or non-linear signalling behaviour, however, the binding reactions may be difficult to disentangle from the response.

(b) Radio-derivatives of a ligand are utilized to indicate binding (particularly using so-called 'ligand-blot' techniques, see e.g. ref. 2) although low-resolution kinetic information is possible to obtain in this manner, real-time kinetics are very difficult to measure and in most cases quite inaccessible.

(c) A covalent chromophoric derivative of the ligand may be used to indicate binding through spectral or intensity (quantum yield) changes. Some elements of doubt inevitably become cause for concern by such covalent modifications to the ligand, however, as they may conceivably interfere with the receptor interactions, particularly if the ligand is small and the chromophoric moeity is large.

A number of specific solutions to these rather generalized problems involve some of the more ingenious techniques as evidenced by the present volume. Of all spectroscopic techniques that are available to study ligand–receptor interactions, however, fluorescence probably offers the greatest versatility of applications, most likely because of its inherent sensitivity and simplicity of use. It is a phenomenon that involves the emission of electromagnetic radiation ca. a few nanoseconds after absorption by an appropriate chromophore centre. This is a convenient time period for a number of molecular processes to take place that can affect the emission properties of the chromophore (see below). Fluorescence, however, is not a feature of all biological molecules and although many proteins, nucleotides, etc. inherently possess this property (i.e. intrinsic) it is sometimes not appropriate to make use of it. It is often necessary, therefore, to incorporate (usually covalently) one of a number of different kinds of fluorescent label (such as fluorescein or rhodamine) in order to visualize a particular reaction but as mentioned above, this may lead to perturbation of the process under investigation. Helpful and comprehensive reviews of the very many applications of fluorescence spectroscopy to biological problems may be found in Lakowicz, and Brand and Johnson (3, 4).

In the present review, we try to emphasize generic applications of fluorescence technology with a number of examples and we outline two general types of study that are as about as non-invasive as it is possible to achieve with spectroscopic studies of biological samples. In the first, we demonstrate that use may be made of the intrinsic fluorescence of tryptophan within a protein (serum albumin) to indicate ligand binding and ligand-dependent conformational equilibria. In this example, the analysis is aided by the fact that the serum albumin utilized (human) possess only a single tryptophan and so we can be confident of the origins of the fluorescence signals. This state of affairs is not typical and often methods must be devised to discriminate the fluorescence originating from several tryptophan sources within a protein (3, 4). More recently it has been possible to engineer proteins to possess desirable intrinsic fluorescence properties and less ambiguity in the origins of the fluorescence.

The main content of the present article, however outlines a generic fluorescence technique that does not involve labelling either the ligand or the receptor with a fluorescent indicator but does involve labelling the matrix upon which the intermolecular associations take place (i.e. membranes). In the latter technique, virtually all ligand–receptor interactions that take place with membrane proteins may be studied in a practically non-invasive manner. In order to describe the methodology for this technique, it is necessary first to outline briefly the electrical properties of membranes as the technology involves measurements of very small changes of the membrane surface electrostatic potential as the result of ligand binding. This potential is one of three separate electrical potentials associated with membranes. Application of the technology to measurements of ligand–receptor interactions are illustrated with a number of ligands in several different receptor systems. In order to maintain a connecting theme within this chapter, however, we have included some of our other work with albumin to illustrate how implementation of the technology reveals something of the nature of the albumin binding reactions with its cell surface receptor. The final development of the same technology is used to make localized measurements of ligand binding on a single cell surface using high-resolution fluorescence microscopy. Albumin is used to illustrate single cell imaging and may also be employed to determine the spatial localization of the ligand (albumin)–receptor interactions on the endothelial cell surface.

2 Ligand–protein interactions revealed by intrinsic (tryptophan) fluorescence

Intrinsic fluorescence from particular amino acids has been found to yield valuable information about their local environment within proteins and has found extensive application as a probe of protein structure and dynamics (5–8) via lifetime and anisotropy decay measurements. Tryptophan in particular offers a powerful intrinsic probe of environments within proteins and of the interactions of peptides with other structures (e.g. membranes).

The ground to first excited state transition of tryptophan involves two closely lying singlet states assigned as ^1L_a and ^1L_b (9, 10) with approximately orthogonal transition dipole moments (10, 11). The polarization of tryptophan fluorescence is strongly dependent upon the relative contributions of these two transitions in both the absorption and emission processes. Studies by Valeur and Weber (9) indicate that in hydrophilic solvents the ^1L_a transition predominates at wavelengths above 295 nm.

A nice example of how these properties may be used is exemplified with human serum albumin (HSA). Albumin is involved in solute transport within the circulatory system (12). The precise mechanisms of transport and targeting are still unclear but there is evidence that albumin may target ligands to particular tissues (12, 13). Fatty acids such as oleic acid is one of a number of important such ligands for their aqueous solubilities are very low and they appear to be carried mostly by albumin (12, 14, 15).

In HSA the presence of multiple tyrosine residues gives rise to an additional absorption route and although the resulting fluorescence is relatively weak there is the potential complication of energy transfer to tryptophan (16). At wavelengths above 290 nm tyrosine absorption, however, is minimal and usually neglected (16). Thus, by exploiting the fact that HSA possesses a single tryptophan (Trp 214) it is feasible to study the accessibility of this moiety by the phenomenon of fluorescence quenching (17) and acrylamide (15, 18, 19).

3 Tryptophan fluorescence measurements to demonstrate ligand–protein interactions

3.1 Steady state tryptophan fluorescence quenching

By offering an alternative route for the depopulation of excited states, certain molecules (often relatively small and electron rich) are able to reduce the fluorescence yield of a given fluorophore provided they are able to move into close proximity of the fluorescence centre. In the present example we have used the aqueous solute, acrylamide, to quench tryptophan fluorescence. The accessibility of these so-called 'collisional' quenchers to the fluorescent moiety is one parameter that dictates their effectiveness at quenching; behaviour that is characterized by the Stern–Volmer equation (3, 15, 16):

$$\frac{F_0}{F} - 1 = K_{sv}Q \qquad [1]$$

where F_0 and F are the fluorescence intensities before and after addition of quencher (concentration Q). K_{sv} is the effective quenching constant, and is dependent on both the lifetime before addition of quencher (τ_0) and the bimolecular rate constant of the quenching reaction (k_q).

According to *Equation 1* a plot of the quencher concentration together with the extent of quenching (shown in *Figure 1*) is linear with a slope of K_{sv}; this latter quantity is known as the effective quenching constant and is made up of the terms shown in *Equation 2*.

$$K_{sv} = k_q \tau_0 \qquad [2]$$

The quenching constant is enormously useful for indicating how fluorescent groups on or within large molecules such as proteins are exposed to the surrounding solvent. And, because exposure or the geometry of such groups may be altered following conformational changes of the protein, a Stern–Volmer quenching analysis before and after such behaviour can be quite revealing of both the properties of such structural changes as well as ligand binding itself. Fortunately, such quenching experiments may be carried out relatively easily with most standard benchtop spectrofluorimeters (e.g. in our laboratory we use Aminco-Bowman series 2 luminescence spectrometers) by measuring the effects of titrating a quencher into the protein sample and measuring steady state fluorescence intensity; then analysed according to the Stern–Volmer relation.

3.2 Time-resolved fluorescence lifetime and anisotropy decay measurements of albumin

Fluorescence lifetime and anisotropy decays measurements may be used to identify changes of the environment of tryptophan following ligand binding to albumin. The experimental difference between this kind of study and the foregoing 'steady-state' analysis is that time-resolved measurements require much more sophisticated instrumentation but yield valuable complementary information including that of the fluorescence lifetime (*Equation 1*) (15, 16). In the present example, (as outlined in *Protocol 1*, performed with Dr Angus Bain and

Dr Neil Chadborn presently at University College London), we used a time-resolved analysis of the changes of polarization over the picosecond time domain. Polarization analysis of the emission can be undertaken using a computer controlled analysing polarizer to resolve the vertical, horizontal, and 'magic angle' polarization components of the emission. The time evolution of each component may be built up using standard time correlated single photon counting (TCSPC) methods (20).

Excitation of an isotropic sample (with an ultrashort excitation pulse) the relationship between the fluorescence intensity detected an analyser angle of β with respect to the excitation polarization direction is given by *Equation 3*:

$$I(t,\beta) \propto N(t)\left[1 + (3\cos^2\beta - 1)R(t)\right] \tag{3}$$

where $R(t)$ is the degree of alignment of the emitting transition dipole moment in the laboratory frame and $N(t)$ is the excited state population. For an analyser angle of $\beta = 54.7°$ (the 'magic angle') the dipole alignment contribution to the emission intensity vanishes yielding *Equation 4*:

$$I(t,\beta = 54.7°) \propto N(t) \tag{4}$$

The fluorescence anisotropy that is experimentally determined is defined by *Equation 5*:

$$R(t) = \frac{I(t,\beta = 0°) - I(t,\beta = 90°)}{I(t,\beta = 0°) + 2I(t,\beta = 90°)} \tag{5}$$

and is a direct measure of the time dependent alignment of the emission transition dipoles of the fluorescing population. TCSPC anisotropy measurements yield an initial anisotropy that reflects the averaging of other faster processes. For tryptophan within HSA, the observed anisotropy should yield double exponential decays consistent with the evolution of two independent diffusive motions with a significant separation in their respective correlation times according to *Equation 6*.

$$
\begin{aligned}
R(t) &= \exp\left(-\frac{t}{t_2}\right) \times \left(A_1 \exp\left(-\frac{t}{t_1}\right) + A_2\right) \\
&= \left(A_1 \exp\left(-\frac{t}{t_1}\right) + A_2 \exp\left(-\frac{t}{t_2}\right)\right), (t_2 \gg t_1)
\end{aligned} \tag{6}
$$

Rotational diffusion (isotropic) of the protein corresponds to the slow correlation time; t_2. Restricted rotational diffusion of the tryptophan residue within the protein is represented by an exponential decay to a constant value; A_2, with a faster correlation time; t_1. The initial fluorescence anisotropy therefore corresponds to the sum of A_1 and A_2. The ratio of A_1 to A_2 however is determined by the degree of orientational averaging within the protein matrix, this has been interpreted in terms of diffusion within a cone of semiangle α which is related to the amplitude of the two decay rates by *Equation 7* (21, 22):

$$A_2 / (A_1 + A_2) = \left[\frac{\cos\alpha(1 + \cos\alpha)}{2}\right]^2 \tag{7}$$

173

Fluorescence lifetimes (I(t, = β = 54.7°) measurements) can be analysed using a standard deconvolution procedure with the response of the apparatus to scattered pump light (5, 23) and the fluorescence anisotropy decays analysed using a least squares fit of the form (24) of *Equation 8*:

$$R(t) = \sum_i A_i \exp\left(-\frac{t}{t_i}\right) \qquad [8]$$

On this basis, time-resolved techniques allow us to be more certain that we are observing true quenching from any lifetime changes that may occur. In addition by implementing the polarization of fluorescence we also have a measure of any attendant structural changes that accompany changes of the Stern–Volmer constant as illustrated in *Figure 1*.

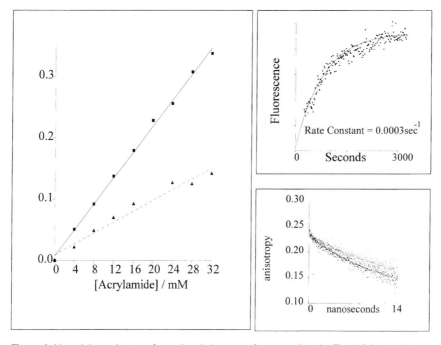

Figure 1 Ligand-dependent conformational changes of serum albumin. The left-hand side plot indicates a Stern–Volmer quenching analysis of Trp 214 of human serum albumin (10 μM) in the absence (■) and bound with oleate (treated for 3 h with a 9:1 ratio of oleate—▲). The upper right-hand side plot indicates the effect of adding oleate at a 9:1 ratio in the presence of 32 mM acrylamide to serum albumin (10 μM). The steady-state fluorescence was measured at 345 nm following excitation at 295 nm. Data was collected at 1 min intervals with the lamp shutter closed to avoid the possibility of photobleaching. The increase of fluorescence indicates that the accessibility to the fluorescence centre (i.e. Trp 214) is being more hindered due to conformational changes of the albumin following ligand binding and at the quencher is less able to quench the fluorescence at the indicated concentration. The lower right-hand side plot indicates the effect on the fluorescence anisotropy measured using TCSPC for defatted albumin (solid line) and albumin ligated with oleate (dotted line) as above.

4 The surface electrostatic potential: one of the trinity of electrical membrane potentials

Membrane potentials are manifest in a number of forms which have quite different origins and properties. Unfortunately, there seems to be considerable confusion associated with these potentials in terms of their origins, identity, and not least, their nomenclature, as well as with their influence on a number of important biological processes. Consequently, it is not unexpected that a clear understanding of the roles some of these processes play in biology are not widely appreciated. Three types of electrical potential are present in membranes (e.g. 30). The 'membrane potential' which has received the most experimental attention and certainly, the best-documented, is associated with gradients of electrical charge across the membrane bilayer. These gradients are engineered and maintained by the translocation of charged solutes (e.g. ions) or electron transport across the membrane and to be established they require an input of energy in one form or another (light, redox, or ATP).

The second *membrane potential* of major significance is known as the *membrane surface potential*. Its existence has been established for some time and the underlying physical theory is well-developed and robust; reviews by McLaughlin (25) and Cevc (26) offer sound and eloquent expositions of the nature of this potential. Similarly, there is much compelling experimental evidence that indicates the surface potential is a major player in membrane biology. The final manifestation of a '*membrane potential*' involves the contribution that electrical dipoles make to the properties of biological membranes. The *Dipole potential* (Ør) is thought to have its origins in the molecular polarizations associated with the carbonyl group and the oxygen-bonded phosphates components of most membranes. In addition water molecules appear to adopt an organized structure along the membrane surface by virtue of their permanent molecular dipoles and so may also make a contribution (27). The Dipole potential, however, is both the least well-documented and the least well-understood of all of these so-called membrane potentials. The appearance and influences that it may play in particular biological processes, therefore, remains somewhat obscure. Evidence is accumulating, however which indicates, that this manifestation of a *membrane potential* is likely to be at least as influential as the others (28, 29).

5 Membrane electrostatics: implemenation of *fluoresceinphosphatidylethanolamine* (FPE) to measure ligand–receptor interactions of membranes

The classical Gouy–Chapman–Stern model may be applied to biological membranes to relate the electrostatic surface potential (Ψ) and the excess surface charge density (σ) (*Equation 10*) (25, 26, 30). The model predicts the variation of the surface potential with the distance from the bilayer surface in the manner shown in *Figure 2*, where the influence of the electrolyte concentration of the solution surrounding the membrane on the magnitude of the surface potential is illustrated.

The concentration of ions in the aqueous phase at the surface of the membrane phase is a function of the surface potential and is described by the Boltzmann equation (*Equation 9*). The potential at the membrane surface (Ψ) causes an uneven distribution of ions in the surface environment. Ions of the same charge to that of the surface will be repelled; ions of the opposite charge will be attracted. The concentration of ions in the aqueous phase at the surface of the membrane phase (C_s), therefore, is a function of surface potential and is described by the Boltzmann equation as follows:

$$C_S = C_B \exp\left(-z\, e\, \Psi / k_B T\right) \qquad [9]$$

where C_B = concentration of a specified ion in the bulk solution, C_S = concentration of a specified ion on the membrane surface, e = elementary charge, k_B = Boltzmann constant, T = absolute temperature (K), and z = positive or negative, indicating the number of charges on the appropriate ion.

A more complete description of the membrane–solution interface is given by a combination of the Poisson equation which relates the electrical field vector to the surface charge density (σ), and the Boltzmann equation which with suitable boundary conditions yields the Stern equation.

$$\sinh\left[e\, \Psi / (2k_B T)\right] = 1 / (8\, N\, \varepsilon_r \cdot \varepsilon_0\, k_B T)^{1/2} \cdot \sigma\, \sqrt{CN_B} \qquad [10]$$

In this relation, N represents Avogadro's number, ε_r is the dielectric permittivity of the medium and ε_0 is the dielectric permittivity of free space. *Equation 10* suggests that changes in the ionic concentration (or the valency of the electrolyte) of the bulk phase ($C.N_B$) or the surface fixed charge density will promote changes in the electrostatic surface potential.

In terms of the responses of an acidic group located on a surface environment it is helpful to rearrange the Boltzmann relation in a logarithmic form (*Equation 11*) and combine it with the Henderson–Hasselbach equation to yield *Equation 12* as follows:

$$\log\left[C_S / C_B \right] = \Psi \cdot 0.059\ \text{Volts} \qquad [11]$$

$$pK_s = pK_B + \Psi \qquad [12]$$

where C = concentration of protons at the s = surface and $_B$ = bulk phase environments, respectively.

This indicates that the pK of an acidic group on an electronegative membrane surface differs from the same group in the bulk phase. Thus, for a fluorescent pH indicator such as the fluorescein moiety of fluoresceinphosphatidylethanolamine (FPE—structure shown in *Figure 2*) located at the membrane surface, changes of the electrostatic surface potential will affect the protonation state of fluorescein leading to changes of fluorescence. In other words, changes in the electrostatic surface potential will promote a change in the apparent pK of FPE.

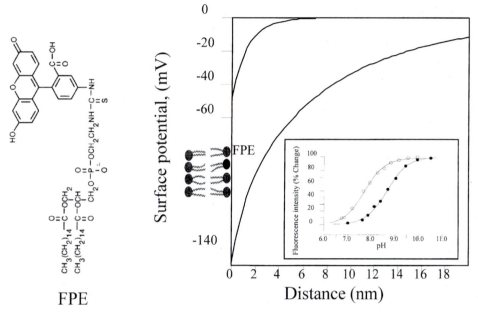

FPE

Figure 2 The position of FPE in the phospholipid membrane and the computed electrical potential with respect to distance from the membrane surface. The left-hand side drawing shows the structure of fluoresceinphosphatidylethanolamine (FPE). The right-hand side illustrates the position of FPE within the phospholipid bilayer and how the electric potential of the membrane surface (sensed by FPE) decays away from the surface. The electric potential is computed according to the doctrines of electrostatic theory outlined in *Equations 9–12*.

This is illustrated in *Figure 2* (inset) where the titration of FPE-labelled erythrocytes ghosts in relatively high and low bulk electrolyte concentration (i.e. leading to two different levels of Ψ) reveals the effect on the experimentally determined (i.e. apparent) pK.

Any changes in the number of surface charges at the membrane, such as the binding of an inorganic ion or a charged oligopeptide, will cause an alteration in Ψ. Thus, it is possible to monitor the interaction of any kind of charged molecule with a membrane or a membrane receptor by observing the fluorescence changes of FPE. Interactions that lead to a more electronegative surface potential (i.e. addition of negative charge to the membrane surface) cause a decrease in the intensity of the fluorescence of FPE whereas interactions causing a more electropositive Ψ (i.e. addition of positive charge to the membrane surface) lead to a fluorescence increase. Furthermore, as the protonation-deprotonation reactions of FPE are not rate-limiting then the kinetics of the molecule–membrane interactions may also be measured. Thus stopped-flow mixing (see Chapter 5) of membranes with molecules offer the possibility of determining the complete time-evolution of their intermolecular interaction (31). Initially we illustrate the techniques using simple membranes made up of well-defined phospholipids prepared as monodisperse, unilamellar phospholpid vesicles (100 nm diameter). This work is followed by more generally interesting studies with living cells. In all cases it is possible to rapidly mix the membrane preparations and the ligand of interest to interrogate the early interactions of the ligand with the membrane. In our experience even stop-flow mixing (see Chapter 5) human cells causes no physical side-effects and does not interfere with fluorescence measurements.

5.1 Labelling human cells with FPE

We have previously developed strategies to label the cell surface with appropriate fluorophores in preparation for studies of either the interactions of molecules with a population of cells in suspension or for single-cell fluorescence imaging for human cells or micro-organisms such as bacteria or fungi. For the present study, human cell lines were labelled with FPE. We have placed some emphasis on doping membranes with only a very small amount of FPE so that the technique is effectively as non-invasive as possible. In typical imaging experiments there was only about 1 (or ca. up to 5) molecule(s) of FPE per 1000 phospholipids. Similarly, self-quenching may be a problem with membrane-located fluorophores but at these concentrations it is not evident.

5.2 Location of FPE in the membrane

The fluorescence moeity of FPE occupies a location as part of the molecular surface of the membrane in contact with the aqueous solution, in other words at the membrane–solution interface as illustrated in *Figure 2*. The experimental criterion used to determine the degree of orientation of the membrane-bound FPE relies on the effects of addition of an impermeable salt to the suspension media. This causes changes in the fluorescence of FPE as expected on the basis of the effect on ψ. By the addition of an appropriate ionophore such as A23187 or ionomycin, the plasma membranes are rendered permeable, facilitating the equilibration of the calcium ions across the membrane. This may result in a further increase of fluorescence if there is any probe present on the internal bilayer leaflet facing the cytoplasm. The ratio of the fluorescence before and after the addition of the calcium ionophore, therefore, yields the ratio of the cell surface probe to that facing the cytoplasm. Depending on the labelling protocol, FPE was found evenly distributed on either side of the membrane bilayer or more typically for binding experiments was located exclusively in the outer bilayer leaflet of the membrane. Under the conditions of the cellular labelling protocol, a significant proportion (ca. 30–70%) of the added FPE finds its way exclusively into the outer bilayer leaflet with none appearing in the inner bilayer leaflet. Similarly, transmembrane 'flipping' over the time scales required for data acquisition for the imaging experiments was not evident.

In order to visualize the spatial disposition of the membrane surface electrostatic potential, it is necessary that some knowledge of the location of the various flurophores utilized as indicators is obtained. In the case of FPE, it has already been demonstrated in earlier publications that the conditions necessary to place it in the outer bilayer leaflet of the membrane. Nevertheless, all preparations of cells were assayed (in the manner described briefly above) for any inner-bilayer leaflet labelling, unless required, as this may compromise the interpretation.

Under some circumstances, it was desirable to have FPE located on the cytoplasm face of the plasma membrane. This was found to occur serendipitously, if the addition of the alcoholic-FPE solution corresponded to a larger percentage of the total aqueous volume than described above or the plasma membrane phospholipid:FPE ratio much smaller than that found to yield successful labelling exclusively of the outer bilayer leaflet. There are also a number of other labelling conditions (e.g. use of other organic solvents such as chloroform) that result in the same effect but it may be considered undesirable to expose cells to these solvents. In the event that the cells are labelled with FPE on both sides of the plasma membrane, active label solely on the inner bilayer leaflet may be achieved by treatment with a rabbit anti-fluorescein antibody (RAF). Re-assay for labelling asymmetry then indicates that no active label remains on the outer bilayer leaflet. In this case, no immediate fluorescence changes occurred due to the addition of calcium ions. Following the ionophore addition, however, the fluorescence change was the same as that measured in the absence of the RAF. Thus, the remaining fluorescence after the RAF titration was found to originate solely from the cytoplasmic facing phospholipid leaflet and indicates that all FPE in the outer bilayer leaflet of the plasma membrane was quenched.

6 Digital imaging: using CCD fluorescence microscopy and confocal scanning laser fluorescence microscopy (CSLM) of ligand binding reactions of living cells

CCD-based fluorescence imaging involved recordings taken with an inverted fluorescence microscope (e.g. Diaphot; Nikon Inc. Melville, NY) using a 50 W mercury lamp and fitted with appropriate filters, e.g. for FPE, a 500 nm cut-off filter and a dedicated dichroic mirror. The objective was equipped with × 40–100 lens and was used under phase contrast. Images were acquired from the microscope system by attaching a peltier cooled, slow-scan charge coupled device (CCD) camera (Wright Instruments, Enfield, UK) to the video port with images focused onto a EEV CCD 02–06–1206 back-illuminated detector. Depending on the level of luminescence sensitivity or spatial resolution (or both) required, an on-chip binning technique was sometimes utilized. This involves collecting charge from a number of neighbouring pixels before conversion to a digital signal. Each analogue to digital conversion adds noise to the signal but because on-chip binning reduces the number of conversions per total chip readout, it increases the signal to noise, this improvement, however, is gained at the expense of spatial resolution. In the case of weak signals, overall the image noise is dominated by the readout noise component so that the signal to noise is increased by the number of pixels per bin. Thus, a 3 by 3 bin produces the same improvement in the signal to noise as a 9-fold increase in the exposure time without binning. Increasing the binning factors decreased the readout time because fewer values have to be converted in the camera control module.

In a 256 colour mode, the 32 K intensities produced by the camera are converted into 256 colours of the bitmap using a simple linear conversion algorithm specified by a minimum and a maximum count and a minimum and maximum colour. The relation of count to colour is a straight line that at 8-bit resolution extrapolates to colour saturation values of 0–255. The minimum and maximum count values required for the conversion could be set to the minimum or maximum count of the image. The minimum and maximum counts are very sensitive to noise (e.g. particularly from cosmic rays). Thus, one solution was to set the minimum and maximum values to the nth percentiles of the distribution of the number of pixels with each possible count. In this case the definition of percentiles is that exactly n percent of the pixels of the image have a count equal to or less than the value of the nth percentile.

In our case, fluorescence images have been recorded in a binary free-format but for additional archiving and presentation, they were converted into .tif, .gif or .jpg formats although in the latter there is often distortion at the image edge that renders some error and so were not used for further analysis. The successful incorporation of FPE into the outer bilayer leaflet of the plasma membrane of various cell types was determined using the response of the membrane to the addition of calcium ions in preparation for quantitative fluorescence imaging. The resulting images indicated that the membrane-bound FPE appears to be dispersed about the membrane surface in a manner suitable enough to study interactions of a number of molecular species with the membrane.

Confocal laser scanning microscopy (CSLM) was undertaken with a Bio-Rad MRC600 Confocal laser microscope typically at $\times$ 600 magnification with oil immersion. Autofluorescence was occasionally present but was easily corrected. Cells were viewed in the bright field at $\times$ 400, $\times$ 600, and $\times$ 1000 magnification, followed by fluorescence image acquisition at the same magnifications.

7 Fluorescent images corrected for probe disposition and photobleaching

In order to achieve a satisfactory image that solely represents the spatial profile of the electrostatic potential on the cell surface and any subsequent changes due to ligand binding to receptors, it is necessary to have equivalent knowledge of the location of the fluorophores. This problem has been considered previously for FPE in a related context as its role as an indicator of macromolecular binding to populations of cells (32). In the present case, however, in addition, to a requirement for the same knowledge of the probes location (i.e. it is important that they face only one side of the phospholipid bilayer), it is also imperative that, knowledge of the spatial disposition of the labelling density of the FPE is also available. The former is easily accessible using the techniques described previously (33, 32) and briefly outlined above but the latter requires further consideration.

This latter complication is a two-dimensional analogue of the problems encountered during the spatial imaging of ionic distributions within cells and may be solved in a similar manner (34, 35). Thus, by making use of the potential dual-excitation properties of FPE over certain ranges, as pointed out previously (32), a probe-concentration independent 'ratio' image of the electrostatic potential may be obtained. Rapid switching between two excitation cassettes is a supplementary feature of many commercial confocal microscopes such as a Bio-Rad MRC600 (used in the present study) or other CCD-based fluorescence microscopes such that dual-excitation images may be acquired reasonably promptly. Whilst it is feasible to modify or purchase fluorescence microscopes equipped with dual-excitation/emission facilities, it is not absolutely necessary as in the present context, it is also possible by additional methods to correct the images for probe disposition. These treatments yield images that are essentially equivalent to 'ratio' images, some of which are outlined as follows: acquisition of images following labelling with FPE, treatment of cells with saturating concentrations of calcium ions at constant pH, leads to the maximum fluorescence intensity all over the cell surface. This image may be utilized, therefore, as the internal standard or is equivalent to illumination at a pseudo-isosbestic excitation/ emission wavelength. Thus, the maximum pixel intensity at any particular location on the cell surface in this image is set to depend only on the local probe density and may be used as a reference image to correct the 'experimental' images. In the event that calcium ions are used to attain the maximum local pixel intensity, e.g. by perfusing with media containing elevated levels of calcium ions (i.e. ca. 5 mM) to obtain the image followed by perfusion with normal media to remove the calcium in preparation for experiments.

A second method to correct images for probe disposition and related to the previous strategy makes use of a RAF antibody although by definition it must be dedicated to images acquired with FPE. The antibody binds only to fluorescein and once visualized, provides a measure of the probe disposition. Visualization is achieved using a non-fluorescein fluorescent covalent derivative of the RAF antibody, in other words as the antibody disposition is visualized, it reflects the FPE disposition. RAF however, is rather expensive and in contrast to the calcium treatment cannot be reversed so this method is retrospective as it can only be applied at the end of experimentation.

Under the majority of the conditions employed for the present studies and in accordance with ratio-fluorescence imaging studies, photobleaching was not found to compromise seriously, image acquisition provided account was taken of the probe disposition. In other words photobleaching is analogous to a reduction in the active probe concentration but does not affect the relative probe disposition.

8 Spatial imaging of the cell surface to identify localized ligand binding

The spatial disposition of the electrostatic potential of the cell surface of a number of different human cell types may be visualized using FPE. Digital fluorescence images were recorded using either a cooled, slow-scan CCD camera or with a laser confocal microscope. The spatial disposition of the electrostatic potential of the surface of all eucaryotic cells types examined except erythrocytes was seen to be highly heterogeneous with delineated regions of differing electrical potential ranging from around 100 nm (or less) to regions covering square micron areas. Differences were also observed between the inner and outer-bilayer leaflets. By addition of different proteins such as serum albumin, the profile of electrostatic potential on the cell surface were seen to change and were interpreted to reflect binding of these proteins. In view of the relationship between the electric potential, surface charge density and fluorescence intensity and provided the net charge of the protein is known, it is possible to deduce the number of protein molecules that become bound to the cell surface as well as their spatial disposition. This analysis was explored with albumin in which the number of protein molecules bound to particular sites on the cell surface are described.

9 Experimental protocols

As a guide to the general user, here are a series of protocols we followed in a series of fluorescence related experiments.

Protocol 1

Ligand binding to serum albumin revealed by tryptophan fluorescence quenching and lifetime studies

Equipment and reagents

- Benchtop fluorimeter or time-resolved laser system
- Cuvette
- Oleic acid (1 M in ethanol)
- Albumin
- Acrylamide

Method

1 The HSA–oleate complex was prepared by mixing a 9:1 molar ratio of oleic acid (1 M in ethanol) with the albumin. The mixture was then gently agitated for 1 h at room temperature.

2 To measure the rate of binding, acrylamide (final concentration 16 mM) was added to protein in a stirred cuvette. 3 µl of oleic acid in ethanol was added (final ethanol concentration was below 0.1%, v/v) and the fluorescence was measured every 60 sec. Between measurements the shutter was closed, to avoid photobleaching.

3 Albumin as either the defatted form or the oleate-bound form, was placed in a clean quartz cell in a buffered aqueous solution at $\sim$ 10 μM. The tryptophan spectra and steady-state fluorescence intensity at a suitable intensity peak (e.g. 345 nm) may then be measured following excitation at 295 nm. The quenching data are then analysed using the Stern–Volmer expression (*Equation 1*).

4 Fluorescence lifetime and anisotropy decays measurements were performed using a CW mode locked Nd:YAG pumped, cavity dumped dye laser (Coherent Antares 76-S and 7220D) running with rhodamine-6G and frequency doubled in β-barium borate (Photox). This provides a 4 MHz train of ca. 7 ps pulses at 297 nm which may be used to excite the tryptophan $S_1 \leftarrow S_0$ transition in albumin samples using a collinear excitation-detection arrangement. In isotropic media the cylindrical symmetry of the excitation process with respect to the (vertical) laser polarization vector ensures the equivalence of collinear and right-angled excitation-detection geometries (36). Fluorescence was filtered with a 70 nm band pass interference filter centred at 400 nm (Corion P70–400).

5 Data is collected over a period of about 20 min and analysed to yield changes of polarization or intensity over the picosecond-nanosecond domain.

Protocol 2

Preparation of model membranes labelled with FPE

Equipment and reagents

- Pressure extruder bomb (Lipex BM Inc., Vancouver, Canada)
- Polycarbonate filters with a 100 nm pore size (Nucleopore Filtration Products, Pleasonton, CA, USA)
- Rotary evaporator
- Eppendorf tubes
- Sephadex PD-10 column
- 100 mg/ml solution of phosphatidylcholine
- 100 mg/ml solution of phosphatidylserine in chloroform/methanol
- Liquid nitrogen
- 2 mg/ml FPE solution
- 95% ethanol

A Preparation of large unilamellar phospholipid vesicles (PLVs) 100 nm diameter made up of phosphatidylcholine/phosphatidylserine (9:1 molar ratio)

1 90 μl from a 100 mg/ml solution of phosphatidylcholine plus 10 μl from a 100 mg/ml solution of phosphatidylserine in chloroform/methanol were mixed in a round-bottomed flask and dried under a stream of oxygen-free argon gas by rotary evaporation until a thin film is formed. The film when fully dried by further exposure to the stream of gas is essentially solvent-free and ready for rehydration with aqueous media.

Protocol 2 continued

2 The lipid film is rehydrated with 1 ml of an appropriate aqueous buffer at room temperature.

3 The resulting solution is frozen and thawed using liquid nitrogen and hot tap-water five times and results in an highly heterogeneous multilamellar vesicular (MLVs) suspension.

4 After the freezing–thawing procedure the MLV suspension is extruded ten times using a pressure extruder bomb through polycarbonate filters with a 100 nm pore size. This results in a monodisperse, unilamellar suspension of unilamellar phospholipid vesicles (PLVs) of a specified size that may be stored under a nitrogen atmosphere for several days at 4 °C before use.

B Labelling of the outer bilayer leaflet of PLV with FPE

1 15 μl of a 2 mg/ml FPE solution is placed in a 1.5 ml Eppendorf tube and dried under a stream of argon gas, followed by resolution with 10 μl of 95% ethanol.

2 1 ml of the PC/PS (9:1) PLVs suspension (10 mg/ml lipid) is added into the Eppendorf tube and the mixture is incubated at 37 °C in the dark for 45–60 min.

3 Unincorporated FPE is removed by size exclusion chromatography using a Sephadex PD-10 column. Labelled vesicles are typically diluted twice following this treatment.

4 The final 5 mg/ml PC/PS FPE-labelled (0.25 mol% FPE) suspension is stored under nitrogen atmosphere at 4 °C and used within 48 h.

C Labelling of both membrane bilayer leaflets of PLVs with FPE

1 90 μl of a 100 mg/ml PC solution plus 9.5 μl of a 100 μl PS solution plus 15 μl of a 2 mg/ml FPE solution in chloroform methanol are mixed.

2 Follow the procedure explained in part A for the preparation of PLVs.

3 Follow the procedure explained in part B, steps 3 and 4.

Protocol 3

Labelling of the outer bilayer leaflet of the plasma membranes of lymphocytes with FPE

Equipment and reagents

- Haemocytometer
- Centrifuge
- Argon
- Jurkat, MOLT4, or RAJI lymphocytes
- Standard culture medium such as RPMI 1640 medium with NaHCO$_3$ (Sigma)
- 10% (v/v) fetal calf serum
- 2 mM glutamine
- 2% (v/v) bash
- Iso-osmotic buffer
- Trypan blue
- FPE in chloroform-methanol
- 95% ethanol

Method

1 Jurkat, MOLT4, or RAJI lymphocytes are cultured in a standard culture medium such as RPMI 1640 medium with $NaHCO_3$, supplemented with 10% (v/v) fetal calf serum, 2 mM glutamine, and 2% (v/v) bash at 37 °C and 5–7% CO_2.

2 Cells are harvested and washed by two centrifugation steps at 2500 g for 5 min in an iso-osmotic aqueous buffer.

3 The viable cells number density is determined by counting with a haemocytometer and the Trypan blue exclusion technique.

4 A volume of FPE in chloroform-methanol at a ratio of 10 µg FPE: 3.10^6 cells is placed in an acid-washed dry test-tube and the organic solvent evaporated under a stream of argon gas (as *Protocol 2*, step 1) followed by resolvation with 15 µl of 95% ethanol.

5 The cell suspension is then added to the FPE suspension and the mixture is gently agitated before incubation in the dark for 45–60 min at 37 °C with occasional, gentle agitation.

6 Unincorporated FPE is removed by two serial centrifugations at 2500 g for 5 min in the appropriate iso-osmotic buffer.

7 The cells were stored at 4 °C and used within 24 h.

Protocol 4

Labelling of erythrocyte membranes with FPE

Equipment and reagents

- Centrifuge
- Argon
- Whole blood

- Buffer pH 7.5
- FPE
- 95% ethanol

Method

1 Whole blood from donors (always with informed consent) is drawn into a tube containing 1 mg K_3EDTA per 1 mg of blood.

2 Erythrocytes were isolated by centrifugation at 2500 g for 10 sec at 4 °C. The plasma and lymphocyte layer (buffy coat) were removed by aspiration.

3 Packed erythrocytes were washed three times with buffer at pH 7.5. After the final wash the cells were left packed (i.e. 100% hematocrit).

4 50 µg of FPE is prepared by removal of the chloroform-methanol solution under a stream of argon followed by resuspension in 15 µl of 95% ethanol.

5 500 µl of buffer at pH 7.5 is added to the ethanolic FPE suspension followed by the addition of 100 µl of packed erythrocyte.

6 The mixture is gently agitated before incubating at 37 °C for 45–60 min in the dark.

Protocol 4 continued

7 Unbound FPE is removed by repeated centrifugation at 5000 g for 5 sec until the supernatant appears clear.

8 Labelled cells were stored at 4 °C under argon and used within 48 h.

Protocol 5

Labelling of B12 rat glial cells plasma membranes with FPE

Equipment and reagents

- See *Protocol 3*
- Glass coverslips
- Haemocytometer
- Eppendorf tubes
- Centrifuge
- Flasks
- B12 cells
- 10% fetal calf serum (FCS)

- Dulbecco's modified Eagle's medium (DMEM)
- 200 mM glutamine
- Trypsin/EDTA
- Sucrose buffer: 280 mM sucrose, 10 mM Tris pH 7.5
- Collagenase

Method

1 B12 cells were cultured on glass coverslips in Dulbecco's modified Eagle's medium (DMEM) supplemented with 10% fetal calf serum (FCS) and 200 mM glutamine. Cells were harvested using trypsin/EDTA and counted in a haemocytometer chamber. They were replated at 10^6 cells/flask.

2 After harvesting the cells were resuspended in medium (DMEM; 10% FCS). Aliquots of 1.5×10^6 cells were transferred to Eppendorf tubes and centrifuged at 1000 g for 8 min. The supernatant was aspirated off and the cells resuspended in 140 μl sucrose buffer. Cells may also be left on coverslips for either fluorescence microscopy (see *Protocol 7*) or the ensemble of adhered cells could be measured for FPE fluorescence using either custom-made or commercially available attachments such as the SLM-Aminco variable-angle coverslip-holder.

3 The cells were digested with collagenase at a working concentration of 0.3 mg/ml. The samples were incubated for 90 min at 37 °C. The cells were washed with 200 μl of the sucrose-based medium and centrifuged twice at 11 600 g for 5 sec to remove cell debris and the enzyme.

4 Collagenase-stripped cells were resuspended in a total volume of 1 ml of sucrose buffer and labelled with FPE as indicated in *Protocol 3*, steps 4–7.

Protocol 6

Preparation of FPE-labelled gp60 vesicles and control liposomes
(Procedure for the purification of membrane receptor, gp60, from crude cell extract from ECV304 endothelial cells)

Equipment and reagents

- 225 cm^2 flasks
- Centrifuge
- WGA-affinity column: wheat germ agglutinin from *Triticum vulgaris* cross-linked to 4% beaded agarose
- Centricon-10 ultrafiltration units
- Sonicator
- ECV304 endothelial cells
- Triton X-100

- PBS: 10 mM sodium and potassium phosphate buffer containing 2.7 mM KCl, 137 mM NaCl pH 7.4
- Ethanol
- Sodium dodecyl sulfate (SDS)
- *N*-acetylglucosamine
- 2% sodium cholate
- Phosphatidylcholine (PC)
- Phosphatidylserine (PS)

A Preparation of a membrane receptor

1 ECV304 endothelial cells were grown to confluence in 225 cm^3 flasks then harvested and stored in their culture medium at $-18\,°C$. Approx. 1 billion cells were collected and washed with 30 ml PBS twice by precipitation at 4500 g for 20 min at 4 °C. Cells (approx. 4 ml) were resuspended in 6 ml PBS containing 1% (v/v) Triton X-100 and then gently stirred on ice for ca. 60 min.

2 Centrifugation of this mixture at approx. 3000 g for 5 min produced a pellet of 3–4 ml and 6 ml of supernatant. The supernatant was rapidly added to 45 ml absolute ethanol at $-18\,°C$. The precipitate that formed was allowed to settle out overnight at $-18\,°C$.

3 The protein precipitate in 88% cold ethanol and was precipitated at 3000 g for 5 min. The ethanol supernatant was removed and 16 ml of 70% ethanol at $-18\,°C$ was added to the pellet, gently resuspended, and precipitated by centrifugation at 3000 g for 5 min. The pellet was then gently stirred with 8 ml PBS on ice for ca. 60 min.

4 Further centrifugation at approx. 3000 g for 5 min produced a pellet of 1–2 ml. The supernatant (10 ml) was adjusted to 1% Triton X-100 and 0.2% SDS. Extraction of the pellet with PBS was repeated and a second aliquot of 10 ml of the supernatant adjusted to 1% Triton X-100 and 0.2% SDS.

5 An affinity column was prepared as follows: wheat germ agglutinin from *Triticum vulgaris* cross-linked to 4% beaded agarose (1 ml bed volume) was regenerated by washing with PBS containing 0.5 M NaCl, 1 mM MgCl$_2$, 1 mM MnCl$_2$, 1 mM ZnCl$_2$, and 1 mM CaCl$_2$. This solution was removed using 15 ml PBS containing 1% Triton X-100 and 0.2% SDS.

6 About 20 ml of the cell extract was passed down the WGA-affinity column at a flow rate of approx. 0.2 ml/min then washed with 15 ml PBS containing 1% Triton X-100 and 0.2% SDS.

7 The gp60 was released from the affinity column by eluting it with PBS buffer containing 1% Triton X-100, 0.2% SDS, and 0.5 M N-acetylglucosamine. Six 1 ml aliquots were collected from the column and the N-acetylglucosamine-containing buffer replaced by PBS buffer containing 0.1% Triton X-100, 0.02% SDS. This was carried out using Centricon-10 ultrafiltration units that were centrifuged at 4500 g.

8 Fraction 2 was found to contain significant amounts of gp60 as indicated by ligand blot analysis using I^{125}-HSA.

B Reconstitution of gp60 into membranes and labelling with FPE

1 A sample of gp60 in PBS buffer containing 1% Triton X-100, 0.02% SDS which had been stored in liquid nitrogen is thawed and dialysed against approx. 300 volumes of PBS containing 2% sodium cholate pH 7.4 overnight.

2 A dispersion of phosphatidylcholine (PC) and phosphatidylserine (PS) is prepared by drying a mixture of 17 mg PC and 2.8 mg PS under argon, 'swishing' with 0.5 ml ethanol, aspirating the solvent, and then redrying under argon. The phospholipid mixture is then dispersed in 1 ml PBS containing 2% sodium cholate using a whirli-mix (5–10 min). The sample is subjected to sonication using a microtip sonicator probe at power setting 4, 30% duty cycle for 10 min. This procedure is carried out at 4 °C. After sonication the sample is centrifuged at approx. 2000 g for 5 min. Two separate samples of phospholipids were prepared in this manner so that one could be used in the preparation of the control liposome sample.

3 0.9 ml of the gp60 sample is mixed with 1 ml of the PC/PS dispersion and transferred to benzoylated visking tubing (1 kDa cut-off). The samples were then dialysed against:

(a) 250 ml of 250 mM sucrose, 10 mM Hepes (Na$^+$), 0.68% sodium cholate pH 7.4 for 4 h.

(b) 250 ml of 250 mM sucrose, 10 mM Hepes (Na$^+$), 0.3% sodium cholate for 4 h.

(c) Finally 250 ml of 280 mM sucrose, 10 mM Tris–HCl pH 7.5 overnight (for at least 16 h).

Control liposomes in the absence of the receptor were prepared by a similar procedure.

4 gp60 vesicles and control liposomes prepared as described above were labelled on their outer leaflet with FPE at a concentration of approx. 0.2% as described in *Protocol 2*, part B.

Protocol 7

Spatial imaging of the cell surface electrostatic potential; use to identify localized binding/membrane insertion interactions

Equipment and reagents

- See *Protocols 3* and *5*
- Fluorescence microscope
- Adherent cell line (e.g. human umbilical vein endothelial cells—HUVECS)

Method

1   An appropriate adherent cell line was labelled essentially as described for the lymphocytes or B12 microglial (but without the need for collagenase treatment). In this case cells were left on the coverslips for microscopy.

2   Cells checked for their calcium response—i.e. relabelled or discarded if the response was too small, i.e. in terms of their fluorescence increment.

3   Labelled cells on a coverslip were placed under a fluorescence microscope equipped for either cooled CCD fluorescence imaging or for CLSM. Cells were irradiated with appropriate light and images collected.

4   Addition of appropriate ligand solution and images collected over time.

5   Images corrected for probe disposition and localized ligand binding identified.

---

10 Case studies

Protocols 1–7 outlined above have been chosen to illustrate a number of types of interaction between molecules of biological interest with their target protein (e.g. HSA) including integral (receptor) proteins within membranes (i.e. FPE studies). These could be seen as generic methods for implementation with virtually any molecule that possess suitable fluorescent properties such as tryptophan or in the case of the membrane ligand–receptor interactions, an electrical charge. We have tried to illustrate the techniques using work ongoing in our laboratory as representative of a variety of properties that may have some solution by bringing to bear fluorescence-based technologies.

10.1 Ligand (oleate) binding to HSA revealed by tryptophan fluorescence quenching

It was found that both the steady state and time-resolved Stern–Volmer collisional-quenching studies of Trp 214 with acrylamide pointed to the existence of an oleate-dependent structural transformation as illustrated in *Figure 1*. The bimolecular quenching rate constant of defatted HSA; 1.35×10^9 M^{-1}s^{-1}, decreased to 0.84×10^9 M^{-1}s^{-1} after incubation with oleic acid (9:1). Time-resolved fluorescence anisotropy measurements of the Trp 214 residue yielded information of motion within the protein together with the whole protein molecule. Characteristic changes of these motions following the binding of oleate to albumin were observed. The addition of oleate was accompanied by an increase in the rotational diffusion time of the albumin molecule from ca. 22 nsec to 33.6 nsec. Within the body of the protein, however, the rotational diffusion time for Trp 214 exhibited a slight decrease from 191 psec to 182 psec and was accompanied by a decrease in the extent of the angular motion of Trp 214, indicating a transition to a more spatially restricted but less viscous environment following oleate binding.

Binding sites have been found to exist on each of the three domains of HSA (12). The closest fatty acid binding site to Trp 214, however, has been reported to involve Lys 351, which is on the opposite side of the structural domain (12). This suggests that changes revealed by the emission spectra, as well as the quenching data, are not due to direct interactions between the oleate moiety on the Trp 214, but instead, are likely to be caused by a change in the conformation of the HSA after binding oleate, resulting in an altered environment of Trp 214.

10.2 Kinetics of oleate-dependent fluorescence changes of HSA

Oleate has been observed to promote a decrease in the Stern–Volmer quenching constant (*Figure 1*, left-hand side plot); such experiments were performed by adding the quenching agent to the fluorophore allowing the establishment of equilibrium and then recording the level of fluorescence. On the other hand, if oleate is added to albumin already in the presence of the quencher, however, changes of fluorescence intensity would be anticipated to indicate changes of the quenching constant. This phenomenon was indeed observed and illustrated in *Figure 1* (upper right-hand plot). These data were found to be best described by a single exponential process, with a rate constant of 8×10^{-4} sec^{-1}. The rate of change in quenching may arise as a direct influence of the presence of oleate, or as a result of conformational changes induced after oleate binding which may occur immediately or subsequently much more slowly. The rate constant for the binding of oleate to HSA is reported to be of the order of 3.2 sec^{-1} (37). In view of the fact that this value is three orders of magnitude larger than that found from our quenching study, the most likely explanation of the observed change in quenching is that it reflects a slow conformational change subsequent to the more rapid ligand binding.

10.3 Interaction of a signal peptide (p25) with model membranes

The interaction of the 25 residue signal sequence of subunit IV of cytochrome *c* oxidase, known as p25, with FPE labelled PC vesicles is shown in *Figure 3*. After each peptide addition there is an increase of the fluorescence intensity of the initial binding of each peptide which is consistent with the interaction of the positive charges on the peptide with the membrane surface causing the surface

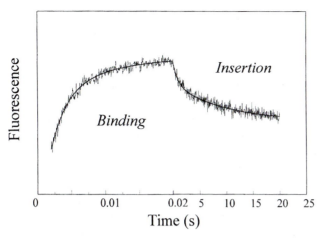

Figure 3 The time-evolution of the interactions of p25 with membranes as revealed with the electrostatic sensing technique. p25 (0.6 μM) was stopped-flow mixed with phospholipid vesicles (200 μM phospholipid) labelled with FPE (0.2 mole%) and the fluorescence at 520 nm following excitation at 490 nm recorded. Two time scales are shown each with 1000 points: 0–20 msec and 20 msec–20 sec. The solid lines through the experimental points represents the best-fit achieved using an iterative least-squares analysis of the data points to a double exponential expression of the form: $y = y_{01} \exp -(k_1 t) + y_{02} \exp -(k_2 t) + c$, where y is the fluorescence intensity, y_{01} and y_{02} are the initial fluorescence signals determined at $t = 0$, for rates k_1 and k_2 respectively, t is the time, and c the value of any signal offset.

potential to become slightly more electropositive. After this initial increase, there is what appears to be a slower excursion phase. We have shown that the latter represents the insertion of the peptide or charged limb of the peptide into the phospholipid membranes since a decrease in fluorescence is consistent with loss of positive surface charge. The FPE-based technique, therefore also offers information on the insertion of the peptides into and possibly across the membrane. In this way, Golding *et al.* (31) show that using a simple fluorimetric titration of the peptide to a population of FPE-labelled vesicles, a binding curve may be constructed and at the point of half saturation a relative affinity may be determined. The using least-squares iteration the experimental data may be fitted to a binding model. Using this technique different membrane compositions can be tested (33) and rapid mixing (stopped-flow) methods can be applied in order to resolve the binding and insertion events earlier than 1–5 sec. Using this approach (31) p25 is observed to bind to model membranes with rate constants up to about 700 s^{-1} and then insert into the membrane with rate constants on the order of 0.4 s^{-1}. Comparison of these processes with similarly time-resolved experiments performed with a stopped-flow CD spectrometer revealed that the p25 sequence first inserts into the membrane and then folds forming an α-helix. The technique has been successfully used in the study of the several other peptides such as melittin (38), fertilin (39), and viral fusion peptides (29) with membranes as well as a number of proteins such as apocytochrome *c* (40).

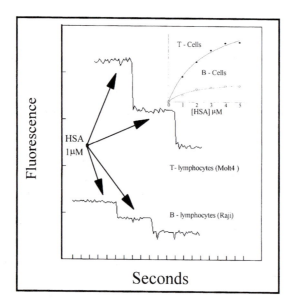

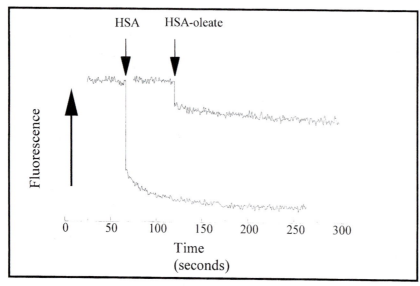

Figure 4 The interactions of serum albumin with lymphocytes. The top illustration shows B and T lymphocytes labelled with FPE at a concentration 1.2×10^5/ml suspended in 280 mM sucrose, 5 mM Tris pH 7.5 and maintained at 20 °C. Fluorescence was recorded continuously at 518 nm following excitation at 490 nm and albumin added at the times indicated by the arrows, the ticks represent 20 second intervals. The inset indicates complete HSA titration of each lymphocyte population at similar cell numbers, the ordinate indicates the change of fluorescence at each corresponding HSA concentration. The lower part of the figure illustrates the difference in response of MolT4 cultured T lymphocytes labelled with FPE following addition of 2 μM HSA or HSA-oleate at the indicated times (the latter has been offset by about 100 sec for clarity).

10.4 The interactions of serum albumin with B and T lymphocytes

Fluorescence changes associated with the binding of albumin to the cell surface of B and T lymphocytes are illustrated in *Figure 4*. The fluorescence is observed to decrease consistently with the addition of negative charge to the membrane surface as the albumin molecule overall is known to be negatively charged (12). At this stage, the *electrostatic sensing* does not discriminate between the addition of electric charge to the membrane as the result of binding to a receptor or the membrane lipid. The overall change of the fluorescent signals, however, exhibits all the characteristics of receptor binding (see below). These signals may be plotted cumulatively as shown in *Figure 4*, to estimate the affinity of albumin for the plasma membrane 'receptor'. Both white cell types exhibit saturation of binding with similar affinity constants (ca. 4 μM) but the relative abundance of the putative albumin receptor is ca. five times greater with T cells. Thus, as the number of cells in each illustrated experiment is the same, the extra binding capacity implies a greater abundance of receptors on the T cells.

Similar studies performed with albumin fully ligated with oleate also shown in *Figure 4*, indicate that the relative affinity of both types of lymphocyte is very much smaller than for the defatted albumin molecule. Additions of oleate alone caused no such signal changes. Similarly, a number of cell types such as endothelia, fibroblasts, and microglia have also been studied in an analogous manner and yielded results similar to those of the lymphocytes (data not shown). Studies have in the past been performed with erythrocytes as outlined in ref. 32. Erythrocytes, however are the only other cells apart from the B12 microglial cells line (see below) that proved to require some additional complication in their treatment to achieve results that could be interpreted simply. Although secure labelling with FPE is achieved using *Protocol 4*, the binding of macromolecules appears complicated (32). In this case gentle neuraminidase treatment proved necessary in order to measure the binding of added macromolecules to the *plasma membrane*. Without this treatment some macromolecules were observed to bind to the extensive sialic acid groups attached to glycophorin. Thus, depending on what is required this may be seen as a virtue if the study of macromolecular interactions with the 'extracellular' matrix/glycocalyx is sought.

10.5 Receptor-mediated membrane binding of albumin

A much more compelling and direct demonstration of the receptor-mediated membrane biding of albumin can be achieved studying the interaction of the protein with the recently discovered plasma membrane albumin receptor gp60 (41, 42) reconstituted into FPE-labelled vesicles (*Figure 5*). As with cells (*Figure 4*), the net fluorescence is ultimately diminished following the addition of albumin to gp60. Over the millisecond time-domain, however, the fluorescence increases rapidly with a rate approaching the resolution of the stopped-flow mixing technique (> 1000 s^{-1}).

Such rates are so rapid that they are unlikely to represent processes other

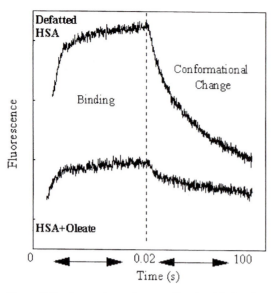

Figure 5 The interactions of serum albumin with the albumin receptor gp60, reconstituted into model membranes. Stopped-flow mixing studies (see e.g. Chapter 5) were carried out with a SX-17 MV Applied Photophysics spectrofluorimeter apparatus equipped with a 150 W xenon arc lamp. 20 μM aliquots of HSA or HSA-oleate solutions were mixed with equal volumes of gp60 vesicles with ca. 20–40 gp60 receptors per vesicle or vesicles without gp60 incorporation, typically this was 100 μl, and the resultant mixture illuminated at 490 nm. Fluorescence was collected above 520 nm continuously after mixing and 2000 data points assigned to each of the indicated time scales (0–20 msec and 20 msec–100 sec). The solid lines through the experimental points represents the best-fit achieved using an iterative least-squares analysis of the data points to a double exponential expression of the form shown in the legend to *Figure 3*. The instrumental dead time was estimated to be between 1–2 milliseconds. All experiments were carried out in 280 mM sucrose, 10 mM Tris–HCl pH 7.5. The concentration of gp60 vesicles was 200 μM on a total lipid basis after mixing. The ambient temperature was maintained at 22 °C.

than binding as in the event, a binding reaction with unrealistic rates would have to be postulated to prelude this. These early encounters leading to the binding of albumin to gp60, therefore, must first involve the addition of positive charges to the membrane. Albumin is known to possess a number of specific domains or regions on its surface which are heavily positively or negatively charged (12) and although not unexpected, the gp60–albumin-binding domain would seem to interact initially with one of these regions.

Following the initial increment over the milliseconds time domain, it is clear in *Figure 5*, that the fluorescence signals then decrease more slowly over a period of seconds. This behaviour has been observed with other membrane–protein interactions and indicates that conformational changes take place that reveal the more dominant negatively charged groups and/or part of the gp60/albumin complex has become more deeply imbedded within the membrane bilayer (31).

Studies of the interactions of purified gp60 with albumin ligated with oleate also shown in *Figure 5*, likewise were consistent with the observations made with the lymphocytes (*Figure 4*) and the other cell types. In particular the gp60–[albumin/oleate] interaction is much reduced and by the same extent for both phases of the receptor interaction as compared to the defatted albumin. Control experiments were performed with the same membranes in the absence of gp60 (data not shown) and although there is a small interaction between the albumin and the membrane it is quite different from that observed in the presence of gp60.

The implication of these data is that the membrane interactions may be regulated by whether or not albumin is carrying certain ligands. The molecular basis of this molecular selectivity seems most likely to reside in the ligand-dependent conformational changes exhibited by albumin, recently identified using steady-state and time-resolved (picosecond) fluorescence anisotropy, lifetime and quenching measurements (summarised in *Figure 1*) (15), and on the basis of X-ray crystallographic studies (43). We found that ligands such as oleate promote changes in both the shape and overall size of the albumin molecule and underlie recognition by gp60. An understanding of this behaviour may contribute to a mechanistic understanding of the very many anecdotal effects of the presence of albumin on a large and growing number of specific cellular processes some associated with uptake and secretion and others associated with cell surface behaviour (12). Of these, the effects on secretion of monoclonal antibodies by hybridoma cells and lymphocytes seems especially relevant (44).

10.6 Influence of interleukin-8 (IL-8) on the interaction of the gp41 fusion domain of HIV with T lymphocytes

It is known that to enter the host cell, HIV requires the use of two cell-membrane receptors, CD4 and a integral membrane protein co-receptor. In the case of HIV-1, the co-receptors are the chemokine receptors CCR5 and CXCR4 (45). The particular determinants on the HIV envelope responsible for the interaction with the chemokine receptors at the time of writing this chapter are not totally clear but it is clear that the five variable regions designated in gp120 as V1–V5 (46) are involved in promoting viral infection.

We have explored a possible influence of IL-8, a chemokine known to interact with receptors CRC5 and CXCR4 present in the lymphocyte cellular membrane, on the interaction of the fusion peptide of the HIV gp41 envelope protein. *Figure 6* illustrates how the positively charged fusion domain of the gp41 protein of HIV (29, 47) interacts with FPE labelled T lymphocytes, generating an increase in the fluorescence of the dye. This increase, however, is much reduced when the sample is pre-incubated with interleukin-8 (IL-8). The result of the experiment suggests that the 16-residue fusion peptide interacts as well with these receptors. The fluorescence increase measured when the cells have been pre-treated with IL-8 could conceivably reflect more non-specific binding interactions of the fusion peptide to the lipidic bilayer. Thus the possibility of an interaction between the fusion peptide of gp41 and a membrane receptor is striking but additional corroborating experiments are necessary to test this

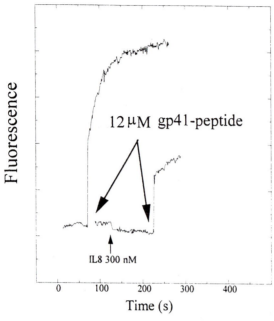

Figure 6 The interactions of the HIV fusion domain, gp41, with T lymphocytes; effects of pre-treatment of IL-8. 10^5/ml Molt4 T lymphocytes were suspended in 280 mM sucrose, 10 mM Tris–HCl pH 7.4 and fluorescence recorded at 522 nm following excitation at 490 nm. Two experiments are shown and offset in the time dimension for clarity. Peptides were added as indicated.

hypothesis. The example illustrates well the applicability of the technique for the study of the interaction of competing ligands with a membrane receptor.

10.7 The interaction of Alzheimer's peptides with B12 microglial cells

Initial experiments were performed with the microglial cell line B12 cultured from rat nervous tissue, to identify the possibility of labelling the cells with FPE in order to carry out studies on the interaction of the amyloid peptide (βA 1–40) involved in Alzheimer's disease with the cellular membrane.

The simplest method to check the efficacy of the incorporation of FPE into the outer bilayer leaflet involves challenging the membranes to an increase of the extracellular calcium concentration. Thus, adsorption of the Ca^{2+} ions onto the cell surface results in a decrease of the electronegative potential and therefore leads to an increase of the FPE fluorescence. In the case of B12 cells it proved necessary to subject the extensive extracellular matrix of the B12 cell line to limited proteolysis. Amongst the many measures explored, the most productive appeared to be the treatment with collagenase (from *clostridium histolicum*). The difference in FPE-labelling of treated and untreated cells is shown in *Figure 7*. It is clear from the figure, that the treatment with collagenase led to a much better incorporation of FPE into the B12 rat glial membrane as

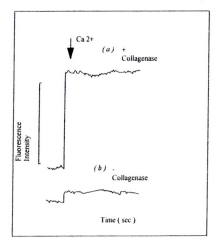 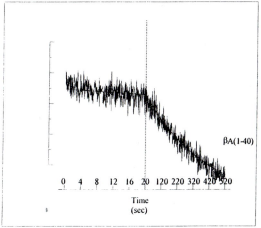

Figure 7 The interactions of Alzheimer's amyloid peptide with microglial cells. The left-hand side figure illustrates the fluorescence response at 518 nm following excitation 490 nm of collagenase treated and untreated B12 microglial cells labelled with FPE in suspension (ca. 10^5/ml) before and after addition of 4 mM calcium ions. The right-hand side figure indicates the time-evolution of the B12 microglial cell line labelled with FPE after treatment with the 800 nm βA(1–40) amyloid peptide. Cells were bathed in 280 mM sucrose/10 mM K-Hepes pH 7.4 at 23 °C.

indicated by the response to Ca^{2+}. It is worth emphasizing that following the addition of Ca^{2+} and the immediate increase in fluorescence no further significant changes were observed. This also indicated that the cellular membrane remained intact during the enzymatic digestion as a decrease in the fluorescence intensity following the addition of calcium ions is not observed. Other studies designed to assess the viability of the cells following the collagenase treatment also indicated that the cells remained a viable condition (data not shown).

The successful incorporation of FPE into B12 cell membranes, facilitated by collagenase treatment, allowed the binding interaction between neuronal cells and peptides to be monitored. The effects of addition of βA(1–40) on the fluorescence intensity of FPE-labelled B12 rat glial cells is also shown in *Figure 7*. Further additions of amyloid peptide led to an immediate decrease in fluorescence intensity which thereafter remained constant until a further addition is made. Saturation is reached at very much lower concentrations of βA(1–40) (ca. 1 μM) than observed with the artificial membrane systems and no further changes in fluorescence intensity of FPE were detected. The relationship between the extent of the binding interaction to amyloid peptide concentration indicated that the cumulative extent of the reduction of fluorescence corresponding to the peptide additions follows a hyperbolic profile and according to Wall *et al.* (33) is consistent with a binding reaction. Subsequent fitting of these data to a binding equation (33) indicated that the dissociation constant (K_d) of the peptide for the B12 plasma membrane is about 1.9 μM (Senyah *et al.*, in preparation).

197

10.8 Quantitation of spatially localized interactions of ligands (albumin) with living cells

The final application of the fluorescence-based ligand binding technology described in the present chapter involves the spatial imaging of the ligand binding phenomena on the cell surface. It is evident that many plasma membrane receptors appear to be recruited to so-called raft structures which are phase-separated patches of more viscous regions of membrane (48). This also appears to be the case for the albumin receptor, gp60, isolated from endothelial cells to illustrate that specific ligand–receptor interactions may also be assayed using the fluorescence electrostatic-sensing technology (*Protocol 7*). Instead of averaging the fluorescence output of an ensemble of cells, however, with the availability of modern imaging technology (i.e. CCD cameras and CLSM), it is feasible to visualize the disposition of fluorescence on the surface of a single cell. Thus, any changes of fluorescence following the addition of charged ligands indicates the interaction of the ligand in exactly the same manner as for the population studies. The virtue of this latter approach, however, is that spatial information of the ligand binding properties are also accessible.

In the case illustrated in Plate 7, the left-hand image shows the fluorescence difference image of albumin interactions with HUVECs. This is a 'false' colour image corrected for the probe disposition and so represent the 'hot' and cold spots of albumin binding. There appears to be localization of binding. By relating the fluorescence intensity difference to the change of electrostatic potential (i.e. modified from *Equation 12*) and then to the change of surface charge density, dividing by the net charge on the bound ligand yields an estimate of the number of molecules bound to a particular region of the cell surface. This is shown as the upright ordinate of the right-hand side plot in Plate 7. With the latest cooled CCD cameras, it has proved possible to visualize single such ligand molecules. The conditions that affect binding and even the binding domain of the ligand can be interrogated in this manner. Thus the whole repertoire of ligand binding is now accessible using these techniques.

11 Conclusions

The techniques outlined in the present chapter are for the most part generic in that they may be applied to a number of quite different systems. We have tried to illustrate this versatility using some of the experimental work ongoing in our laboratory. In the case of the tryptophan fluorescence studies, however, it is also worth emphasizing, that the position of tryptophan within the body of a protein does not always offer revealing information about the properties of a ligand–protein interaction. In the case of albumin it was fortunate that fluorescence changes were observed that correlated with the ligand-dependent conformational equilibria but not with the initial ligand binding.

The membrane sensor technology we have developed particularly with FPE, however, is much less exclusive in the sense it ought to be appropriate with virtually every ligand provided it possesses an electric charge. In this case we would like to emphasize that this is a truly generic technology. In addition the technique does not even depend upon FPE, i.e. other probes may be more appropriate to use. Thus, whilst FPE is more or less optimal for mammalian cells (32), coumarins have been used with micro-organisms such as bacteria and fungi (49–51). Similarly, if a quite different pH range were desired to be used a dye could be used with a more appropriate pK within the range of the experimental pH.

Acknowledgements

We are grateful to the many graduate research students, post-doctoral researchers, and academic visitors who have passed through our *Cell Biophysics Laboratory* and have contributed to the work outlined in this chapter. In particular, we would like to thank the following for their respective contributions to the particular examples and other work outlined in *Protocols 1–7* and the corresponding discussion as follows: [1] Neil Chadborn, Jason Bryant, and Angus Bain. [2–4] Jon Wall and Fayad Ayoub. [5] Nancy Senyah. [6] Roy Harris, Gareth Jones, Viril Patel, and Neil Chadborn. [7] Neil Chadborn, George Georgiou, Fayad Ayoub, and Jon Wall.

References

1. Eftink, M. R. (1997). In *Methods in enzymology* (Brand, L. and Johnson, M. L. (ed.)), Vol. 278, p. 221. Academic Press, London.
2. Dunbar, B. S. (1994). *Protein blotting: a practical approach*. Series 140. IRL Press, Oxford, UK.
3. Lackowicz, J. (1987). *Principles of fluorescence spectroscopy*. Plenum Press, New York, USA.
4. Brand, L. and Johnson, M. L. (ed.) (1997). *Methods in enzymology*, Vol. 278. Academic Press, London.
5. Hochstrasser, R. M. and Negus, D. K. (1984). *Proc. Natl. Acad. Sci. USA*, **81**, 4399.
6. Ross, J. B. A., Rousslang, K. W., and Brand, L. (1981). *Biochemistry*, **20**, 4361.
7. Ludescher, R. D., Volwerk, J. J., de Haas, G. H., and Hudson, B. S. (1985). *Biochemistry*, **24**, 7240.
8. Gratton, E., Alcala, J. R., and Marriott, G. (1986). *Biochem. Soc. Trans.*, **14**, 835.
9. Valeur, B. and Weber, G. (1977). *Photochem. Photobiol.*, **25**, 441.
10. Callis, P. R. (1997). In *Methods in enzymology* (Brand, L. and Johnson, M. L. (ed.)), Vol. 278, p. 113. Academic Press, London.
11. Yamamoto, Y. and Tanaka, J. (1972). *Bull. Chem. Soc. Jpn.*, **65**, 1362.
12. Peters, Th. Jr. (1996). *All about albumin*. Academic Press Inc., London, UK.
13. Weiseger, R., Gollan, J., and Ockner, R. (1981). *Science*, **211**, 1048.
14. Rose, H., Conventz, M., Fischer, Y., Jüngling, E., Hennecke, T., and Kammermeier, H. (1994). *Biochim. Biophys. Acta*, **1215**, 321.
15. Chadborn, N., Bryant, J., Bain, A., and O'Shea, P. (1999). *Biophys. J.*, **76**, 2198.
16. Boaz, H. and Rollefson, G. K. (1950). *J. Am. Chem. Soc.*, **72**, 3425.
17. Lakowicz, J. R. and Weber, G. (1973). *Biochemistry*, **12**, 4171.
18. Eftink, M. R. and Ghiron, C. A. (1976). *Biochemistry*, **15**, 672.

19. Eftink, M. R. and Ghiron, C. A. (1977). *Biochemistry*, **16**, 5546.

20. O'Connor, D. V. and Philips, D. (1984). *Time correlated single photon counting*. Academic Press, London.

21. Kinosita, K. Jr., Kawato, S., and Ikegami, A. (1977). *Biophys. J.*, **20**, 289.

22. Ameloot, L., Hendrickx, H., Herreman, W., Pottel, H., Van Cauwelaert, F., and Van Der Meer, W. (1984). *Biophys. J.*, **46**, 525.

23. Janes, S. M., Holtom, G., Ascenzi, P., Brunori, M., and Hochstrasser, R. M. (1987). *Biophys. J.*, **51**, 653.

24. Munro, I., Pecht, I., and Stryer, L. (1979). *Proc. Natl. Acad. Sci. USA*, **76**, 56.

25. McLaughlin, S. (1989). *Annu. Rev. Biophys. Biophys. Chem.*, **18**, 113.

26. Cevc, G. (1990). *Biochim. Biophys. Acta*, **1031**, 311.

27. Brockman, H. (1994). *Chem. Phys. Lipids*, **73**, 57.

28. Cladera, J. and O'Shea, P. (1998). *Biophys. J.*, **74**, 2434.

29. Cladera, J., Martin, I., Ruysschaert, J. M., and O'Shea, P. (1999). *J. Biol. Chem.*, **274**, 29951.

30. O'Shea, P. (1988). *Experientia*, **44**, 684.

31. Golding, C., Senior, S., Wilson, M. T., and O'Shea, P. (1996). *Biochemistry*, **35**, 10931.

32. Wall, J., Ayoub, F., and O'Shea, P. (1995). *J. Cell Sci.*, **108**, 2673.

33. Wall, J., Golding, C., van Veen, M., and O'Shea, P. (1995). *Mol. Membr. Biol.*, **12**, 183.

34. Grynkiewicz, G., Poenie, M., and Tsien, R. Y. (1985). *J. Biol. Chem.*, **260**, 3440.

35. Montana, V., Farkas, D. L., and Loew, L. M. (1989). *Biochemistry*, **28**, 4536.

36. Bain, A. J. and McCaffery, A. J. (1984). *J. Chem. Phys.*, **80**, 5883.

37. Schieder, W. (1980). *J. Phys. Chem.*, **84**, 925.

38. Wolfe, C., Cladera, J., and O'Shea, P. (1998). *Mol. Membr. Biol.*, **15**, 221.

39. Wolfe, C., Cladera, J., Lahda, S., Senior, S., Jones, R., and O'Shea, P. (1999). *Mol. Membr. Biol.*, **16**, 257.

40. Bryson, E. A., Rankin, S. E., Carey, M., Watts, A., and Pinheiro, T. J. (1999). *Biochemistry*, **38**, 9758.

41. Schnitzer, J. E. and Oh, P. (1994). *J. Biol. Chem.*, **269**, 6072.

42. Tirrupathi, C., Song, W., Bergenfeldt, M., Sass, P., and Malik, A. B. (1997). *J. Biol. Chem.*, **272**, 25968.

43. Curry, N., Mandelkow, H., Brick, P., and Franks, N. (1998). *Nature Struct. Biol.*, **5**, 827.

44. Wall, J. S., Ayoub, F., and O'Shea, P. (1996). *Front. Biosci.*, **11**, 46.

45. Clapham, P. R. (1997). *Trends Cell Biol.*, **7**, 264.

46. Hoffman, T. L. and Doms, R. (1999). *Mol. Membr. Biol.*, **16**, 57.

47. Martin, I., Schaal, H., Scheid, A., and Ruysschaert, J. M. (1996). *J. Virol.*, **70**, 298.

48. Simons, K. and Ikonen, E. (1997). *Nature*, **387**, 569.

49. Davies, K., Afolabi, P., and O'Shea, P. S. (1996). *Parasitology*, **113**, 1.

50. Jones, L., Hobden, C., and O'Shea, P. S. (1995). *Mycol. Res.*, **99**, 969.

51. Hobden, C., Teevan, C., Jones, L., and O'Shea, P. (1995). *Microbiology*, **141**, 1875.

Chapter 5
Stopped-flow techniques

John F. Eccleston and Jon P. Hutchinson
National Institute for Medical Research, The Ridgeway, Mill Hill, London NW7 1AA, UK.

Howard D. White
Department of Biochemistry, East Virginia Medical School, Norfolk, Virginia 23507, USA.

1 Introduction

There are many methods for measuring the equilibrium binding constant of a protein with a ligand, where the ligand can be a small molecule or another macromolecule such as a protein or nucleic acid, as described in other chapters in this book. However, many biological processes are controlled not by equilibrium constants but by the kinetics of the interaction between protein and ligand and for this reason it is necessary to measure the association and dissociation rate constants in order to understand a system fully. Although this can be done using surface plasmon resonance techniques (see Volume 1, Chapter 6), and in certain cases by NMR measurements (see Chapter 10), rapid-reaction methods are generally the method of choice if a suitable intrinsic optical probe exists in the system or an extrinsic probe can be introduced.

Rapid-reaction methods include stopped-flow techniques in which the solutions are rapidly mixed, or relaxation methods in which an equilibrium mixture is perturbed by rapid-changes in temperature (T-jump) or pressure (P-jump). This chapter is solely concerned with rapid mixing techniques in which reactions can be followed in the millisecond to second time scale. The need to make measurements over this time scale can be seen if the association rate constant of a ligand is assumed to have a value of 10^6 M^{-1} sec^{-1}. At a concentration of ligand of 10 μM, the observed first-order rate constant will be 10 sec^{-1} which corresponds to a half-time of reaction of 69 msec. If the concentration of ligand is 100 μM, then the rate constant will be 100 sec^{-1}, giving a half-time of 6.9 msec. The rationale for these arguments is described in more detail below. Obviously, the actual rate constants of a protein–ligand interaction vary depending on the system and it is the aim of this chapter to describe how these measurements can be made.

2 First- and second-order reactions

Since this chapter is mainly concerned with a practical approach to stopped-flow measurements, the following theory section is necessarily brief. However, since an understanding of the basic theory of first- and second-order reactions is necessary for the correct interpretation of data, this is given below. For a comprehensive discussion of the analysis of kinetic data, the reader is referred to books by H. Gutfreund (1), A. Fersht (2), and a review by K. A. Johnson (3).

2.1 First-order irreversible reactions

Although this chapter by definition concerns second-order reactions where the reaction rate is controlled by the concentrations of two species, it is appropriate first to describe the behaviour of a first-order reaction for reasons which will become obvious.

In the reaction:

$$A \xrightarrow{k} B$$

the rate of loss of A is given by:

$$\frac{-d[A(t)]}{dt} = k[A(t)] \qquad [1]$$

Integrating gives:

$$\ln [A(t)] - \ln [A_o] = -kt \qquad [2]$$

where $[A_o]$ is the initial concentration of A, and $[A(t)]$ is the concentration of A at time t. This can be rearranged to give:

$$\ln [A(t)] = \ln [A_o] - kt \qquad [3]$$

or

$$[A(t)] = [A_o]e^{-kt} \qquad [4]$$

Experimental data can be fitted to the linear equation (*Equation 3*) by plotting ln $[A(t)]$ against t which gives a straight line of slope k. However, the use of non-linear least squares fitting procedures allow the data to be fitted directly to *Equation 4*.

Two consequences of the properties of first-order reactions follow from these equations. First, since the value of kt is proportional to the natural logarithm of the ratio of $[A_o]$ to $[A(t)]$, absolute concentration units for $A(t)$ do not need to be measured, only some property which is proportional to the ratio of $[A_o]$ and $[A(t)]$. Secondly, the time taken to complete a definite fraction of the reaction is independent of the initial concentration of A. For example, for the reaction to proceed to 50% completion (i.e. when $[A_o]/[A(t)] = 2$) then substituting for $[A_o]/[A(t)]$ in *Equation 2* gives:-

$$t_{1/2} = \frac{\ln (2)}{k} = \frac{0.69}{k} \qquad [5]$$

where $t_{1/2}$ is the time taken for half the reaction to occur. This provides a rapid method for evaluating the value of k from the half-time of a first-order reaction.

2.2 Second-order irreversible reactions

Second-order reactions are best studied under one of two special conditions. The first of these is when the concentration of one of the reactants is in a large excess over the other and so does not effectively change over the time course of the reaction. This is called a pseudo first-order reaction. (What constitutes a large excess in practical terms is discussed in Section 7.1.)

For the reaction

$$A + B \xrightarrow{k} C$$

the rate of formation of C is given by:

$$\frac{d[C(t)]}{dt} = k\,[A(t)]\,[B(t)] \qquad [6]$$

where $[A(t)]$ and $[B(t)]$ are the concentrations of A and B at time t.

If $[A_o] \gg [B_o]$, $[A(t)]$ can be regarded as a constant and equal to $[A_o]$, so the rate equation is:

$$\ln [B(t)] - \ln [B_o] = -([A_o]\,k)t \qquad [7]$$

This has the same form as *Equation 2* where $k_{obs} = [A_o]\,k$ with units of sec^{-1}.

If the observed rate constant is determined with respect to $[A_o]$, a plot of k_{obs} against $[A_o]$ gives a straight line of slope k with units of $M^{-1}\,sec^{-1}$ (if $[A]$ is measured in M, i.e. mol/L).

The second condition under which a second-order reaction is often measured is when the concentrations of the two reactants are equal. *Equation 6* then becomes:

$$\frac{d[C(t)]}{dt} = k\,[A(t)]^2 \qquad [8]$$

which on integration gives:

$$\frac{1}{[A(t)]} - \frac{1}{[A_o]} = kt \qquad [9]$$

and the experimental data can be fitted to this equation. It should be noted that the initial concentrations of reactants need to be known since the second-order constant involves a concentration term. In this case, at half completion of the reaction:

$$[A(t)] = [A_o]/2 \qquad [10]$$

so

$$\frac{2}{[A_o]} - \frac{1}{[A_o]} = kt_{1/2} \qquad [11]$$

or

$$t_{1/2} = \frac{1}{k\,[A_o]} \qquad [12]$$

Unlike first-order reactions, the half-time of the reaction is dependent on the concentration of A_o and so the half-time progressively increases as the reaction proceeds.

If neither of the two special conditions can be used, the general equation for a second-order reaction is:

$$kt = \frac{1}{[A_o] - [B_o]} \cdot \ln \frac{[B_o][A(t)]}{[A_o][B(t)]} \tag{13}$$

2.3 First-order reversible reaction

The above discussions have assumed that the reactions under study are irreversible. This is usually not true in practice. If a first-order reaction is reversible:

$$A \underset{k_{-1}}{\overset{k}{\rightleftharpoons}} B$$

At equilibrium:

$$K = \frac{[B_{eq}]}{[A_{eq}]} = \frac{k_1}{k_{-1}} \tag{14}$$

where K is the equilibrium constant for the reaction and $[A_{eq}]$ and $[B_{eq}]$ are the concentrations of A and B at equilibrium. At equilibrium:

$$k_1[A_{eq}] = k_{-1}[B_{eq}] \tag{15}$$

The rate of approach to equilibrium is given by:

$$k_1([A_o] - [B_{eq}]) - k_{-1}[B_{eq}] = 0 \tag{16}$$

$$[A_o] = \frac{k_{-1}}{k_1}[B_{eq}] + [B_{eq}] \tag{17}$$

$$\frac{d[B(t)]}{dt} = k_1([A_o] - [B(t)]) - k_{-1}[B(t)] \tag{18}$$

$$\frac{d[B(t)]}{dt} = k_{-1}[B_{eq}] + k_1[B_{eq}] - k_1[B(t)] - k_{-1}[B(t)] \tag{19}$$

$$= (k_{-1} + k_1)([B_{eq}] - [B(t)]) \tag{20}$$

This has the same form as an irreversible first-order reaction (*Equation 1*) except that the observed rate constant is the sum of k_1 and k_{-1}. When k_{-1} is reduced to zero, this equation becomes equivalent to *Equation 1*.

Knowledge of the equilibrium constant allows individual values of k_1 and k_{-1} to be determined from *Equation 14*.

2.4 Second-order reversible reactions

As with irreversible second-order reactions, these are best done under pseudo first-order conditions for ease of analysis of the data.

In the scheme:

$$A + B \underset{k_{-1}}{\overset{k_1}{\rightleftharpoons}} AB$$

k_1 is the second-order association rate constant and k_{-1} is the first-order dissociation rate constant. If [A] is much larger than [B], its concentration is effectively constant during the time course of the reaction.

If $[B(t)]$ is the concentration of B at time t and $[B_{eq}]$ is the concentration of B at equilibrium, $[B_o]$ is the concentration of B at time zero, and similar notations are used for the concentrations of A and AB.

At equilibrium:

$$k_1 [A][B_{eq}] = k_{-1} [AB_{eq}] \qquad [21]$$

Therefore

$$k_1 [A][B_{eq}] - k_{-1} [AB_{eq}] = 0 \qquad [22]$$

Since [A] remains constant throughout the reaction, $k_1[A]$ is a constant and is defined here as k'. Therefore:

$$k' [B_{eq}] - k_{-1} [AB_{eq}] = 0 \qquad [23]$$

since:

$$[B_{eq}] = [B_o] - [AB_{eq}]$$

$$k' ([B_o] - [AB_{eq}]) - k_{-1} [AB_{eq}] = 0$$

$$k' [B_o] - k' [AB_{eq}] - k_{-1} [AB_{eq}] = 0$$

$$[B_o] = \frac{k' [AB_{eq}] + k_{-1} [AB_{eq}]}{k'} \qquad [24]$$

$$= \frac{k_{-1}}{k} [AB_{eq}] + [AB_{eq}] \qquad [25]$$

The rate of formation of AB is given by the equation:

$$\frac{d[AB(t)]}{dt} = k' [B(t)] - k_{-1}[AB(t)] \qquad [26]$$

Since:

$$[B(t)]] = [B_o] - [AB(t)]$$

$$\frac{d[AB(t)]}{dt} = k'([B_o] - [AB(t)]) - k_{-1}[AB(t)] \qquad [27]$$

$$= k' ((k_{-1}/k')[AB_{eq}] + [AB_{eq}] - [AB(t)]) - k_{-1} [AB(t)]$$

$$= k_{-1}[AB_{eq}] + k' [AB_{eq}] - k' [AB(t)] - k_{-1} [AB(t)]$$

$$= (k_{-1} + k') ([AB_{eq}] - [AB(t)]) \qquad [28]$$

This has the same form as a first-order reaction shown in *Equation 1*. A plot of $\ln([AB_{eq}] - [AB(t)])$ against time will therefore yield an observed first-order rate constant, k_{obs}, which is equal to $k' + k_{-1}$. Since $[AB_{eq}] = [B_o] - [B_{eq}]$ and $[AB(t)] = [B_o] - [B(t)]$, a plot of $\ln([B(t)] - [B_{eq}])$ will yield the same information. Again, concentrations need not be in any specific units and any parameter proportional to concentration of the species can be used.

If the reaction is performed over a range of concentrations of A, the dependence of k_{obs} on $[A_o]$ can be measured. Since:

$$k_{obs} = k_1[A_o] + k_{-1} \qquad [29]$$

a plot of k_{obs} against $[A_o]$ will give a straight line of slope k_1 (second-order rate constant; units, M^{-1} sec^{-1}) and intercept on the ordinate of k_{-1} (first-order rate constant; units, sec^{-1}). Hence k_1 and k_{-1} can be determined together with the value of the equilibrium dissociation constant (k_{-1}/k_1) for the reaction.

3 Principle of operation of the stopped-flow instrument

The stopped-flow technique was originally developed by Chance and Gibson (4) from the continuous flow technique. Continuous flow was used by Hartridge and Roughton (5) to study the binding of oxygen to haemoglobin. Solutions of haemoglobin and oxygen were rapidly mixed together and passed down a glass tube (*Figure 1a*). The age of the reaction solution at any given point along the tube is defined by the flow rate and the volume between mixing and that point. Therefore, by making measurements along the length of the tube, the time course of the reaction could be followed. For example, if the mixed solution has a linear flow rate of 10 metres/sec, making measurements every 1 cm gives the reaction profile at 1 msec time points. In this way, even though the measurements took several minutes to make at each point, reactions could be followed on the millisecond time scale. However, this technique is very expensive with regard to solution volumes and can only be used where large amounts of protein and ligand are readily available. The advent of photomultiplier tubes in the 1940s allowed light to be monitored on a rapid time scale and led to the development of the stopped-flow method, although use has been recently made of the continuous flow method which has allowed reactions on the sub-millisecond time scale to be measured (see Section 9.6).

In the stopped-flow technique, two solutions are rapidly mixed together and at any given point the age of the solution is again defined by the flow rate and the volume between mixing and observation. However, by the use of a back syringe, the flow of mixed reactants is suddenly stopped and the reaction is followed in real time with a suitable detection system (*Figure 1b*). *Figure 2* shows a typical stopped-flow trace which illustrates the processes occurring in the instrument where the recording device is storing data from 50 msec before the point when flow stops. The reaction is that of N-acetyl tryptophanamide with excess N-bromosuccinimide (6) which results in the loss of the N-acetyl tryptophanamide fluorescence. Initially we see the signal from the end of the previous

a)

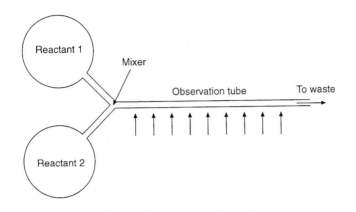

b)

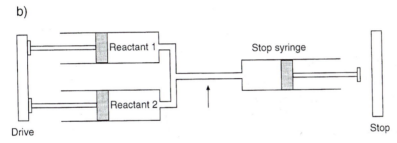

Figure 1 Diagrammatic representations of (a) the continuous flow and (b) the stopped-flow methods. The arrows represent the points of observation.

run on the instrument which has negligible fluorescence. When the reactant syringes are driven forward, the new solution displaces the old solution from the observation cell. Then a period of constant signal is seen. Over this time, the reactant solution has a constant age as described for the continuous flow method. This age is defined as the dead-time of the instrument and is the shortest time from which reactions can be monitored. When the flow stops, the reaction is followed with time.

In the above example, the reaction was followed by monitoring fluorescence. Owing to the inherent sensitivity of this method, it is widely used for stopped-flow studies and is mainly used in the examples in this chapter, although absorbance measurements are also discussed. Other specialized methods of detection that have been used include CD, NMR, FTIR, and calorimetry which are discussed elsewhere in these two volumes.

In a typical stopped-flow experiment, equal volumes of the two reactant solutions are mixed. The concentrations in the observation cell are therefore half of those in the syringe and it is important to define which concentration is being quoted.

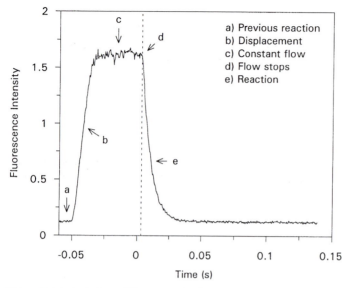

Figure 2 A typical stopped-flow trace showing the different phases of the observed signal. In this example, the recording was commenced at 50 msec before flow stopped in order to illustrate all of the phases.

4 Testing the performance of a stopped-flow instrument

4.1 The mixing process

The performance of the stopped-flow instrument depends crucially on rapid mixing of the two solutions. The main aim is that the solutions are mixed with such geometry and linear flow rate that turbulent flow of the mixed reactants occurs. This should allow complete mixing of the solutions before they reach the observation point of the reaction typically in 1–6 msec. As described above, the dead-time is related to the flow rate and distance of observation of the reaction from the point of mixing to the point of observation. It can therefore be reduced by increasing the flow rate or decreasing this distance. However, a very short dead-time is not necessarily a good thing since mixing might not then be complete.

Although the user of commercial equipment has little control over the above factors, the mixing efficiency and dead-time of the instrument should be measured, particularly when very fast reactions are being studied.

4.2 Mixing efficiency

It is important that the solutions are thoroughly mixed when observation commences so that the observed rate of the reaction is the true one and is not limited by the mixing process. This can be demonstrated by performing a reaction that is instantaneous on the stopped-flow time scale (7). If any sign of a reaction can be seen, then this must be due to incomplete mixing. Protonation reactions occur with rate constants much faster than the stopped-flow time scale so that if an indicator at a pH above its pK_a is mixed with buffer below its pK_a, the reaction should be completed within the dead-time of the instrument. *Protocol 1* describes such an experiment for a stopped-flow instrument in the fluorescence mode. A similar experiment can be designed for absorption mode measurements using phenol red (λ_{max}, 560 nm; ε, 54 000 $M^{-1}cm^{-1}$).

Protocol 1

Testing the mixing efficiency of a stopped-flow instrument

Equipment and reagents

- 0.1 M sodium pyrophosphate adjusted to pH 8.7 with HCl
- 0.1 M sodium pyrophosphate adjusted to pH 6.2 with HCl
- 100 μM 4-methylumbelliferone in water

- Stopped-flow instrument operated in the fluorescence mode with excitation at 366 nm and emission monitored with 399 nm cut-off filter

Method

1 Dilute 10 μl of 100 μM 4-methylumbelliferone into 1 ml of 0.1 M sodium pyrophosphate pH 8.7. Place into one syringe of the stopped-flow apparatus. Fill the other syringe with 0.1 M sodium pyrophosphate pH 8.7.

2 Do several pushes to make sure that the new solution fills the observation cell of the instrument.

3 Increase the high voltage on the photomultiplier to give a suitable signal (typically 4 V).

4 Dilute 10 μl 4-methylumbelliferone into 1 ml of 0.1 M sodium pyrophosphate pH 6.2. Place into one syringe of the stopped-flow apparatus. Fill the other syringe with 0.1 M sodium pyrophosphate pH 6.2.

5 After several pushes, without altering the high voltage, note the value of the signal at the recorder. It should be ~ 0.1 V. The signal therefore changes from 4 V at pH 8.7 to ~ 0.1 V at pH 6.2.

6 Fill one syringe with 10 μM 4-methylumbelliferone in 0.1 M sodium pyrophosphate pH 8.7 and the other with 0.1 M sodium pyrophosphate pH 6.2. Do several pushes. Then record the signal after a push collecting data over a 50 msec time scale.

7 Determine that after the flow has stopped, there is a constant signal over this time range so that complete mixing is occurring in the dead-time of the instrument.

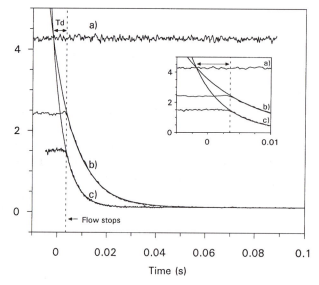

Figure 3 Measurement of the dead-time of a stopped-flow instrument. The reaction conditions were 10 μM N-acetyl tryptophanamide in one syringe and in the other (a) buffer, (b) 600 μM N-bromossuccinimide, (c) 300 μM N-bromossuccinimide. These correspond to final concentrations of 5 μM N-acetyl tryptophanamide, and 300 μM or 150 μM N-bromosuccinimide. The fitted lines are to a single exponential and extrapolated back to the start of the reaction shown in (a). Inset shows the process on an expanded time scale. The figure is discussed more fully in the text.

4.3 Measurement of dead-time

Having proved that efficient mixing is occurring, the dead-time of the instrument should be measured. This can be done by extrapolating the time course of a reaction back to its known starting point. This is illustrated in *Figure 3*, again using the reaction of N-acetyl tryptophanamide with N-bromosuccinimide under pseudo first-order conditions. See *Protocol 2*.

Protocol 2

Measurement of the dead-time of a stopped-flow instrument in the fluorescence mode

Equipment and reagents

- Stopped-flow instrument operated in the fluorescence mode with excitation at 290 nm and emission monitored through a WG 320 cut-off filter

- 10 μM N-acetyl tryptophanamide in 0.1 M sodium phosphate pH 7.5
- 600 μM N-bromosuccinimide in 0.1 M sodium phosphate pH 7.5

Method

1   Push equal volumes of 10 μM N-acetyl tryptophanamide in 0.1 M sodium phosphate

pH 7.5 against 0.1 M sodium phosphate pH 7.5, in the stopped-flow apparatus. After several pushes, increase the high voltage in the photomultiplier to give a suitable signal (typically 4 V).

2 Record a trace over 100 msec.

3 Push equal volumes of 10 μM N-acetyl tryptophanamide in 0.1 M sodium phosphate against 600 μM N-bromosuccinimide. Record the signal from the reaction over 100 msec time scale. An exponential decrease in signal should be seen with a rate constant of approximately 200 sec^{-1} (at 20 °C) with an end-point at close to zero.

4 Fit the data to a single exponential and extrapolate the fitted line back to the signal obtained when N-acetyl tryptophanamide was pushed against 0.1 M sodium phosphate pH 7.5.

5 Determine the dead-time T_d of the instrument as shown in *Figure 3*.

6 Repeat using 300 μM N-bromosuccinimide.

The dead-time of the instrument is the difference in time between the observed start of the reaction and the intercept of the extrapolation with the starting point. Typically this is in the region of 1–6 msec. It should be noted that the time labelled in the record as zero time is not necessarily the point at which flow stops and from when the reaction is monitored (*Figure 3*). This is because it is difficult to co-ordinate the precise stopping time with the trigger signal to within less than a few milliseconds.

For the determination of the dead-time of instruments operating in the absorption mode, the alkaline hydrolysis of 2,4-dinitrophenylacetate can be measured. This has a second-order rate constant of 45 M^{-1} sec^{-1} so that mixing a solution of 2,4-dinitrophenylacetate with 1 M NaOH (final concentration) will give a reaction with a rate constant of 45 sec^{-1} ($t_{1/2}$ = 15.4 msec). See *Protocol 3*.

Protocol 3

Measurement of the dead-time of a stopped-flow instrument in the absorbance mode

Equipment and reagents

- Stopped-flow instrument operated in the transmission mode with incident light at 360 nm
- 5 mM HCl in water
- 2 M NaOH in water

- 12.5 mM 2,4-dinitrophenylacetate (preferably recrystallized from ethanol) in propan-2-ol; just prior to use dilute 10-fold with 5 mM HCl

Method

1 Push water through the observation cell and then set the high voltage in the

Protocol 3 continued

photomultiplier to give a suitable signal (typically 4 V). Set this value as 100% transmission.

2   Close the shutter in the photomultiplier and set the signal as 0% transmission.

3   Push equal volumes of 1.25 mM 2,4-dinitrophenol acetate in 5 mM HCl against 5 mM HCl.

4   Record data over 100 msec time scale.

5   Push 1.25 mM 2,4-dinitrophenol acetate against 2 M NaOH and record the reaction over 100 msec time scale. An exponential decrease in transmission occurs with a rate constant of 45 sec^{-1} (20 °C).

6   Convert traces obtained in steps 4 and 5 from percentage transmission to absorbance.

7   Fit the data from the reaction (in step 5) to a single exponential. Extrapolate the exponential back to the value of the absorbance at the start of the reaction (in step 4).

8   Determine the dead-time of the instrument as shown in *Figure 3*.

5 Optimization of instrument

The main aims of optimizing instrument conditions are to maximize the size of the signal and reduce the noise of the signal. The actual solution conditions affect these factors but first the variables that can be changed on the instrument are discussed.

5.1 Lamp selection

For fluorescence experiments, a high light intensity is required for excitation and for this reason the best light source is either a xenon, mercury, or xenon/mercury arc lamp. Xenon arc lamps have a relatively smooth emission spectrum whereas mercury or xenon/mercury lamps have a series of very intense emission bands which, depending on the excitation spectrum of the fluorophore, may be used to advantage. A disadvantage of arc lamps is that their emission is relatively noisy compared to deuterium or quartz halide lamps. Deuterium and quartz halide lamps have lower intensity outputs and are suitable for absorbance measurements which have a much lower requirement for light intensity. Quartz halide lamps can also be sometimes used for fluorescent excitation in the visible region if the fluorophore has a high extinction coefficient and quantum yield. Manufacturer's information on the dependence of light intensity on wavelength should be consulted.

5.2 Monochromator

The only two aspects in the operation of the monochromator over which the user has control are the correct alignment of the light source with the monochromater and the slit widths of the monochromator. Alignment should be done according to the manufacturer's instructions but the main aim is that the light beam is focused on the inlet slit and that the diverging beam inside the monochromator does not overfill the first mirror. If the latter does occur, then the amount of stray light will increase. With regard to the monochromator slits, this is a balance between spectral purity of the exciting light and the intensity of light transmitted. For fluorophores with a large Stokes shift (i.e. the wavelength difference between the excitation and emission maxima) a large slit width can be used to increase light intensity. For fluorophores with a short Stokes shift, the monochromator slit widths may need to be reduced to enhance the exclusion of scattered light from the photomultiplier. Alternatively, the excitation light may be set at shorter wavelengths than the maximum. These decisions need to be made in conjunction with the choice of emission conditions (see Section 5.3).

A further use of the monochromator slit widths is the possibility of reducing the light intensity to avoid photobleaching of the fluorophore, although this could be done equally well using a neutral density filter. The main aim of the optical system of a fluorescent stopped-flow instrument is to have the intensity of light reaching the sample as high as possible. Photobleaching is not usually a problem on the time scale of < 1 sec but over longer periods may be significant. Photobleaching causes a slow decrease in fluorescence emission intensity. It can be quantified by pushing the fluorophore alone against buffer and recording the decrease in fluorescence intensity. This control can be used in the analysis of reaction traces.

5.3 Choice of emission filter

The emitted light is usually detected by a photomultiplier after it has passed through a suitable optical filter. The aim of the filter is to pass only fluorescence emission light and not any of the exciting light which may be scattered by the solution.

Figure 4 shows the variation with wavelength of the light intensity detected at right angles when either a solution of water or 1 μM *N*-acetyl tryptophanamide is irradiated with light at 280 nm. The fluorescence emission of *N*-acetyl tryptophanamide occurs with a maximum wavelength at 350 nm. However, with this solution, and with water alone, light is detected at 280 nm and at 560 nm. This is due to Rayleigh scattering of the exciting light (280 nm) and of the first harmonic of the exciting light (560 nm). Light is also detected at 309 nm resulting from Raman scattering by the water. The wavelength of the Raman peak depends on the wavelength of the incident light and can be calculated from the equation $1/\lambda_R = 1/\lambda_{EX} - 0.00034$. As the excitation wavelength is increased, the first harmonic scattered will progressively move to the red and will become less of a problem as the photomultiplier will be less responsive to these longer wavelengths.

Filters are chosen to maximize the fluorescence signal and minimize these other signals. This can be done by using either a cut-off and/or band pass filter.

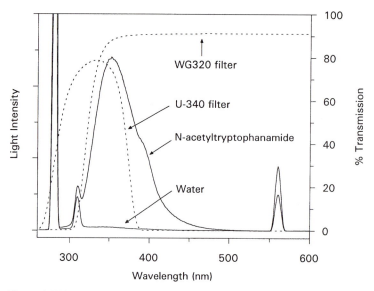

Figure 4 Light intensity observed orthogonal to the exciting light with respect to wavelength, when water or 1 μM *N*-acetyl tryptophanamide is irradiated at 280 nm. The transmission spectra of a WG 320 cut-off filter and Hoya U-340 band pass filter are also shown.

The transmission characteristics of two such filters are also shown in *Figure 4.* The WG 320 cut-off filter blocks light in the UV region and passes light at higher wavelengths with 50% transmission at 320 nm, whereas the Hoya U-340 band pass filter passes light between 270 nm and 390 nm. It can be seen that the WG 320 filter will totally block the 280 nm and most of the 310 nm scattered light but transmit the light from fluorescence emission. However, the WG 320 filter will pass the 560 nm scattered light. This can be blocked by the addition of the U-340 filter although the fluorescence signal will be attenuated.

The above discussion is based on a change of tryptophan fluorescence intensity without a significant red or blue shift. This is the case for ATP binding and hydrolysis by rabbit skeletal myosin subfragment 1 and can be measured with a good signal to noise ratio using solutions containing as little as 100 nM (12 μg/ml protein) (8). If there is a significant shift in the emission maximum of the fluorescence with ligand binding, appropriate band-pass or sharp cut-off filters should be selected. For example, calcium binding to whiting parvalbumin induces a large blue shift in the emission spectrum of a single tryptophan residue (9) which can be observed with either a 323, 340, or 380 nm interference filter or a 360 nm cut-off filter. No change in fluorescence is observed with a UG3 band pass filter as the small change in the quantum yield is cancelled by the spectral response of the filter.

5.4 Stop volume

The considerations discussed above are concerned with optimization of the optical system. A final consideration is to decide on the volumes of reactants to

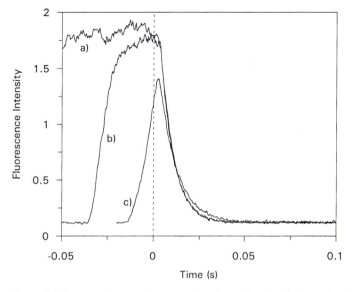

Figure 5 Effect of volume pushed on a reaction profile. A solution of 5 μM *N*-acetyl tryptophanamide was mixed with 300 μM *N*-bromosuccinimide and tryptophan fluorescence observed. The total volume pushed was (a) 150 μl, (b) 100 μl, (c) 50 μl.

be mixed. This can usually be done by adjusting the position of the back-stop of the stopping syringe. When flow is activated, the two solutions are mixed at increasing speed until they reach a constant flow rate governed by the resistance of the flow circuit. Enough solution needs to be mixed to achieve this and so have a short dead-time without wasting often valuable material. The optimal volume to be pushed can be determined as shown in *Figure 5*. Here, the reaction of *N*-acetyl tryptophanamide with excess *N*-bromosuccinimide is done by pushing total volumes of mixed solution of 50 μl, 100 μl, and 150 μl. At both 150 μl and 100 μl pushed volume, constant flow is achieved but not at 50 μl. In the last case, the amplitude of the observed process is reduced and the reaction appears to be slightly slower, possibly due to incomplete mixing of the reactants.

5.5 Time constant

Electronic noise on the signal is limited by the response time or time constant of the electronics. The noise is inversely proportional to the square root of the time constant. Therefore increasing the time constant by 100-fold decreases the noise ten-fold. The value of the time constant should be such that it is long enough to reduce noise but not to affect the rate of the process being observed. This is illustrated in *Figure 6*. It can be seen that with no time constant the signal is very noisy but increasing the time constant reduces the noise without affecting the observed rate of the reaction. However, at longer time constants, the observed process becomes progressively slower until in the extreme case, the rate of the observed process it solely determined by the time constant (*Figure 6*

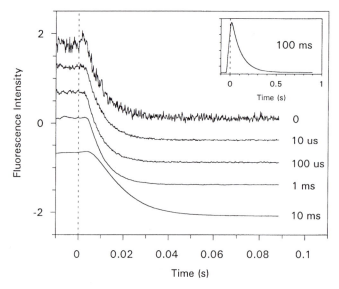

Figure 6 Effect of the time constant on the observed process. The reaction shown is that of *Figure 5b* with time constants used as indicated. The traces are shown with a vertical displacement for the sake of clarity.

inset). It is best to use a time constant that is < 10% of the half-time of the fastest process being observed.

The signal to noise ratio can also be increased by operating the instrument in the 'oversampling' mode so that the data is collected at the fastest rate possible by the instrument and blocks of data are averaged to give single time points. For example, if the computer collects data with a frequency of 1 MHz, samples can be collected at 1 μsec intervals and averaged in suitable size blocks. The signal to noise ratio increases with the square root of the number of samples averaged.

6 Probes to monitor stopped-flow reactions

A signal is required which changes on ligand binding which can be monitored with time. This may be intrinsic to the system, or one of the components of the system is modified either covalently or non-covalently with an extrinsic probe.

6.1 Intrinsic probes

Tryptophan fluorescence in proteins is commonly used as an intrinsic probe. Large changes in fluorescence may occur on interaction of a protein with a ligand. For example, on the interaction of myosin subfragment 1 with ATP, there is a 20% enhancement of tryptophan fluorescence which has been extensively used for studies aimed at understanding the mechanism of the myosin ATPase (8). NADH is also fluorescent and has been used for stopped-flow studies on many NADH dependent systems (10). Both tryptophan and NADH may also undergo absorption changes on protein–ligand interactions but the effects are generally much smaller than the fluorescence changes. Other naturally occurring chromophoric cofactors and coenzymes include flavins, pyridoxal phosphate, haem, coenzyme B_1, folates, type 1 blue copper centres, and Ni (II) centres (11).

6.2 Extrinsic probes

An extrinsic probe can be introduced into the system by covalently binding a probe to the protein or ligand. The choice of probe is determined first by the need for high environmental sensitivity of the probe so that it gives large changes of signal during a binding process. Secondly, since fluorescence intensity is proportional to absorption coefficient and quantum yield, it is preferable to work with fluorophores with high values of both these properties. These two factors may not necessarily be compatible. For example, the N-methylanthraniloyl moiety has an extinction coefficient at 354 nm of 5700 M^{-1} cm^{-1} and quantum yield 0.2 in water, whereas fluorescein has an extinction coefficient of 80 000 M^{-1} at 490 nm and a quantum yield of > 0.9. Therefore, fluorescein fluorescence can be detected at much lower concentrations than the N-methylanthraniloyl fluorophore but the latter is more sensitive to its environment.

The probe can be attached to either the protein or ligand. There is a wide range of fluorescent probes available commercially that react with functional groups of biomolecules, usually with thiol or amine groups. It is easier to modify the ligand if it is a small molecule because it will generally be easier to produce a well characterized product and the fluorophore will be close to the binding site in the complex and therefore more likely to give fluorescence intensity changes. Many small biomolecules have been modified in this way including nucleotides and peptides (12, 13).

Covalent modification of proteins has also been extensively used although, because more than one site may be labelled, it can be difficult to obtain a reproducible product. Ideally, the modified protein should be characterized by a combination of mass spectroscopy, and proteolysis experiments, and the modification shown not to affect the biological activity of the protein. Examples of such studies on calmodulin (14) and phosphate binding protein (15) should be consulted for more detailed information.

If a protein contains a single cysteine residue, covalent labelling at a single site may be readily achievable. However, it may be well removed from the ligand binding site and thus less likely to be sensitive to the ligand binding process. This problem has been overcome with phosphate binding to phosphate binding protein. This protein contains no cysteine residues so Brune *et al.* genetically engineered single cysteine residues into the protein based on a knowledge of the structure of the protein. Cysteine residues were introduced at positions where the label would be expected to be perturbed by phosphate binding. By a combination of different residues and fluorophores, they produced one modified protein that showed a 13-fold increase in fluorescence on the binding of phosphate (15).

Similarly, genetic engineering may be used to produce proteins with a single tryptophan moiety. Recently, analogues of tryptophan have been introduced into proteins. These analogues have longer wavelength excitation and emission maxima than tryptophan itself and therefore can be detected separately from tryptophan. In this way the interaction of a protein containing a single tryptophan analogue with a multi-tryptophan containing protein can be monitored (16).

7 Measurement of association and dissociation rate constants

7.1 Initial consideration

As described in Section 2.2, second-order reactions are best studied under conditions where the concentrations of the reactants are such that one is in a large excess over the other or that the concentrations of both reactants are equal. This makes the analysis easier although the advent of global analysis methods as described later (Section 10) reduces these restrictions. However, in the following discussion, use is made of pseudo first-order conditions in which there is excess ligand over protein. The question arises of what exactly a large excess means. Gutfreund (1) simulated a ligand binding reaction at different ratios of reactant concentrations and then fitted the simulated data to single exponentials. He showed that analysing the data over 97% of the theoretical time course, a three-fold excess of one reagent gave a 11% error in the fitted rate constant whereas a ten-fold excess gave only a 3% error. However, it should be noted that in practice when the concentration of the excess reactant approaches that of the other, the reaction becomes more second-order, with a long 'tail' and so the end-point may not be well defined, thus increasing the error. It is important that the end-point of the reaction is well defined. If it is chosen before the real end of a first-order reaction, the rate constant derived will be faster than the actual rate constant, and if chosen before the end of a second-order reaction may show a perfectly good fit to a first-order reaction.

For excess reactant conditions (pseudo first-order), it is best to have the fluorophore attached to the low reactant concentration species, since in the opposite case, the background signal will be increased as the concentration of excess reactant increases. However, in certain cases this may be overcome by making use of fluorescence energy transfer which enables the bound form of the fluorophore to be excited selectively. For example, Woodward *et al.* (17) studied the interaction of myosin subfragment 1 with 2'-O-(3'-O)-N-methyl-anthraniloyl ATP by exciting the tryptophans of the subfragment 1 and monitoring the fluorescence of the N-methylanthraniloyl moiety. In this way they were able to observe the interaction at 1 μM subfragment 1 with up to 100 μM ATP analogue and still have an observable signal.

Before making kinetic measurements it is preferable to perform a steady-state spectral titration of the protein and ligand (e.g. see Section 2). In addition to giving information about the molar dissociation constant, K_d of the interaction, it indicates wavelengths where the signal is largest and these can then be used in conjunction with the variables described in Section 5 to optimize the optics of the system.

In the following example it is assumed that the ligand is in excess of the protein although the experimental details and analysis are identical, if the situation is reversed. It is also assumed that the signal arises from the protein. The concentrations are described as a guide for making initial measurements and may need to be modified in the light of these.

7.2 Association reactions

First, a solution of 1 μM protein is pushed against buffer. The high voltage on the photomultiplier should then be adjusted to give a suitable signal in the recording device. The manufacturer's information should be consulted for exact details but generally the voltage should be high enough so that the photomultiplier responds linearly to fluorescence intensity but not exceed a value where it may be damaged. The buffer is then replaced by a solution of 10 μM ligand and the reaction is recorded, typically over the range of 0–50 msec to 0–500 msec. The time range should be chosen such that the exponential change is well defined as is the end-point. A good guide is to record over 10 times the half-time of the reaction. The reaction is then repeated with increasing concentrations of ligand. As wide a range of concentrations as possible should be used but this may become limited by the reaction getting too fast to measure or by the availability of material. The data should then be fitted to single exponentials. This can either be done on individual traces or after averaging a number of traces at any given ligand concentration. The signal to noise ratio is proportional to the square root of the number of traces averaged so that averaging four times will give half the noise of a single trace.

If the protein–ligand interaction is a simple second-order binding reaction, a plot of the observed rate constant against ligand concentration will give a straight line with the slope giving the second-order association rate constant and the intercept giving the first order dissociation rate constant (see Section 2.4).

The dissociation rate constant may not be well defined in the above experiment if the intercept is close to the origin. It can be independently measured by a displacement experiment.

7.3 Displacement reactions

In a displacement reaction, a solution containing protein (P) and fluorescent ligand (fL) is mixed with an excess of non-fluorescent ligand (nfL). The following scheme applies:-

$$\text{P.fL} \underset{k_1}{\overset{k_{-1}}{\rightleftharpoons}} \text{P} + \text{fL} \overset{\overset{\displaystyle \text{nfL}}{\diagdown}}{\underset{k_{-2}}{\overset{k_2}{\rightleftharpoons}}} \text{P.nfL}$$

The concentration of fL is chosen such that a reasonable saturation of the protein is achieved. A high concentration of nfL is required such that once fL has dissociated from the protein, it cannot reassociate and that the rate constant of the observed process is determined solely by k_{-1}. Depending on the relative values of the other rate constants, this may not necessarily be true and the observed rate constant may be a complex function of all four rate constants. Chock and Gutfreund (10) have done computer simulations of displacement reactions with varying concentrations of displacing ligand and have shown that observed rate constants may be faster or slower than the true dissociation rate constant. The best test to show that the true dissociation rate constant is being measured is to show that it is independent of the concentration of the displacing ligand.

If the protein–ligand interaction is a simple second-order reaction the value of the dissociation rate constant obtained from the displacement experiment should agree with the value obtained from the intercept of the binding experiment. Furthermore, a value for the K_d from the interaction can be calculated from the relationship: $K_d = k_{off}/k_{on}$ where k_{on} is the second-order association rate constant and k_{off} is the first-order dissociation rate constant and should agree with the K_d determined from equilibrium titration measurements.

7.4 Example of determination of association and dissociation rate constants

A typical example for characterizing the interaction of two proteins is shown in *Figure 7* where the interaction of a domain of the Raf protein with Ras.mantGTP is shown. The Raf domain is in excess and mant fluorescence is observed. It can be seen that on mixing Ras.mantGTP with excess Raf, an exponential increase in fluorescence intensity occurs (*Figure 7a*). The observed rate constant of this process is linearly dependent on the Raf concentration and gives a second-order binding rate constant of 8.3×10^7 M^{-1} sec^{-1} (*Figure 7b*). The intercept is too close to zero to define k_{-1} with any certainty but it is less than 20 sec^{-1}. A displacement experiment in which the Ras.mantGTP is displaced from Raf by Ras.GTP gives an observed rate constant of 2.9 sec^{-1} (*Figure 7c*). From these values of association and dissociation rate constants, the value of the K_d for the interaction of Ras.mantGTP with Raf can be calculated as 35 nM.

7.5 Kinetics of interaction of a non-fluorescent ligand with a protein

In certain cases, it is possible to measure the interaction of a non-fluorescent ligand with a protein by competitive binding studies with a fluorescent ligand. If the protein is mixed with the fluorescent and non-fluorescent ligands premixed together the following processes occur:

$$\text{P} + \text{fL} \underset{k_2}{\overset{k_1}{\longrightarrow}} \text{P.fL}$$

$$\text{P} + \text{nfL} \longrightarrow \text{P.nfL}$$

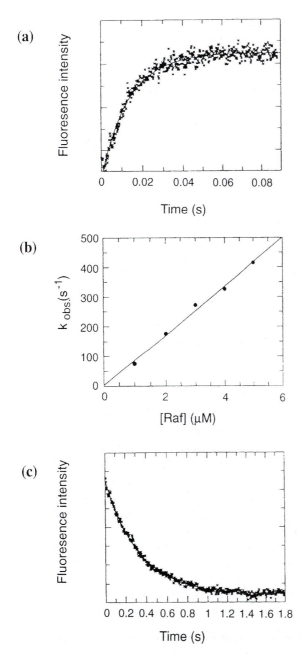

(a)

(b)

(c)

Figure 7 The interaction of Ras.mantGTP with Raf. (a) A solution of 0.25 μM Ras.mantGTP was rapidly mixed with a solution of 1 μM ras-binding domain of Raf (final concentration). The solid line is the best fit of the data to a single exponential giving a rate constant of 75 sec^{-1}. (b) A plot of k_{obs} from (a) against concentration of Raf. The straight line gives a second-order binding constant of 8.3×10^7 M^{-1}sec^{-1}. (c) A solution of 0.25 μM Ras.mantGTP and 0.50 μM Raf was rapidly mixed with a solution of 10 μM Ras.GTP. The solid line is the best fit of the data to a single exponential giving a rate constant of 2.9 sec^{-1}.

221

where fL is the fluorescent ligand and nfL is the non-fluorescent ligand. The observed rate constant for the formation of P.fL is:

$$k_{obs} = k_1 \, [\text{fL}] + k_2 \, [\text{nfL}] \qquad [30]$$

Therefore, a fixed concentration of fluorescent ligand is first mixed with the protein. Then an increasing concentration of non-fluorescent ligand is added to the fluorescent ligand solution and the observed first-order rate constant is measured. A plot of k_{obs} against the concentration of nfL gives a straight line of slope k_2 and intercept $k_1[\text{fL}]$.

The dissociation rate constant of the non-fluorescent ligand can be determined by displacement with a fluorescent ligand as described earlier, although the high fluorescent background will reduce the observed signal considerably.

8 Complex binding reactions

In the above discussion and example the binding reaction can be interpreted as a simple second-order reaction. However, data for even apparently simple binding reactions may show evidence for complex behaviour. For example, binding reactions may show as biphasic processes characterized by two exponentials. Alternatively, even though the data can be fitted to a single exponential, the observed rate constants show a hyperbolic dependence on ligand concentration. Such data can be explained by a two-step binding mechanism which involves isomerization of the protein or ligand before they interact, or an isomerization of the protein–ligand complex after binding, or the formation of two different protein–ligand complexes (7). The type of experimental data obtained from these mechanisms depends on which step of the binding mechanism the signal change is monitoring. The situation becomes more complex if a catalytic step can follow the initial binding steps since catalytic steps, isomerization steps and product release steps also need to be considered. The traditional method of analysing these reactions is to develop an analytical solution in which the observed rate constants are related to the elementary rate constants of the mechanism. The data are then fitted to this equation over a wide range of ligand concentrations. This approach is not always possible and also simplifying assumptions may have to be made which may not all be valid. An alternative approach is to use 'Global Analysis' methods (see Chapter 4). A comparison of these two methods is given in Section 10.

9 Specialized techniques

The above discussion involved the mixing of two equal volumes of solution and the fluorescence intensity, or absorption, of the solution was recorded with time. There are several variations on this method which can give additional information on the system under study.

9.1 Double mixing

Instead of simply mixing two reactant solutions and observing the reaction, some instruments allow the two solutions to be mixed and after a pre-set time, the mixed solution is then rapidly mixed with a third solution. In this way the interaction of a short-lived intermediate with a third reactant can be studied. This technique is not generally useful for simple protein–ligand interactions but can provide important information for more complex systems such as studies on enzymic mechanisms. It is important that the delay time before the second mixing be chosen to maximize the formation of the short-lived intermediate, C, and to avoid artefacts from unreacted starting materials A and B in the final reaction. In general the rate of formation k_1 of the intermediate must be much greater than the rate of its decomposition k_2:

$$A + B \xrightarrow{k_1} C \xrightarrow{k_2} D$$

For the simplest case shown above, in which there is irreversible formation and breakdown of C, the time course for the formation and decay of C is described by:

$$d[C(t)]/dt = k_1/(k_1 + k_2)\,(e^{-k_1 t} - e^{-k_2 t}) \qquad [31]$$

and the time at which the formation of C is maximal is equal to:

$$t_{max} = \ln \frac{k_1/k_2}{k_1 - k_2} \qquad [32]$$

An example is shown in *Figure 8*. In this experiment, a solution of 4 μM myosin subfragment 1 is initially mixed with 2 μM ATP and after either 0.5 sec or 5 sec, this solution is mixed with 30 μM actin. All the solutions contain the fluorescently labelled phosphate binding protein described earlier so that phosphate release from the hydrolysis of ATP is measured. It can be seen that there is a bi- or triphasic release of phosphate. At 0.5 sec not all of the ATP has bound to the myosin. The faster component represents phosphate release from the bound ATP and the larger slow component is due to the binding and hydrolysis of ATP which was not bound at 0.5 sec. In contrast, 5 sec after mixing, there was an instantaneous release of P_i from hydrolysis of ATP by subfragment 1 in this time, followed by a biphasic release process after mixing with actin. Now the amplitude of the slow phase is reduced since almost all of the ATP has bound and the two phases are controlled by the rate constants of ATP cleavage and phosphate release. For a detailed explanation of the mechanisms of the processes involved, the reader is referred to White *et al.* (18).

9.2 Variable volume mixing

Instead of mixing equal volumes of two solutions, some instruments allow variable ratios of reactant to be mixed. This technique is not generally useful for protein–ligand binding studies and is mainly used for protein folding studies where large changes in denaturant concentrations need to be made by rapid dilution.

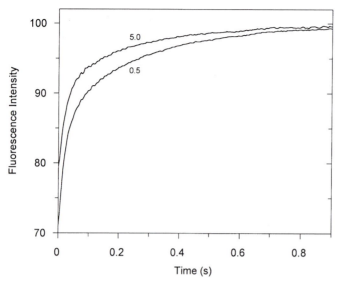

Figure 8 An example of a double mixing experiment. 4 μM myosin subfragment 1 was mixed with 2 μM ATP and this reactant solution mixed with 30 μM actin after either 0.5 sec or 5 sec. Phosphate release was monitored with the fluorescent phosphate binding protein. Webb, M. R. and White, H. D., unpublished data.

9.3 Logarithmic time base

Spectroscopic data are often collected over a linear time scale. However, in complex systems reaction processes may occur with very different rate constants. Separate reactions can be performed collecting the data over different time scales, but for economy of reactants it is preferable to do a single push and collect data over two different time scales or collect data with a logarithmic time base (19). The latter method allows data points to be collected at short intervals at the start of the reaction and at progressively longer time intervals as the reaction proceeds. In this way sufficient data points are collected to define processes with large differences in rate constants. Examples of logarithmic time base traces are shown in *Figure 9* where a simulated single exponential process, a double exponential process with both signals in the same direction, and a double exponential process with signals in the opposite direction are shown in linear and logarithmic time bases. It should be noted that if the data are collected with a logarithmic time base in the 'oversampling' mode (Section 5.5), the signal to noise ratio will increase over the course of the reaction since progressively more samples will be averaged for each time point.

9.4 Diode array detection

The availability of relatively low cost multichannel detectors such as diode array devices for the observation of complete UV/vis spectra with respect to time, together with the advances in computer technology, has given rise to a very

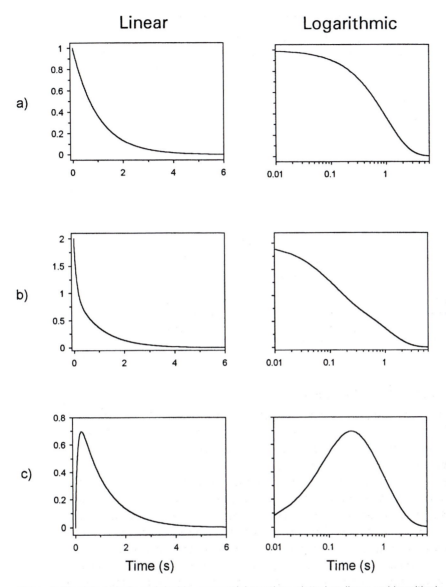

Figure 9 Simulated single and double exponential reactions plotted on linear and logarithmic time scales. (a) $k = 1$ sec^{-1} (amplitude 1.0). (b) $k_1 = 1$ sec^{-1} (amplitude 1.0), $k_2 = 10$ sec^{-1} (amplitude 1.0). (c) $k_1 = 1$ sec^{-1} (amplitude 1.0), $k_2 = 10$ sec^{-1} (amplitude -1.0).

effective method of acquiring and analysing kinetic spectroscopic data. With scan repetition rates (integration times) as short as 1.5 msec, these detectors provide adequate temporal resolution for many stopped-flow applications. Their sensitivity requires a high light level, and at present are only suitable for absorption measurements.

The power of diode array detection is its ability to expose kinetic data to more rigorous analysis than that at a single (or a few) wavelengths. In many cases the overall reaction can only be characterized by a range of wavelengths where minor intermediates are detectable and reaction phases clearly separated. Furthermore these large homogeneous 3D blocks of data can be achieved in a single, or maybe a few pushes thereby using a minimum amount of solution.

With systems exhibiting relatively complex mechanistic behaviour with spectrally distinct components, these collected spectra represent an over-determined system of equations that describe both the individual component spectra and their concentration profiles. It is now possible to access computational methods that can fit multivariate observations to a mechanistic model for a kinetic system with commercially available software (20). Through the use of factor analysis it is possible to determine the *minimum* number of coloured components contributing to kinetic spectral scans, independently of any model. An important method is singular value decomposition (SVD) (20). This performs a mathematical decomposition of a data matrix into linearly independent components ordered in terms of their significance. The output from SVD is a series of basis spectra and a corresponding series of time dependent amplitudes.

With the use of theoretical models to describe concentration profiles, the results from factor analysis can be fed into a non-linear least squares fitting routine to obtain *globally optimized* parameters and calculate component spectra.

An example of this method of detection is shown in *Figure 10*. Although this is not a protein–ligand interaction, it demonstrates the type of data obtained and the analysis of these data. The reaction is that of 5,5'-dithiobis (2-nitrobenzoic acid) (DTNB) with excess thioglycerol (RS). This is an irreversible two-step process:-

$$RS^- + DTNB \xrightarrow{\ k_1\ } TNB_1 + RS\text{-}TNB$$

$$RS^- + RS\text{-}TNB \xrightarrow{\ k_2\ } TNB_2 + RSSR$$

Figure 10a shows the absorbance spectra during the course of the reaction. 90 spectra were recorded over a time scale of 18 sec although fewer are shown here for clarity. Fitting the absorbance at 440 nm against time shows a biphasic formation of TNB, with fitted rate constants of 5.3 sec^{-1} and 0.25 sec^{-1} (*Figure 10b*).

However, by analysing the data over the whole wavelength range with time using a multivariant global analysis programme it is possible to obtain the spectrum of the short-lived RS-TNB intermediate (*Figure 10c*).

9.5 Fluorescence anisotropy

This—along with other fluorescence based methods—has already been covered in the previous chapter. Here we focus on its particular relevance to stopped flow measurements. In fluorescence anisotropy measurements a fluorophore is excited with vertically polarized light and the intensity I of the emitted light

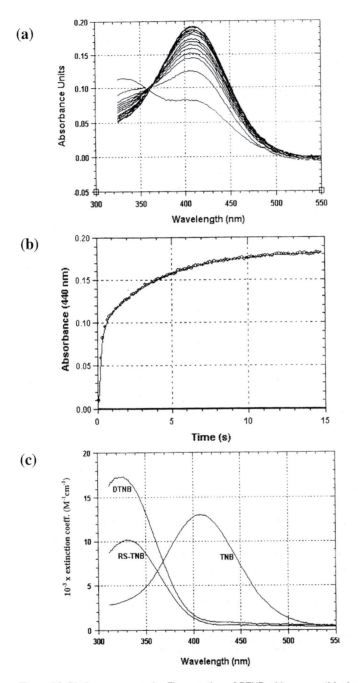

Figure 10 Diode array example. The reaction of DTNB with excess thioglycerol monitored with a diode array detector. A solution of 20 μM DTNB was rapidly mixed with a solution of 0.5 mM thioglycerol. (a) Absorbance scans were measured over the time period of 0–18 sec. (b) Absorbance at 440 nm plotted against time. The solid line is the best fit to two exponentials giving rate constants of 5.3 sec^{-1} and 0.25 sec^{-1}. (c) The computed spectra of DTNB, RS-TNB, and TNB. Reproduced with permission from ref. 21.

227

polarized parallel (//) and perpendicular (⊥) to the plane of the exciting light is measured. Fluorescence anisotropy (r) is calculated from the relationship:-

$$r = \frac{I_{//} - I_{\perp}}{I_{//} + 2I_{\perp}} \qquad [33]$$

The anisotropy is related to the rotational correlation time of the fluorophore by the equation:-

$$r = r_o / [1 + (\tau/\tau_c)] \qquad [34]$$

where r_o is the limiting anisotropy of the fluorophore, τ is the excited state lifetime, and τ_c is its rotational correlation time.

There are two advantages of making fluorescence anisotropy measurements over intensity measurements. First, since the anisotropy of a fluorophore is related to its rotational correlation time, changes in anisotropy give structural information. For example, a fluorophore bound to a protein will show an increase in anisotropy if that protein binds to a second protein, or a decrease in anisotropy if it is displaced from the protein. Secondly, fluorescence anisotropy changes may occur during a binding process which are not accompanied by an intensity change.

It should be emphasized that the above discussion is a very simplified description of fluorescence anisotropy. For example, it assumes that the fluorophore is rigidly attached to the protein and only undergoes global movement, whereas local motion of the fluorophore may contribute significantly to the anisotropy. A more rigorous treatment applied to equilibrium measurements is given by Jameson and Sawyer (22).

For stopped-flow fluorescence anisotropy measurements, the instrument first needs to be equipped with a polarizer filter on the excitation path which can be rotated through 90° to give light polarized either vertical or horizontal to the laboratory axis. It is best to make measurements in the 'T' format with two photomultipliers at right angles to the incident light, one with a polarizer filter monitoring light polarized parallel and one polarized perpendicular to the exciting light.

Since the two photomultipliers will respond differently to the intensity of the parallel and perpendicular light, it is first necessary to normalize them. This is done by exciting the fluorophore with horizontally polarized light. The amount of light depolarized to either the parallel or perpendicular planes to the photomultipliers will then be equal and so the high voltage on each photomultiplier is adjusted to give the same output signal.

The sample is then irradiated with vertically polarized light and the stopped-flow measurements made, recording the output from each photomultiplier simultaneously where $I_{//}$ is the signal at the parallel photomultiplier and $I_{\perp}$ is the signal at the perpendicular photomultiplier. Anisotropy is then calculated from the relationship given above and the total intensity can be determined from the relationship:

$$I = I_{//} + 2I_{\perp}$$

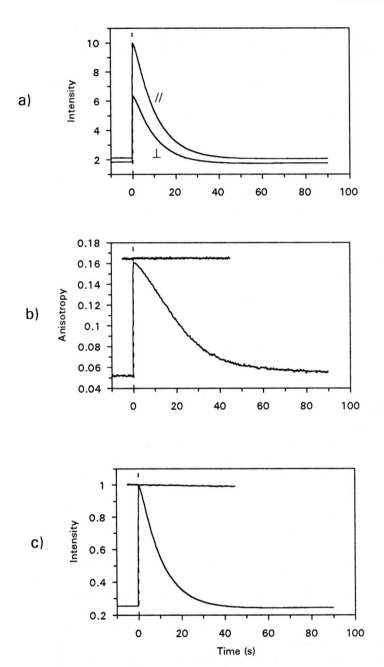

Figure 11 Fluorescence anisotropy stopped-flow experiment. A solution of 1 μM rho.mantGDP was mixed with 250 μM GDP in the presence of 40 mM EDTA and 200 mM ammonium sulfate. The sample was excited with vertically polarized light and light emitted parallel and perpendicular were both monitored. (a) Photomultiplier signals from parallel and perpendicular polarized light detectors. (b) Record of anisotropy from $r = (I_{//} - I_{\perp}) / (I_{//} + 2I_{\perp})$. (c) Record of intensity from $I = I_{//} + 2I_{\perp}$. The horizontal traces in (b) and (c) are data obtained on mixing 1 μM rho.mantGDP with 40 mM EDTA and 200 mM ammonium sulfate.

229

An example of such a measurement is shown in *Figure 11* where the displacement of the fluorescent nucleotide, mantGDP, from the small G-protein, RhoA, by GDP in the presence of EDTA and ammonium sulfate is measured. *Figure 11a* shows the record of parallel and perpendicular polarized light signals. These have been converted to fluorescence anisotropy (*Figure 11b*) and total intensity (*Figure 11c*). It can be seen that the anisotropy of the mantGDP decreases from 0.16 when bound to protein to 0.05 after displacement.

The time courses of the reaction monitored by intensity and anisotropy can be seen to be markedly different. This arises because the fluorescent intensity of mantGDP is directly proportional to the relative concentrations of bound and free mantGDP and so follows an exponential decrease in intensity. However, the observed anisotropy is determined by the fractional intensity of light from each species as well as their individual anisotropies.

Therefore, at any given time the observed anisotropy in a reaction of A converting to B will be:

$$r_{obs}(t) = f_a(t) \, r_a + f_b(t) \, r_b \qquad [35]$$

where $f_a(t)$ and $f_b(t)$ are the fractional intensities of A and B respectively at time t, and r_a is the anisotropy of A and r_b is the anisotropy of B.

$$f_a(t) = \frac{[A(t)]}{[A(t)] + D[B(t)]} \text{ and } f_b(t) = 1 - \frac{[A(t)]}{[A(t)] + D[B(t)]} \qquad [36]$$

where D is the factor by which B is more fluorescent than A on a molar basis.

Substituting the values of f_a and f_b into equation [35] shows that the anisotropy at any given time is given by:

$$r_{obs}(t) = \frac{[A(t)] \, (r_a - r_b)}{[A(t)] + D[B(t)]} + r_b \qquad [37]$$

If A is decaying to B exponentially, then the concentrations of A and B at any given time are:

$$[A(t)] = [A_o] \, e^{-kt}$$
$$[B(t)] = [A_o] - [A(t)] \qquad [38]$$

where $[A_o]$ is the initial concentration of [A].

Substituting these equations into *Equation 37* gives the observed anisotropy for the process at any given time as:

$$r_{obs}(t) = \frac{(r_a - r_b)}{(1 - D) + D \, e^{kt}} + r_b \qquad [39]$$

By fitting the anisotropy data to this equation where D is determined from the relative values of the intensity at the start and end of the reaction, values of the observed rate constant from the anisotropy data agree with the intensity data. It should be noted that from inspection of *Equation 39*, the closer D is to unity, the closer the anisotropy data will be to a single exponential.

A fuller description of stopped-flow fluorescence anisotropy measurements is given by Otto *et al.* (23), and methods are presented where both the intensity and anisotropy data can be combined in a single analysis.

9.6 Modern continuous flow techniques

There have been several recent developments using continuous flow methods which can measure reactions on an even faster time scale than stopped-flow measurements. Such instruments are not available commercially yet but two are briefly described here.

Shastry *et al.* (24) have described an instrument in which the two solutions are mixed in a novel mixer involving co-axial glass capillaries with a small platinum sphere at the tip of the inner capillary. The mixed solution flows down a fixed quartz observation cell and the fluorescence emission along the cell was monitored with a CCD detector. After correcting for spatial variations in incident light intensity by filling the cell with a constant fluorescent solution and for background light, a plot of fluorescence against distance along the cell could be obtained which corresponds to fluorescence against time. In this way mixing occurred within 15 μsec and the dead-time of the instrument was 50 μsec.

Even faster times have been claimed by Knight *et al.* (25). In their instrument, by miniaturizing the mixing and observation cell, they can rely solely on diffusion mixing rather than turbulent mixing. Mixing and detection is done on a silicon chip, and fluorescence intensity along the observation cell is measured using confocal scanning microscopy. Mixing times below 10 μsec were obtained.

10 Global analysis

Traditional approaches to the analysis of kinetic data usually involve fitting the time dependence of an observed signal, obtained at various reactant concentrations, to one or more exponential terms, some examples of which have been given in Sections 2 and 8. In favourable cases rate constants can then be assigned based upon the concentration dependence of these observed rates (1–3, 26). A weakness of such analyses is that the amplitudes of the observed exponential processes, which contain important information about the reaction, are often ignored. Furthermore, the requirement for an explicit mathematical solution to the rate equations describing the reaction can necessitate simplifying assumptions and place constraints on the experimental conditions (Section 7.1). The availability of PC-based software which can solve differential equations by numerical methods (for example *FITSIM* (27); *SCIENTIST*, MicroMath Inc.; *MODEL-MAKER*, Cherwell Scientific) has opened up the possibility of directly fitting kinetic data to an appropriate mechanism.

Global analysis of kinetic data involves the simultaneous fitting of multiple sets of data, obtained at various concentrations of reactants. Certain variable parameters, for example rate constants in a kinetic mechanism, are common to all the data sets. The simultaneous analysis of data obtained under different experimental conditions has the potential to achieve better definition of these variable parameters (28).

10.1 Strengths and limitations

Traditional methods of kinetic analysis should be applied first (Section 2). The derived rate constants, in conjunction with simulations of the observed processes, can then be used to establish a plausible minimal kinetic scheme for the process under investigation. This is the starting point for global analysis, which may be used as a test of the consistency of the proposed mechanism and to 'polish' the estimated rate constants.

In the most favourable cases global analysis may be used to distinguish between different kinetic mechanisms or to place limits on the rate constants of kinetic steps that were previously not attainable by traditional methods of analysis. Another strength of global fitting is that data obtained by several different methods (e.g. stopped-flow, steady state kinetics, rapid chemical quench, equilibrium binding) can in principle be used to provide a rigorous test for relatively complex kinetic mechanisms. Examples are those encountered in actomyosin ATP hydrolysis (29), kinesin-microtubule ATP hydrolysis (30), and G-protein nucleotide exchange mechanisms (31).

Global fitting of kinetic data should be performed with extreme care as its misuse could easily lead to erroneous results. It is important that the rate constants which are allowed to vary during the fitting are adequately constrained by the data. A wide range of values for a poorly constrained rate constant could give equally good fits of the data. The extent to which an individual rate constant is defined by the data can be tested by simulating the mechanism and varying each rate constant in turn. Rate constants which are poorly constrained should be held constant, either at an estimated value or at one determined separately, for example in a displacement experiment (Section 7.3). Fitting of additional variable parameters, other than rate constants, may also be required. For example, if fluorescence is being measured it may be necessary to have the relative emission intensities of the intermediates of the reaction as variable parameters in the global fitting. In some cases simultaneous multi-wavelength acquisition of spectral data (Section 9.4) can be used to establish the spectroscopic properties of transiently formed intermediates. This is generally of greater use for absorption spectra which have complex signatures, than for changes in fluorescence spectra, which may be relatively featureless.

10.2 Example: myosin-subfragment 1.ADP binding to pyrene-actin

One of the authors has used global fitting to analyse kinetic data obtained from stopped-flow studies of myosin-subfragment 1.ADP binding to actin, which was fluorescently labelled at Cys 374 with pyrene-iodoacetamide. The large fluorescent signal from the actin decreases approximately five-fold on binding of subfragment 1.ADP. Kurzawa and Geeves (32) showed that the quenching of the actin fluorescence is proportional to the amount of subfragment 1 bound and that binding at low ionic strength is very tight ($K_d < 100$ nM).

10.2.1 Conventional kinetic analysis

First, the data obtained were analysed by conventional methods. *Figure 12a* shows the time course of pyrene fluorescence when 1 μM pyrene-actin was mixed with 5 μM subfragment 1.ADP. The decrease in intensity was not well described by a single exponential, and showed evidence of a lag phase. Hence the association of pyrene-actin with subfragment 1.ADP must comprise at least two steps (see Section 8). As shown, a better fit was obtained to a sum of two exponentials, where the amplitudes of the two terms are of opposite sign (the arithmetic expression for a lag). Time courses for the reaction of pyrene-actin with subfragment 1.ADP, in the concentration range 2.5 μM to 60 μM, were fitted to two

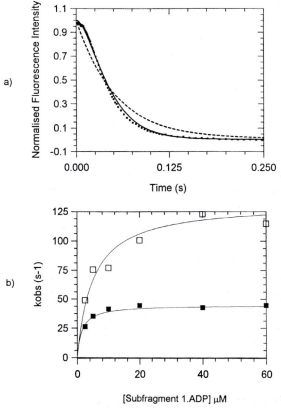

Figure 12 Traditional analysis of the binding of myosin-subfragment 1.ADP to pyrene-actin. (a) 5 μM subfragment 1.ADP was mixed with 1 μM pyrene-actin and pyrene fluorescence was observed. Intensity data were normalized as described in the text and converted to a pseudo-logarithmic time base. The dashed line shows the best fit of the data to a single exponential with $k_{obs} = 21$ sec^{-1}. The solid line shows the best fit to a double exponential with $k_{obs1} = 75$ sec^{-1} and $k_{obs2} = 35$ sec^{-1}. (b) The dependence of k_{obs1} (open symbols) and k_{obs2} (solid symbols) on subfragment 1.ADP concentration. Time courses were obtained for subfragment 1.ADP from 2.4 μM to 60 μM as described above, and fitted to a double exponential. The lines show a fit of the data to a simple hyperbola, giving a maximum value for k_{obs1} of 132 sec^{-1} and k_{obs2} of 45 sec^{-1}.

exponentials to yield observed rate constants for the lag phase and the fluorescence decrease step. *Figure 12b* shows the dependence of these observed rate constants on subfragment 1.ADP concentration. Since the observed rate of the lag phase approached a limiting value, this process cannot represent the bimolecular formation of a collision complex between pyrene-actin and subfragment 1.ADP. However, these results are consistent with a three-step mechanism shown below, where two consecutive isomerization steps follow the initial formation of a collision complex, and the fluorescence change is associated with the interconversion of AM1 and AM2. The lag phase represents interconversion of the collision complex AM and AM1.

$$A + M \underset{k_2}{\overset{k_1}{\rightleftharpoons}} AM \underset{k_4}{\overset{k_3}{\rightleftharpoons}} AM_1 \underset{k_6}{\overset{k_5}{\rightleftharpoons}} AM_2$$

The observed rate of the lag phase at saturation represents the sum of the rate constants for the interconversion of AM and AM1 ($k_3 + k_4 \sim 130$ sec^{-1}), and the observed rate of the intensity decrease phase at saturation represents the sum of the rate constants for the interconversion of AM1 and AM2 ($k_5 + k_6 \sim 45$ sec^{-1}).

10.2.2 Global analysis

Having established a plausible minimum mechanism for the interaction of pyrene-actin with subfragment 1.ADP, global analysis was used to refine the rate constants determined above and provide estimates of some constants in the scheme which were not readily obtained by conventional analysis. Although the overall quality of the data was extremely good, variations of 5–10% in the total amplitude of the signals made it impossible to perform global analysis directly on the raw data. To scale the data, the individual time courses were fitted to a double exponential including an offset, which defines the end-point of the reaction. Each time course was then normalized to give the same total fluorescence change. Determination of kinetic constants by global fitting of stopped-flow data to the mechanism shown in the scheme was performed using a non-linear least squares fitting routine in the program *SCIENTIST*. Experimental data were entered in the spreadsheet window of the program in three columns: time, total subfragment 1.ADP concentration, and normalized fluorescence intensity, using the same three columns for data obtained at different subfragment 1.ADP concentrations. *Figure 13* shows how the mechanism was compiled in the model window of *SCIENTIST*. Time (t) and total subfragment 1.ADP concentration (M_o) were entered as 'independent' variables, i.e. variables whose values do not depend on the values of other variables. The six rate constants in the scheme were entered as 'parameters', which can either be fixed at their initial value or allowed to vary during the fit. Normalized fluorescence (FL) was entered as a 'dependent' variable, values of which can be calculated from the independent variables and parameters. The other components of the model shown in *Figure 13* are listed below.

(a) Differential equations describing the mechanism and defining the individual rate constants.

```
       // 3 Step model of Actin Binding A+M<==>AM<==>AM1<==>AM2
       // NetworkApproach
       IndVars: T,MO
       DepVars: FL,A,M,AM,AM1,AM2
       Params: k1,k2,k3,k4,k5,k6
1)     //Differential Equations
       A'=-ABS(k1)*M*A+ABS(k2)*AM
       M'=-ABS(k1)*M*A+ABS(k2)*AM
       AM'=ABS(k1)*M*A-ABS(k2)*AM -ABS(k3)*AM +ABS(k4)*AM1
       AM1'=ABS(k3)*AM -ABS(k4)*AM1-ABS(k5)*AM1+ABS(k6)*AM2
       AM2'=ABS(k5)*AM1-ABS(k6)*AM2
2)     //Conservation equations
       A=AO -AM -AM1 -AM2
       M=MO -AM -AM1 - AM2
3)     // Fluorescence is the same for A, AM, and AM1
       FL=AO-AM2
4)     // Initial Parameter Values
       K1=1000
       K2=10000
       K3=1000
       K4=1.0
       K5=60
       K6=0.2
5)     //Initial Conditions
       T=0
       M=MO
       A=AO
       AM=0.0
       AM1=0.0
       AM2=0.0
       MO=1.0
       AO=1.0
       ***
```

Figure 13 The *Scientist* routine used to analyse the interaction of subfragment 1.ADP with pyrene-actin according to the three-step mechanism.

(b) Mass conservation expressions, relating total subfragment 1.ADP concentration (M_o) and total pyrene-actin concentration (A_o) to the concentrations of the individual species.

(c) An expression relating the normalized observed fluorescence to the concentrations of the fluorescently active species.

(d) Initial parameter values, which can be fixed or allowed to vary during the fit.

(e) Initial conditions, which are constraints required for finding solutions to the differential equations.

In order to reduce the number of variable parameters, the forward rate constant of the initial binding step (k_1) was fixed at a value of $10^3 \ \mu M^{-1} sec^{-1}$ and the

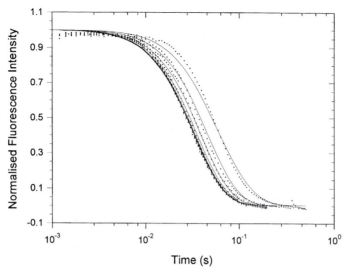

Figure 14 Global analysis of the binding of myosin-subfragment 1.ADP to pyrene-actin. Subfragment 1.ADP (1 μM to 60 μM) was mixed with 1 μM pyrene-actin and pyrene fluorescence was observed. Intensity data were normalized as described in the text and converted to a pseudo-logarithmic time base. Lines through the data show the global fit to the three-step mechanism, where $k_1 = 10^3$ μM^{-1}sec^{-1}, $k_2 = 4710$ sec^{-1}, $k_3 = 80$ sec^{-1}, $k_4 < 1.0$ sec^{-1}, $k_5 = 49$ sec^{-1}, and $k_6 < 0.2$ sec^{-1}.

rate constants k_4 and k_6 were fixed at zero, hence the two isomerization steps were initially considered to be irreversible. The rate constants k_2, k_3, and k_5 were the parameters which were allowed to vary during the fit. *Figure 14* shows the normalized stopped-flow data for the entire range of subfragment 1.ADP concentrations on a logarithmic time base. The solid lines were calculated from the fitted parameters, $k_2 = 4710$ sec^{-1}, $k_3 = 80$ sec^{-1}, and $k_5 = 50$ sec^{-1}. A value for the association constant of the initial complex (k_1/k_2) of 0.2×10^6 M^{-1} can be calculated. Subsequently, upper limits were placed on the values of the reverse rate constants k_4 and k_6 by comparing the experimental data to data simulated using the above rate constants and non-zero values of k_4 and k_6. Any values of $k_4 < 1$ sec^{-1} and $k_6 < 0.2$ sec^{-1} did not affect the quality of the fit, but if these values were exceeded a significant deterioration in the fit was observed. The values of the rate constants obtained by global analysis are consistent with those from the conventional analysis. However, the global method resolved the forward rate constants k_3 and k_5, whereas the conventional analysis gave the sum of the forward and reverse rate constants for the two isomerization steps ($k_3 + k_4$ and $k_5 + k_6$).

Acknowledgements

We would like to thank E. J. King of Hi-Tech Scientific for contributing to the section on diode array detection and M. R. Webb for providing *Figure 8*.

References

1. Gutfreund, H. (1995). *Kinetics for the life sciences*. Cambridge University Press, Cambridge.
2. Fersht, A. (1984). *Enzyme structure and mechanism* (2nd edn). W. H. Freeman & Co., NY.
3. Johnson, K. A. (1992). *The enzymes* (ed. D. S. Sigman), Vol. 20, p. 1.
4. Gibson, Q. H. (1969). In *Methods in enzymology* (ed. K. Kustin), Vol. 16, p. 187.
5. Roughton, F. J. W. (1963). *Rates and mechanisms of reactions*, p. 703. Wiley, NY.
6. Peterman, B. F. (1979). *Anal. Biochem.*, **93**, 442.
7. Bagshaw, C. R., Eccleston, J. F., Eckstein, F., Goody, R. S., Gutfreund, H., and Trentham, D. R. (1974). *Biochem. J.*, **141**, 351.
8. White, H. D. and Rayment, I. (1993). *Biochemistry*, **32**, 9859.
9. White, H. D. (1988). *Biochemistry*, **27**, 3357.
10. Chock, P. B. and Gutfreund, H. (1988). *Proc. Natl. Acad. Sci. USA*, **85**, 8870.
11. Brzovic, P. S. and Dunn, M. F. (1995). In *Methods in enzymology* (ed. K. Sauer), Vol. 246, p. 168.
12. Jameson, D. M. and Eccleston, J. F. (1997). In *Methods in enzymology* (ed. L. Brand and M. L. Johnson), Vol. 278, p. 363.
13. Schmid, D., Baici, A., Gehring, H., and Christen, P. (1994). *Science*, **263**, 971.
14. Török, K. and Trentham, D. R. (1994). *Biochemistry*, **33**, 12807.
15. Brune, M., Hunter, J. L., Corrie, J. E. T., and Webb, M. R. (1994). *Biochemistry*, **33**, 8262.
16. Ross, J. B. A., Szabo, A. G., and Hogue, C. W. V. (1997). In *Methods in enzymology* (ed. L. Brand and M. L. Johnson), Vol. 278, p. 151.
17. Woodward, S. K. A., Eccleston, J. F., and Geeves, M. A. (1991). *Biochemistry*, **30**, 422.
18. White, H. D., Belknap, B., and Webb, M. R. (1997). *Biochemistry*, **36**, 11828.
19. Walmsley, A. R. and Bagshaw, C. R. (1989). *Anal. Biochem.*, **176**, 313.
20. Malinowski, E. R. (19XX). *Factor analysis in chemistry* (2nd edn). Wiley Interscience, NY.
21. King, E. J. (1997). *Hi-Tech Scientific Application Note*, AN.008.S20.
22. Jameson, D. M. and Sawyer, W. H. (1995). In *Methods in enzymology* (ed. K. Sauer), Vol. 246, p. 283.
23. Otto, M. R., Lillo, M. P., and Beechem, J. M. (1994). *Biophys. J.*, **67**, 2511.
24. Shastry, M. C. R., Luck, S. D., and Roder, A. (1998). *Biophys. J.*, **74**, 2714.
25. Knight, J. B., Vishwanath, A., Brody, J. P., and Austen, R. H. (1998). *Phys. Rev. Lett.*, **80**, 3863.
26. White, H. D. (1982). In *Methods in enzymology* (ed. D. W. Frederiksen and L. W. Cunningham), Vol. 85, Part B, p. 698.
27. Zimmerle, C. T. and Frieden, C. (1989). *Biochem. J.*, **258**, 381.
28. Beechem, J. M. (1992). In *Methods in enzymology* (ed. L. Brand and M. L. Johnson), Vol. 210, p. 37.
29. Walker, M., Zhang, X.-Y., Jiang, W., Trinick, J., and White, H. D. (1999). *Proc. Natl. Acad. Sci. USA*, **96**, 465.
30. Gilbert, S. P., Webb, M. R., Brune, M., and Johnson, K. A. (1995). *Nature*, **373**, 671.
31. Klebe, C., Prinz, H., Wittinghofer, A., and Goody, R. S. (1995). *Biochemistry*, **34**, 12543.
32. Kurzawa, S. E. and Geeves, M. A. (1996). *J. Muscle Res. Cell Motil.*, **17**, 669.

Chapter 6

Protein–ligand interactions studied by FTIR spectroscopy: methodological aspects

Michael Jackson and Henry H. Mantsch

Institute for Biodiagnostics, National Research Council Canada, 435 Ellice Avenue, Winnipeg, Manitoba R3B 1Y6, Canada.

1 Introduction

Many techniques exist for the study of protein–ligand interactions, with each technique providing unique and/or complementary information. For example, calorimetric studies provide information relating to the energetics of protein–ligand interaction (Volume 1, Chapters 10 and 11), while X-ray diffraction techniques provide information on the molecular co-ordinates of atoms in the protein–ligand complex (see Chapter 1). Between these extremes lies a bewildering array of methods varying in technical sophistication, cost, and information content.

Determining the technique of choice for any particular experiment can often be difficult for the uninitiated. However, as a starting point it is wise to remember that the technique that should be used to address a particular problem depends upon the nature of the questions being posed. For example one may determine that thermodynamic parameters are the most important information required. In such a case, calorimetric techniques are required. Alternatively, it may be that it is important to have structural and spatial information at the atomic level. In this case, calorimetric techniques are inappropriate and spectroscopic techniques are required. In fact, only diffraction techniques and nuclear magnetic resonance (NMR) spectroscopy can be used to study ligand–protein interactions with such resolution, reducing the choices available to the investigator. However, few proteins can be crystallized readily, and large proteins or membrane-associated proteins are difficult to study by NMR spectroscopy. In practice, the use of diffraction and NMR techniques is therefore restricted.

Circular dichroism (CD), Raman and infrared (IR) spectroscopies are more widely applicable to protein studies, although application of CD spectroscopy to membrane associated proteins is limited by light scattering artefacts. Raman (see Chapter 7) and IR spectroscopy can be applied with equal success to both small and large proteins, and membrane-associated or embedded proteins can be readily studied. In fact, Raman and IR spectroscopy offer the advantage that both the membrane and protein constituents of samples can be probed simultaneously, providing information not only about the protein, but also the supporting milieu. The main disadvantage of Raman spectroscopy is a reduced signal to noise ratio compared to IR spectroscopy, which can complicate analysis. However, resonance Raman spectroscopy is proving increasingly powerful as a tool for the study of enzyme reactions, and may be the technique of choice in some instances.

Clearly, many aspects of protein–ligand interactions may be studied using a variety of methods. However infrared spectroscopy is perhaps the most versatile technique available. High quality spectra may be obtained with small amounts of material (10 µl of sample at 5–10 mg/ml). Both membrane bound/associated and water soluble proteins may be studied, and high molecular weight does not present problems for the technique.

2 Protein structure determination by FTIR spectroscopy

The sensitivity of IR spectroscopy to protein and peptide secondary structure was first demonstrated by Elliot and Ambrose as early as 1950 (1). In this landmark study, it was shown that the frequency of the amide I absorption band, the major absorption band in protein spectra, was determined by the predominant secondary structures present (*Figure 1*). This study and many others have conclusively shown that for proteins with a predominantly α-helical configuration an amide I band maximum occurs at 1653–57 cm^{-1}, in aqueous solution. A similar frequency is observed for proteins that are predominantly unstructured.

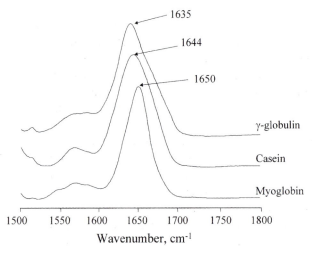

Figure 1 Infrared spectra of proteins with predominantly α-helical (myoglobin), β-sheet (γ-globulin), and unordered secondary structures (casein).

However, upon dissolution in D_2O the amide I maximum for helical proteins is shifted to 1650–54 cm^{-1} while the amide I for unstructured proteins undergoes a much greater shift, to 1640–46 cm^{-1}. Finally, β-sheet proteins exhibit an amide I maximum at 1635–40 cm^{-1}, shifting to 1630–35 cm^{-1} upon dissolution in D_2O.

This brief discussion illustrates two points. First, it is possible to deduce the major secondary structures present in a protein or peptide based upon infrared spectra. Secondly, it is clear that it is advisable never to carry out IR spectroscopic studies of proteins in H_2O alone. To discriminate between helical and unordered polypeptide chains, it is necessary to acquire spectra in D_2O.

The reason for the sensitivity of IR spectroscopy to protein secondary structure becomes apparent when one considers the nature of the vibration giving rise to the amide I absorption and the forces governing protein conformation. The amide I absorption band arises predominantly from stretching vibrations of the C=O moieties of polypeptide amide groups. The frequency of C=O stretching absorptions is strongly dependent upon the strength of any hydrogen bonds formed. A non-hydrogen bonded amide C=O group is typically observed at around 1666 cm^{-1}, with hydrogen bonding causing shifts to lower wavenumber. It is this sensitivity to hydrogen bonding which forms the basis for the sensitivity of IR spectroscopy to protein secondary structure. Secondary structures within proteins and peptides are stabilized by hydrogen bonds between amide C=O and N–H groups. The pattern of hydrogen bonds, i.e. length, linearity, and strength, differ for different secondary structures. In other words, each type of secondary structure is characterized by a unique pattern of hydrogen bonds. This pattern of hydrogen bonds therefore imparts unique spectroscopic properties to each type of secondary structural element (2, 3). Thus, β-sheet secondary structures, which have the strongest hydrogen bonds, exhibit an amide I maximum at a much lower frequency than α-helices.

The amide II absorption band also shows some sensitivity to protein secondary structure. The amide II absorption band arises from the bending vibration of amide N–H groups strongly coupled to the stretching of amide C–N groups. This absorption bands is significantly less 'pure' than the amide I absorption band, and thus is less useful as a predictor of secondary structure. However, as discussed below, the amide II absorption band is very sensitive to deuteration, and can be used as a probe for determining the strength of hydrogen bonds present within secondary structures or for assessing the degree of exposure of structural elements to solvent.

It should be borne in mind that all proteins and most peptides contain more than one type of secondary structural element, and amide I absorption bands are therefore complex, composite absorption profiles resulting from the superimposition of the amide I absorption bands from each secondary structural element present. To determine the frequency of each of the overlapping absorptions comprising the composite amide I or amide II profile band narrowing techniques such as Fourier self-deconvolution or derivation are required (see below).

Clearly infrared spectroscopy may be used to determine the major secondary structural characteristics present within proteins and peptides. However, without a clear understanding of the nature of the information and potential artefacts that can be obtained with the use of infrared spectroscopy, this information cannot be reliably recovered. In the following sections we highlight the experimental aspects which much be considered to allow acquisition and interpretation of data in a meaningful manner.

3 Experimental design

Experimental design is crucial in any study of protein–ligand interactions. The correct choice of experimental technique, experimental conditions, and data manipulation methods are all essential components of good experimental design. Each of these aspects will be discussed in turn.

3.1 Experimental technique

Infrared spectroscopy is unique among spectroscopic techniques, in that methodologies exists to acquire spectra from proteins in any physical state, including crystals, powders, dry films, and aqueous solutions. However, the availability of techniques to sample proteins in such states does not mean that protein–ligand interactions should be studied in these forms. The applicability of each technique depends upon the nature of the question asked.

Much terminology and a few basic principles are common to all of the techniques that may be applied to the analysis of proteins and protein–ligand interactions, and these will be outlined first. Infrared spectra are typically plotted as absorption spectra, with absorbance plotted against wavenumber. Absorbance is calculated as the negative logarithm of the sample spectrum divided by a reference or background spectrum:

$$\text{absorbance} = -\log\left(\text{sample} / \text{reference}\right) \qquad [1]$$

The background and sample spectra are referred to as single beam spectra. A background spectrum is required as the air circulating within the spectrometer and certain instrument components exhibit infrared absorption bands (for example water vapour and carbon dioxide in air and dust on mirrors). The intensity of these features should be constant, and calculating the ratio of the sample spectrum to a background spectrum effectively removes these spectral interferents. The negative logarithm of this ratio is termed the absorbance spectrum, plotted in absorbance units (AU) versus wavenumber.

Clearly it is important, when calculating absorption spectra, that interferences such as water vapour are efficiently removed prior to spectra acquisition, or present at stable levels. Fluctuations will result in inadequate compensation when the background and sample single spectra are ratioed. To ensure high quality spectra, the spectrometer should be well purged with nitrogen or dry air before the background spectrum is acquired. This can be evaluated by acquiring the single beam spectrum and visually inspecting the region between 1500–1900 cm^{-1}. If water vapour is present, then a series of sharp absorption bands

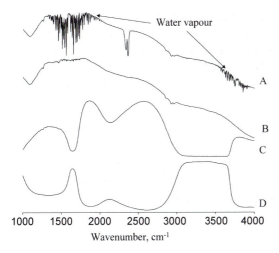

Figure 2 Single beam background spectra with (A) and without (B) water vapour contributions, a single beam sample spectrum (C), and the calculated absorbance spectrum (D).

will be seen (*Figure 2A*). If these absorptions are present, the spectrometer should be purged until these absorption bands are no longer seen or exhibit a stable intensity (*Figure 2B*). It should be noted that it is not always possible to completely purge a spectrometer. In such cases the next best approach is to maintain a stable, low water vapour content in the spectrometer such that water vapour contributions in the background and sample single beam spectra are identical, and so cancel out when the sample spectrum is ratioed against the background.

Once a stable background spectrum has been acquired, then the sample spectrum may be acquired. As this involves opening the spectrometer sample compartment to insert the sample, the sample compartment becomes contaminated with water vapour. Before acquiring a sample spectrum it is therefore important to allow the sample compartment to purge, using oxygen-free nitrogen or dry air, for about 10 minutes. Most modern instruments allow the investigator to monitor the sample during this time, by repetitively acquiring and displaying (but not saving) spectra. This allows the level of water vapour in spectra to be assessed with time. When the purge is deemed adequate, the spectrum may be acquired (*Figure 2C*). The absorbance spectrum may then be calculated (*Figure 2D*).

With these simple precautions, any of the standard methods for spectral acquisition will produce spectra of acceptable quality.

The simplest method of spectral acquisition is to mix a dry (lyophilized) protein with KBr and compress the mixture to form a disk. The disk is then placed in the spectrometer and an absorption spectrum acquired. This process can be repeated for protein lyophilized from a ligand containing solution, and structural differences determined based upon a spectral comparison. The obvious drawback to this approach is that the sample is in a non-physiological (non-hydrated) state. However, in certain circumstances this may not pose a significant problem. For example, many proteins are lyophilized after purification and regain their native structure upon dissolution in aqueous media. In other cases significant activity is lost following lyophilization. For such proteins, activity and thus structural stability can often be retained by lyophilization in the presence of stabilizing materials. In such instances it may be useful to study the conformation of the protein in the presence and absence of stabilizing agents.

In principle, proteins may be studied as solids in a more physiological manner by infrared microscopic analysis of protein crystals (4). While few studies performed on protein crystals have been reported in the literature, this technique may be useful in the study of protein–ligand interactions. An IR microscope functions as a beam condenser, focusing a bright beam of IR light onto a small spot. This allows IR spectra of very small samples (30 μm) to be routinely acquired. Using such instrumentation it should be possible to collect infrared spectra of proteins and protein–ligand complexes in crystalline matrices. The advantage of this approach is that direct correlation may be made between structures determined by infrared microscopy and those determined by techniques such as X-ray diffraction, and comparisons can also be made with structural determinations conducted in solution.

The most physiological method of studying proteins is in solution. Proteins may be dissolved in a range of solvents, including trifluoroethanol, dimethyl sulfoxide, H_2O, and D_2O (see below for a detailed discussion of the merits of each solvent). In addition to a range of possible solvents, spectra of protein solutions may be obtained in a variety of ways. The most widespread and convenient method for acquiring solution spectra is by standard transmission methods. Briefly, the sample solution is placed between two optical windows separated by a known path length. This path length may be introduced by etching one of the optical windows or by placing a spacer (typically composed of an inert material such as Teflon or Mylar) of known thickness between the windows.

Obviously, windows should be chosen which have correct physical and chemical characteristics for the experiment planned. In other words, the windows must be transparent in the region of interest and must be inert and insoluble under the conditions of the experiment. Common window materials include barium fluoride, calcium fluoride, zinc selenide, sodium chloride, potassium chloride, potassium bromide, and silver chloride. Sodium chloride, potassium chloride, and potassium bromide are all water soluble materials, and should not be used with aqueous systems. Barium fluoride, calcium fluoride, zinc selenide, and silver chloride are all insoluble in water. However, zinc selenide and barium fluoride are both soluble in strong acids, and care should be exercised if studies are to be performed at low pH. In addition, zinc selenide has a high refractive index that can result in the presence of interference fringes in spectra unless the windows are machined to ensure that the window surfaces are not parallel. Finally, silver chloride is light sensitive, with silver deposits being formed upon exposure to UV light that darken the window and reduce transmission. In addition, silver chloride reacts with sulfur compounds to produce silver sulfide.

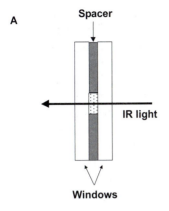

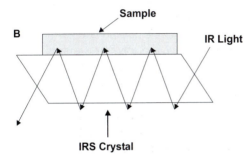

Figure 3 Schematic representation of a transmission (A) and IRS (B) experiments.

Obviously silver chloride windows should not be used with sulfur-containing proteins or peptides. The window material of choice for most experiments involving proteins is either calcium or barium fluoride (depending upon pH), with barium fluoride having better transmission at low wavelengths.

The correct choice of window will reduce the likelihood of spectral artefacts and damage to the windows. Once the window material has been chosen, a small volume of sample (typically 5–10 μl of aqueous solution of protein) is placed between the windows. It is important at this point to ensure that air bubbles are not present in the sample (for example introduced during pipetting), as the refractive index mismatch between air and the window material will result in the appearance of sinusoidal interference fringes in spectra. If no air bubbles are present, the sample is mounted in a cell holder and spectra obtained by passing infrared light through the windows (*Figure 3A*).

Spectra of protein solutions may also be obtained using the more elaborate technique known as attenuated total reflectance (5) or more correctly as internal reflection spectroscopy (IRS). A beam of infrared light entering an infrared transparent crystal will undergo total internal reflection when the angle of incidence at the sample interface is greater than a critical value. This value is determined by the refractive indices of the sample and the crystal. This process is illustrated in *Figure 3B*. The infrared beam enters the crystal, which is typically a parallelogram, trapezoid, or rod. At the interface of the crystal and the sample the light penetrates a small distance into the sample (penetration depth increases with decreasing wavenumber) and is then reflected back into the crystal. This process may be repeated a number of times, depending upon the geometry and length of the crystal. The result is an infrared spectrum resulting from absorption of light each time the infrared beam penetrates the sample.

Protein–ligand binding experiments may be conducted in a number of ways using this technique. For example many IRS accessories are available which consist of a rod-shaped crystal enclosed in a flow-through cell. Protein solutions and ligand solutions may be injected into the flow cell and allowed to mix and spectra acquired. Comparison of spectra with spectra of the pure protein solution will allow characterization of ligand induced changes. Alternatively, a flow-through system may be used to analyse the interaction of membrane embedded or associated proteins with ligands by preparing a film of the membrane dispersion upon a trapezoidal crystal. Flowing buffers with various compositions across the surface of the membrane dispersion will then allow protein structure as a function of ligand binding to be assessed. Importantly, this apparatus allows difference spectroscopy (see below) to be performed, allowing very subtle structural changes associated with ligand binding to be analysed.

While IRS is a powerful measurement tool, it should be stressed that artefacts await the unwary. For example protein adsorption to the IRS crystal can be a problem. In addition, as mentioned above the depth of penetration of the infrared light into the sample is wavelength dependent, and the intensity of absorption bands at low wavenumbers may be preferentially enhanced. Finally, IRS is inherently less sensitive than standard transmission experiments, as the IR light must be focused onto the IRS crystal. This results in significant energy loss, and throughput may be reduced to 30–50%. If high signal to noise ratios are crucial, then this is probably not the technique of choice.

For most experimental situations, the technique of choice for acquisition of spectra of proteins is transmission spectroscopy. This technique is simple, low cost, allows solution spectra to be acquired, and can be used to reliably obtain artefact-free spectra with a few simple precautions.

Protocol 1

Procedure for acquisition of transmission spectra

Equipment and reagents

- Spectrometer
- Protein solution

Method

1   Allow the spectrometer to purge adequately. Monitor the purge efficiency by inspection of the region between 1400–1900 cm^{-1} on a single beam spectrum.
2   When there are no sharp water vapour absorptions between 1400–1900 cm^{-1}, or when their intensity is minimal and stable, acquire a background single beam spectrum.[a]

3 Prepare the protein solution, avoiding the use of organic solvents. If D_2O is the solvent of choice care should be exercised to ensure that atmospheric water vapour does not contaminate this highly hygroscopic solvent. Pipette a small volume (5–10 μl) onto an infrared transparent window on which has been placed a spacer of appropriate dimensions (10 μm or less if H_2O is used, 50 μm if D_2O is used). Ensure that no air bubbles are present to eliminate interference fringes.

4 Allow the spectrometer to purge adequately (approx. 10 min). Monitor the purge efficiency by inspection of the region between 1400–1900 cm^{-1}. Most spectrometers have a 'monitor' feature to enable purge efficiency to be assessed.

5 Acquire a sample single beam spectrum[a] and compute the absorbance spectrum.

[a] It is essential to ensure that background and sample single beam spectra are acquired under exactly the same instrumental conditions, e.g. same resolution, apodization function, etc.

3.2 Experimental conditions: choice of solvent

In general, most experiments involving protein–ligand interactions will be performed in solution. However, the range of solvents available for spectroscopic studies of proteins is large, and a poor choice of solvent will result in the acquisition of non-physiological, and so irrelevant, data. In the following sections we evaluate the major solvents used in infrared spectroscopic studies of proteins.

3.2.1 Aqueous versus non aqueous

The question 'aqueous or non-aqueous?' may seem a strange question to ask when studying protein–ligand interactions. However, the study of protein–ligand interactions can prove difficult with many techniques if membrane bound or membrane associated proteins are of interest. For example, the presence of lipid bilayers causes artefacts due to light scattering when applying circular dichroism spectroscopy, and significant line broadening when using NMR spectroscopy. This has led many investigators to conduct studies of proteins and ligand–protein interactions using a so-called 'membrane mimetic' solvent in place of a lipid bilayer. A number of infrared spectroscopists have followed suite, either to allow direct comparison between data obtained with NMR, CD, and FTIR spectroscopy or because they feel that proteins and peptides in organic solvents are easier to handle than membrane bound or associated proteins.

However, many investigators have performed such studies uncritically, and the suitability of such solvents for studies involving proteins has often been ignored. Although solvents can be chosen to reflect the dielectric constant of the membrane interior, important differences exist. For example membrane lipids are zwitterionic, while organic solvents on the whole are not (although they may carry significant dipoles). In addition, the interior of a lipid bilayer presents a relatively rigid environment, compared to the highly fluid environment of a solvent. Finally, lipid molecules are significantly larger than commonly employed solvent molecules, and may thus be expected to interact differently with proteins. Each of these factors suggests that dissolution of proteins in organic solvents probably results in structures different from those found in lipid bilayers.

A review of the NMR, CD, and IR spectroscopy literature strongly supports perturbation of protein secondary structure by membrane mimetic solvents. For example it is apparent, from a brief survey of the scientific literature, that all studies of proteins and peptides dissolved in pure trifluoroethanol (TFE), a common membrane mimetic solvent, appear to report an α-helical configuration. In contrast, studies using dimethyl sulfoxide (DMSO) as a solvent appear to report a lack of structure. The fact that a variety of proteins and peptides dissolved in a particular solvent are found, by a variety of techniques, to exhibit the same structural characteristics strongly implies a major effect of the solvent.

This effect was demonstrated in a systematic study of the effects of organic solvents upon protein secondary and tertiary structure (6, 7). It was found that when dissolved in DMSO solution all proteins and peptides, irrespective of their native conformation, were indeed completely unstructured. Furthermore, in aqueous/DMSO mixtures all proteins and peptides were found to aggregate. These phenomena were attributed to disruption of hydrogen bond within the protein by the $S=O$ groups of the solvent. In contrast, it was found that halogenated alcohols induced helical structures in all proteins and peptides, again regardless of native structure. As with aqueous DMSO solutions, aqueous alcohol solutions induced aggregation. Apparently, the structures adopted by proteins and peptides in membrane mimetic solvents owe more to intrinsic properties of the solvents than any inherent structural characteristics of the protein.

Clearly, organic solvents induce significant structural perturbations in proteins and peptides. Equally clearly, such solvents should not be used in studies of ligand–protein interactions. While it may be the case that ligand binding is demonstrated and structural alterations can be characterized, it is highly likely that this binding and structural alteration will represent non-physiological phenomena. In fact, it cannot be otherwise if the starting conformation of the protein or peptide is non-native.

In addition to the induction of non-physiological structures, the use of organic solvents as a substitute for the lipid bilayer negates an important advantage of infrared spectroscopy: the ability to study the lipid and protein moieties of membrane systems simultaneously. This ability potentially allows subtle structural perturbations, which perhaps cannot readily be detected, to be monitored through effects on the adjacent lipid molecules. Furthermore, the use of lipid bilayers allows the effect of bilayer composition upon ligand binding to be assessed. For example, studies with reconstituted ion transport systems such as the Ca-ATPase of the sarcoplasmic reticulum can be performed in which the structure of the protein in the presence of ligands (ATP, ADP, vanadate, Ca^{2+}) can be assessed as a function of lipid composition. Such studies should prove valuable given the proven dependence of many transport systems upon membrane composition.

3.2.2 H₂O versus D₂O

The use of deuterium oxide (D_2O or ^2H_2O) as a solvent in place of water offers advantages to the infrared spectroscopists. Most importantly, the use of D_2O removes a strong interfering absorption in the spectral region of most interest in studies of protein–ligand interactions. Water exhibits a series of extremely intense absorption bands in the mid infrared spectral region. The most important of these absorptions is the O–H bending vibration centred at around 1640 cm^{-1}. This absorption is important because it occurs in the spectral region in which the protein absorption bands most useful for protein structural prediction (the amide I) is observed. Unfortunately, the intensity of the O–H bending absorption band of water is much more intense that the amide I absorption bands of proteins. Absorption bands of water therefore dominate spectra of proteins dissolved in water. The intensity of the water absorption bands limits the path lengths which can be used for infrared spectroscopic studies to around 10 μm. Other important absorptions include the broad O–H stretching absorption band at 3000–3400 cm^{-1} and the so-called combination absorption band at 2200 cm^{-1} (*Figure 4A*).

The strong absorption band arising from the O–H stretching vibration of water may be removed with the use of digital subtraction routines (see below), allowing visualization of the underlying protein absorption band. However, digital subtraction of this strong absorption band can be problematic (see below). Problems associated with solvent subtraction are less severe when D_2O is used as a solvent. The difference in mass between the deuteron and proton leads to significant differences in the infrared spectra of the two solvents. The presence of the heavier deuteron shifts the O–D stretching absorption band by approximately 400 cm^{-1} to lower wavenumber, leaving the protein C=O stretching absorption band relatively free from interfering absorption bands (*Figure 4B*). The only interference from the solvent now arises from the much weaker

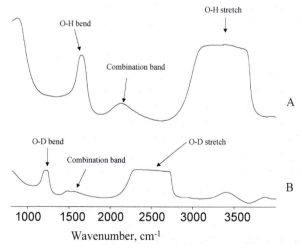

Figure 4 Infrared spectra of H₂O (A) and D₂O (B).

combination absorption band of D_2O, which is also shifted to lower wavenumber. In fact, the reduced intensity of the combination absorption band compared to the O–H bending absorption band means that in most studies the amide I absorption band can be analysed, at least in a preliminary manner, without digital subtraction of the solvent.

The use of deuterium oxide has additional advantages. The reduced intensity of the combination absorption band of D_2O allows significantly greater path lengths to be used when using D_2O as a solvent, which in turn allows lower concentrations of proteins to be studied. Typically, studies in H_2O require concentrations of 20–50 mg/ml. With 50 μm path lengths typically used in experiments utilising D_2O, concentrations as low as 1 mg/ml may be used.

Perhaps more importantly, the use of D_2O allows a more reliable assessment of secondary structure. As discussed above, the amide I absorption bands arising from helical and unstructured peptide chains show significant overlap. However, when dissolved in D_2O, unstructured polypeptide chains show a shift to 1640 cm^{-1}, while helical polypeptide chains exhibit an absorption maximum at 1650–53 cm^{-1}. Thus, the use of D_2O as a solvent allows a more complete description of protein secondary structure to be obtained.

3.3 Solvent and water vapour subtraction

The problems associated with the strong water absorptions in aqueous protein solutions can be alleviated by the use of digital subtraction routines. Digital subtraction routines allow the interactive subtraction of spectra from one another, using the general formula:

$$\text{result} = \text{spectrum} - (\text{reference spectrum} \times \text{factor}) \qquad [2]$$

With this type of subtraction algorithm, the spectrum to be subtracted (in this case the spectrum of D_2O, *Figure 5A*) is multiplied by a factor and subtracted from the aqueous protein spectrum (*Figure 5B*). The factor is adjusted until a visually acceptable resultant spectrum is achieved. When subtracting water spectra, the general consensus is that the most useful assessment of the accuracy of the subtraction is gained by monitoring the 1700–2200 cm^{-1} spectral region. This spectral region is useful as most biological materials do not normally exhibit absorption bands in this spectral region. Thus, in the absence of water or D_2O this spectral region should be relatively flat. The subtraction factor is therefore varied until a flat baseline is achieved between 1800–2400 cm^{-1} (*Figure 5C*).

While subtraction routines offer a powerful aid to the spectroscopist, they are not without pitfalls. Subtle changes in the position of absorption bands can result in the introduction of significant artefacts in difference spectra. For example the O–H bending vibration is extremely sensitive to temperature. Small changes in temperature over the course of an experiment can result in subtle but important shifts in the frequency of this absorption band. Subtraction of a water spectrum acquired at a different temperature from the experimental spectrum will therefore result in the introduction of artefacts in the resultant

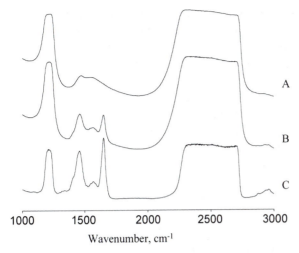

Figure 5 Infrared spectra of D_2O (A), myoglobin in D_2O (B), and the result of interactive subtraction of spectrum A from spectrum B (C).

spectrum. To avoid this problem, spectra of solvents and samples should be acquired at the same temperature using a thermostated cell.

In addition to solvent water absorptions, the infrared spectroscopist is also confronted with spectral interference from atmospheric water vapour. Unfortunately for the biological spectroscopist, a series of narrow, intense absorptions appear between 1500–1900 cm^{-1} which arise from vibrational-rotational modes of gaseous water. This series of absorption bands lies in the spectral region of most use when studying proteins. Effective purging of the spectrometer with dry air or nitrogen is therefore an absolute requirement for obtaining high quality infrared spectra of proteins.

The effect of water vapour on spectra of proteins is illustrated in *Figure 6*, which shows a spectrum of a protein dissolved in D_2O. A series of sharp absorption bands is apparent (*Figure 6A*). The presence of these sharp absorption bands can have a significant impact upon spectral interpretation, particularly if band-narrowing techniques such as derivation of deconvolution are used (see below).

Digital subtraction routines such as those used to subtract solvent water from spectra may be used to remove interfering water vapour absorption bands. In order to perform the subtraction, a water vapour spectrum must be acquired. This is achieved by simply acquiring a spectrum with the spectrometer sample compartment open to the atmosphere and without any sample in the sample compartment (*Figure 6B*). It is advisable to acquire this spectrum at the end of the day, as it may require some time to fully purge the instrument after acquisition of the spectrum. The water vapour spectrum is then interactively subtracted from the protein spectrum. To monitor subtraction the spectral region 1700–1900 cm^{-1} is often used, as this region is devoid of absorptions in most biological systems. The correct subtraction factor is achieved when no sharp absorption bands are seen between 1700–1900 cm^{-1} (*Figure 6C*).

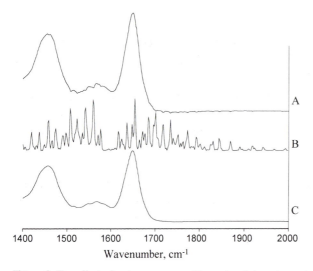

Figure 6 The effect of water vapour on IR spectra. Infrared spectrum of myoglobin (A), the spectrum of water vapour (B), and the result of interactive subtraction of the spectrum of water vapour from the spectrum (C).

It is important to remember that water vapour contributions may be seen in protein spectra for three reasons:

(a) The spectrometer is inadequately purged. This will result in positive water vapour peaks in spectra.

(b) The spectrometer is adequately purged, but was poorly purged when the background was acquired. This will result in the appearance of negative water vapour absorption bands (as there is less water vapour in the sample spectrum than the background).

(c) The buffer spectrum used to subtract water or D_2O absorptions was acquired from a poorly purged spectrometer.

Investigators should be aware that noise in spectra is additive. In other words when a subtraction is performed, the resultant spectrum contains the noise from both spectra. Thus while water vapour contributions may be interactively subtracted from spectra, this operation introduces additional noise. It is therefore preferable to acquire water vapour-free spectra.

Protocol 2

Post-processing of spectra

1 Correction for solvent absorptions is the first step in spectral processing. A reference spectrum must be acquired using the same instrumental parameters (e.g. resolution, apodization function, etc.). Again it is important that the reference spectrum be of high quality, with no contributions from water vapour.

2 The reference spectrum should be interactively subtracted from the sample spectrum. The correct subtraction factor is achieved when a flat baseline is attained in the spectral region 1800–2400 cm^{-1}.

3 The resultant spectrum should then be examined for contributions from water vapour. The spectral region 1700–1900 cm^{-1} should be free from absorptions (unless lipids are present, which will produce an absorption at 1740 cm^{-1}). The presence of water vapour, due to inadequate purging, will be evident from a series of sharp positive or negative peaks in this region.

4 If significant contributions from water vapour are apparent then an interactive subtraction of a spectrum of water vapour should be performed. The subtraction should be judged as adequate when no sharp absorptions are seen between 1700–1900 cm^{-1}.

3.4 Derivation and deconvolution

Proteins and polypeptides usually contain more than one secondary structural motif, and consequently give rise to more than one amide I absorption. Unfortunately for the spectroscopist, the width and separation of these absorptions is such that they overlap considerably and produce a composite, often featureless absorption profile. While some information may be obtained by an inspection of the frequency of the composite amide I absorption maximum and any visible shoulders, changes in the position of such a composite absorption can prove deceptive. Frequency shifts in composite absorption bands may be caused by true frequency shifts of one or more component absorption bands, or by variations in the relative intensities of component bands. Thus a shift to lower wavenumber of a composite absorption may mean that one or more of the underlying absorptions has shifted to a lower frequency. Alternatively, such a shift may be produced by an increase in the intensity of a low frequency component, or a decrease in the intensity of a high frequency component, resulting in a redistribution of intensity without any frequency shifts. Analysis of the composite band alone cannot distinguish between these possibilities and may lead to misinterpretation of spectral shifts. Deduction of structural parameters, and particularly changes in these parameters, from the relatively featureless amide I band alone should therefore only be undertaken with caution.

Clearly infrared spectroscopy would be of limited use if only such limited information could be obtained. To enhance the utility of infrared spectroscopy in this field the user must take advantage of mathematical techniques such as Fourier self-deconvolution (FSD) and derivation (8), which mathematically reduce the width of absorption bands and so allow visualization of overlapping bands. As these techniques allow visualization of overlapping absorption bands they are often referred to as 'resolution enhancement' techniques. However this terminology is misleading, resolution is an instrumental parameter that cannot be increased after a spectrum is recorded. Procedures such as FSD should therefore more correctly be referred to as band narrowing procedures. FSD is the most widely used of these mathematical methods and the general principles and potential pitfalls will be discussed.

An infrared absorption band can be considered to arise from the convolution of a delta function that has position but no width and a Lorentzian that has width but no position to produce a Lorentzian with both position and width. In the Fourier domain, this is expressed as the convolution of the Fourier transform of the delta function (a cosine, the period of which is the frequency of the delta function) with the Fourier transform of the Lorentzian (a decaying exponential, the rate of decay of which is determined by the width of the Lorentzian) to produce an exponentially decaying cosine. The rate of decay of this cosine is determined by the width of the Lorentzian, the wider the Lorentzian, the more rapid the rate of decay. Conversely, narrow bands are characterized by slowly decaying exponential cosines in the Fourier domain. Clearly, it is possible in theory to reduce infrared absorption bands to delta functions which have no width but maintain their frequency characteristics by deconvolving the correct Lorentzian from the absorption profile (hence the term Fourier self-deconvolution, referring to the removal of the intrinsic shape of the absorption band). In practice this is achieved by multiplying the Fourier transform of the absorption band (the exponentially decaying cosine) by the correct increasing exponential, to regenerate the corresponding cosine. The inverse Fourier transform then produces the delta function at the frequency of the original absorption band. Applying this process to a composite absorption such as the amide I absorption should, in principle, result in the appearance of a number of clearly defined delta functions.

Of course in practice we never fully deconvolve the line shapes from our spectra, as we are usually dealing with many decaying cosines superimposed upon each other and decaying at a different rate (it should be remembered that every amide C=O group in a peptide chain gives rise to an amide I absorption band). Rather what is attempted is to reduce the rate of decay of the underlying cosines by multiplication with an increasing exponential, and so reduce the width of the corresponding absorptions.

Unfortunately the choice of the increasing exponential must be determined by the user, and hence is subjective. A poor choice of deconvolution parameters will produce poor results. If the deconvolution parameters are chosen such that the rate of increase of the exponential corresponds to a band width greater than the width of the absorptions being studied, the result will be side lobes at the edges of the absorption bands which can lead to problems with visualization of weaker neighbouring bands. On the other hand, if the chosen band width is to small, the absorption bands will appear unaltered and no additional information will be gained.

In addition to the above caveats, it should be stressed that FSD also reduces the width of water vapour absorptions and enhances noise, producing very sharp peaks. With even minimal deconvolution, noise and water vapour can very

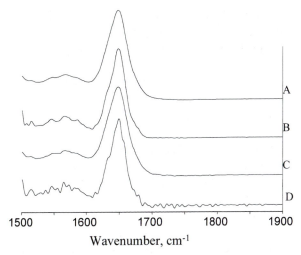

1500   1600   1700   1800   1900

Wavenumber, cm⁻¹

Figure 7 The effect of residual water vapour upon deconvolved spectra. Following application of Fourier self-deconvolution to the relatively featureless absorbance spectrum (A) a number of distinct absorption bands are apparent (B). If even small amounts of water vapour are present (C) deconvolution results in the presence of additional features (D).

quickly become a problem in deconvolved spectra. Thus FSD should only be performed on spectra with a high signal to noise ratio and a low contribution from water vapour.

The effect of water vapour upon deconvolved spectra is shown in *Figure 7*. It can readily be seen that after correct application of FSD (*Figure 7B*) a number of shoulders are apparent which are not apparent in the original absorption spectrum (*Figure 7A*). If the spectrum contains traces of water vapour (*Figure 7C*), then application of the same deconvolution parameters results in a significantly different spectrum (*Figure 7D*) with many more absorption bands apparent. Obviously, assignment of these features to amide modes will result in an incorrect structural determination. As discussed above, a spectrum may be judged to be free of water vapour if no sharp absorptions are seen between 1700–1900 cm⁻¹. Inclusion of at least part of this spectral region in illustrations will aid readers in determining the extent of possible water vapour contributions to deconvolved spectra, and for this reason should be considered good practice for feature-rich spectra.

In addition to FSD, calculation of spectral derivatives is often used to aid in the visualization of overlapping absorptions. The caveats that apply to FSD with respect to noise and water vapour are also applicable to derivation. In addition, the user should be aware that relative integrated intensities are not maintained in derivative spectra. It can be seen that additional care is needed in the interpretation of derivative spectra. Ideally, spectra should be subjected to both derivation and FSD and only features present in both derivative and deconvolved spectra should be assigned in order to avoid artefacts due to data processing.

Protocol 3

Application of band narrowing techniques

1   Due to the composite nature of absorptions band narrowing techniques such as Fourier self-deconvolution or derivation are required to extract meaningful information from protein spectra. Most commercially available deconvolution routines permit interactive deconvolution of spectra, allowing the user to experiment until the correct parameters are chosen. For the purposes of this discussion we will assume the reader is familiar with the widely used deconvolution routine described by Cameron *et al.* (8). This routine requires two parameters, a band narrowing factor and an estimate of absorption band half-width.

2   Select initial deconvolution parameters. Typically, we find that an estimated absorption band half-width of 13–15 cm^{-1} to be useful. The band narrowing factor that may be used depends upon the quality of the spectra. We find a factor of 1.5–1.8 to be an acceptable starting value. If this does not produce adequate band narrowing the factor should be progressively increased until the required narrowing is achieved.

3   Deconvolved spectra should be examined for artefacts. Common sources of artefacts are water vapour, noise, and over-deconvolution. Examination of spectra in such cases reveals the presence of side lobes on absorption bands that may obscure weaker absorption bands. Excessive water vapour or noise are apparent in spectra as sharp features between 1700–1900 cm^{-1}. If such features are visible, the band narrowing factor should be reduced until these features are acceptable.

4   In addition to deconvolution, second derivative spectra should be calculated. Savitsky–Golay second derivatives are the most common type of derivative used. The user is required to enter the number of date points over which the derivative is calculated. The smaller the value, the greater the band separation will be. Larger values lead to smoothing of date and should be used for poorer quality data. Typically, point spacings of 9–15 should suffice. Again the quality of spectra is assessed by visual inspection of the region 1700–1900 cm^{-1}.

5   Deconvolved and derivative spectra should be compared, and only absorption bands visible in both spectra should be assigned. An obvious point, but the investigator should be aware that peaks in second derivative spectra are inverted. Some investigators multiply second derivative spectra by (–1) to produce the more familiar positive bands.

4 Thermal and solvent manipulation techniques for assessing protein–ligand interactions

Detection of ligand-induced changes in protein secondary structure is dependent upon the relative proportion of amino acids involved in the structural rearrangement. As a rule of thumb, to be easily detectable by infrared spectroscopy, a change must involve 5% of the amino acids in the polypeptide chain. However that does not mean that more subtle structural changes cannot be detected. A number of methods exist to elucidate the nature of subtle structural changes induced by ligand binding.

The simplest way of assessing subtle effects of ligand binding is to monitor the rate of proton–deuteron exchange for samples dissolved in D_2O. As discussed above, both the amide I and amide II absorption bands are sensitive to deuteration. In the case of the amide I absorption band this sensitivity arises due to the effect of hydrogen bonds upon the $C=O$ stretching vibration. Substitution of a deuteron for a proton in the amide group, so forming a $C=O$---D–N 'deuterium' bond, affects the frequency of the $C=O$ stretching absorption band, shifting it to lower wavenumbers. The magnitude of the shift will depend upon the number of N–H groups that become deuterated. The rate at which the absorption band shifts will be determined by the strength of the hydrogen bonds involving the N–H groups (strong hydrogen bonds reduce the rate of proton–deuteron exchange) and by the degree of solvent accessibility of the amide N–H groups.

The shift seen in the amide II absorption band upon deuteration is much larger (approximately 100 cm^{-1}), due to the fact that the vibration giving rise to the amide II absorption band involves the N–H group directly. Deuteration of the N–H group to form N–D groups shifts the amide II absorption band to approximately 1400 cm^{-1}. This large shift gives the appearance that the amide II absorption band disappears, and deuteration of proteins is often monitored by evaluation of the 'disappearance' of the amide II absorption band. Again the rate of ' disappearance' of this absorption band is related to the strength of hydrogen bonds involving the N–H groups and the degree of solvent accessibility.

Monitoring of deuteration is straightforward. The protein sample is dissolved in D_2O and placed in a transmission cell. This cell is then mounted inside the spectrometer in a thermostated cell holder. This thermostated cell holder is important as the rate of deuteration is strongly influenced by temperature, and the sample should be maintained at a constant temperature throughout the course of the experiment. Spectra are then acquired from the sample at defined intervals, typically every hour for 12–24 hours. However, it is important to note that deuteration does not occur at a constant rate, and it may be advisable to acquire spectra more frequently at early time points (e.g. every 15 minutes). After all spectra have been acquired, spectra of the buffer acquired at the same temperature should be subtracted from each spectrum. The rate of deuteration may then be expressed in a number of ways. For example deuteration may easily be monitored by plotting the intensity of the residual amide II absorption band at 1550 cm^{-1} as a function of time, or the frequency of the composite amide I absorption band as a function of time. Alternatively, following deconvolution or derivation the frequency of individual amide I absorption band components, corresponding to different secondary structural elements, may be plotted as a function of time. This approach is particularly useful as it allows the kinetics of deuteration of the different classes of secondary structures present to be determined individually.

This procedure should then be repeated in the presence of the ligand of interest. Comparison of the kinetics of deuteration in the presence and absence of the ligand may then reveal information relating to ligand binding. For example, it may be found that the rate of deuteration is retarded in the presence of the ligand. Such a finding may arise if the hydrogen bonding network in sections of the polypeptide chain has been strengthened (reducing proton deuteron exchange) or if the protein has adopted a more compact structure (reducing solvent penetration into the protein). Whichever of the two mechanism leads to the reduced rate of deuteration it is clear that this analysis technique potentially allows very subtle differences in protein structure to be detected.

Thermal stressing of proteins in the presence and absence of ligands can also be used to determine subtle differences in protein structure induced by ligand binding. Thermal denaturation of proteins is routinely studied by infrared spectroscopy. Samples are prepared as described for deuteration experiments. However, rather than keep the sample at a constant temperature, spectra are acquired as the temperature is increased. This is achieved with the use of a thermostated cell holder. Typically, the sample is equilibrated at 20 °C and a spectrum is acquired. The sample temperature is then elevated by a predetermined amount and allowed to equilibrate. A second spectrum is then acquired. This process is repeated until spectra have been acquired over the entire temperature range of interest. Spectra of the buffer solution in which the protein is dissolved are then acquired at the same temperature. As the spectrum of aqueous buffers are extremely sensitive to temperature it is important to ensure that the sample spectra and buffer spectra are well matched with respect to temperature. Spectra of buffer are then subtracted from spectra of the protein acquired at the corresponding temperature.

This process is repeated in the presence of the ligand of interest. The response of the protein to thermal stress in the presence and absence of the ligand can now be assessed. Typically, the response of proteins to such thermal stress is denaturation and aggregation. Denaturation is generally manifest as the disappearance of the characteristic bands associated with helical and β-sheet secondary structures and a broadening of the amide I absorption profile. Aggregation accompanying denaturation is typified by the appearance of characteristic absorption bands at 1625 and 1680 cm^{-1}. It is important that these absorption bands are not confused with β-sheet absorption bands, which unfortunately is common in the infrared spectroscopic literature. These new absorption bands do not arise from secondary structural elements within the protein, but rather arise from quaternary interactions. Thermal stressing of the protein results in unfolding of the native secondary structures present within the protein. Of necessity this involves rupture of the hydrogen bonds stabilising the secondary structures, with concomitant formation of free C=O and N–H groups. These free C=O and N–H groups are energetically unfavourable. The result is that hydrogen bonds form between the C=O and N–H groups of any polypeptide strands with which they come in contact. The consequence of this is that many hydrogen bonds are formed between polypeptide chains in neighbouring protein molecule, forming an aggregate stabilized by very strong intermolecular hydrogen bonds.

The temperature at which this denaturation and aggregation occurs as detected by infrared spectroscopy is characteristic of the protein under investigation and may be influenced by ligand binding. For many proteins, ligand binding is associated with the formation of a more compact and hence more stable tertiary structure. This more compact and stable tertiary structure typically results in an increased temperature of denaturation. Thus, even if no gross changes in secondary structure are detectable upon ligand binding, thermal stressing of proteins may be a useful method to assess the degree of stabilization of the protein secondary structure induced by ligand binding. In principle it is also possible to study preferential stabilization of discrete types of secondary structure based upon the effects of ligand binding upon the denaturation temperatures of discrete secondary structures, as deduced from the behaviour of individual amide I absorption bands in deconvolved or derivative spectra.

A similar approach to assessment of protein stability in the presence and absence of ligands involves structural perturbations by solvents. For example it has been shown that low concentrations of organic solvents such as DMSO and TFE reduce the stability of proteins in aqueous solution (6, 7). This reduced stability may be manifest as an increased rate of proton–deuteron exchange or a decreased thermal stability. The effects of ligand binding can thus be monitored via the effects of solvents upon thermal stability or proton–deuteron exchange of the protein.

Protocol 4

Thermal stressing of proteins

Equipment and reagents
• See *Protocol 1*

Method

1 Prepare samples for analysis and acquire spectra as in *Protocol 1*, ensuring that the sample is adequately thermostated.

2 Increment the temperature by the desired amount and allow the sample to equilibrate. Typically, for initial studies 5 °C increments are used. For more detailed studies, smaller temperature increments may be used. Equilibration periods vary, depending upon experimental set-up and temperature increments. Ideally, temperature should be monitored at the sample, and spectra acquired only when the sample has been at the desired temperature for a few minutes. It is also important to ensure that the cell and cell holder are well insulated in order to allow high temperatures to be achieved.

3 Acquire spectra at temperatures covering the range of interest.

4 Acquire buffer spectra at the same temperatures.

5 Perform buffer subtractions at each temperature as outlined in *Protocol 2*.

Protocol 4 continued

6   Perform deconvolution or derivation as outlined in *Protocol 3*.

7   Repeat this procedure for samples in the presence of the ligand of interest.

5 Difference spectroscopy

In many instances, the differences induced in the secondary structure of proteins upon ligand binding are extremely small. For example rather than producing gross structural changes, ligand binding may result in perturbations of small regions of secondary structural elements or even individual amino acid side chains. Visual comparison of spectra of the protein in the ligand bound and ligand free state may be insensitive to these small changes. Methods are therefore required to enhance the spectroscopic consequences of these changes.

The relative insensitivity of infrared spectroscopy to subtle structural changes arises from the fact that infrared spectroscopy probes all of the amide groups in the protein simultaneously. For example in a protein containing 1000 amino acids, one obtains amide I absorptions from 999 amide C=O groups. If binding of a ligand to this protein results in a structural perturbation of 10 amide groups, then this may be expected to produce a change of only 1% in the amide I absorption band. Such a change will be difficult to detect reproducibly. The answer is simply to remove the amide I components which do not change from the spectrum, leaving only the spectral signatures of the protein components altered by ligand binding. This is achieved by the technique of difference spectroscopy. In practice this involves subtracting the spectrum of the ligand-bound protein from the ligand-free protein. The resultant difference spectrum contains either weak positive features which correspond to new structural elements induced by ligand binding or weak negative features indicative of the loss of structural elements upon ligand binding.

Spectra of the two protein states must be recorded independently under carefully controlled conditions to allow correct subtraction. Interactive subtraction of two spectra matched as well as possible with respect to concentration, temperature, and path length must be performed. Correct subtraction is judged by the appearance of a flat baseline in regions that contain no absorptions (i.e. 1700–2200 cm^{-1}). A sloping baseline indicates over or under subtraction. This approach is obviously subjective but under correctly controlled conditions can provide valuable information.

6 Isotope edited difference spectroscopy

When the ligands under investigation is another protein or peptide, determining the structural perturbations induced by binding presents unique problems. A specialized form of difference spectroscopy known as isotope edited difference spectroscopy may be employed in such circumstances. Isotope edited difference spectroscopy makes use of the spectral shift induced in infrared absorptions when substituting an atom for a heavier isotope. For example substitution of ^{13}C for ^{12}C in amide C=O groups results in a shift of the amide I absorption maximum by approximately 50 cm^{-1} to lower frequency. Practical use is made of this shift in studies of ligand–protein interactions by labelling either the protein of interest or the protein/peptide ligand with ^{13}C. For example in the first demonstration of the utility of this technique calmodulin was completely labelled with both ^{13}C and ^{15}N (9). Thus, isotopic shifts are observed in both the amide I and amide II absorption bands (this is important as the shift in the amide I absorption bands induced by ^{13}C labelling brings the amide I absorption band into the spectral region in which the ^{14}N amide II absorption band is typically observed). The amide I absorption maximum of the labelled calmodulin was shifted from 1643 cm^{-1} to 1589 cm^{-1}. Upon interaction with a small ^{12}C, ^{14}N target peptide the amide I absorption bands from the isotopically labelled calmodulin and the target peptide could therefore be clearly distinguished, and the effects of peptide binding upon the structure of calmodulin investigated. While this technique has yet to find widespread applications (perhaps due to cost) it will surely become an important tool for structural characterization of protein–protein and protein–peptide interactions. The ability to specifically label structures of interests (i.e. binding sites) and determine site-specific changes in conformation as a function of ligand binding will certainly make this new tool attractive.

7 Reaction-induced difference spectroscopy

Subtle changes in protein secondary structure caused by ligand binding may also be detected using the technique of reaction-induced difference spectroscopy, or RIDS. RIDS makes use of the fact that under certain experimental conditions spectra may be acquired from some proteins in the presence and absence of a ligand without having to dismantle the sample cell or remove the cell from the spectrometer. In experiments of this type, a spectrum of the protein in the non-perturbed state is recorded. The same sample of the protein is then perturbed (e.g. by the photolytic release of a caged substrate, by illumination of a light transducing protein or by oxidation or reduction of a redox protein by application of a potential difference) and the spectrum of the perturbed protein is recorded, often as a function of time.

Spectra acquired before and after perturbation of the protein may then be directly compared and differences in structure assessed. However, a more reliable method of assessing the subtle changes produced in such experiments is to calculate the ratio of the two spectra to produce a spectrum that corresponds to the difference between the two states. In essence this corresponds to producing an absorbance spectrum of the structural difference, with the spectrum prior to perturbation corresponding to a background spectrum. A number of good review articles on this topic exist for the interested reader (10–12).

Reaction-induced difference spectroscopy allows extremely small differences in secondary structure to be studied. The first examples of reaction induced difference spectroscopy involved the use of light as a trigger, to probe the differences between structural intermediates of light transducing proteins such as bacteriorhodopsin and rhodopsin. Such studies have been able to identify specific amino acid residues taking part in proton transfer and structural rearrangements of the retinal chromophore in response to illumination (10, 11).

RIDS has also been used to study binding of more conventional ligands to proteins. For example the catalytic cycle of the Ca^{2+}ATPase has been studied by RIDS (12). Such studies have made use of the fact that so-called 'caged' substrates can be produced. Caged substrates are trapped within a matrix that may be degraded, releasing the ligand, by irradiation with UV light. The most common caged ligand is caged ATP. This substrate may be used to assess the conformation changes associated with hydrolysis of ATP by transport proteins such as the Ca^{2+}ATPase.

The experimental procedure is straightforward in principle. A sample of the protein is prepared in the presence of the caged ligand. A single beam spectrum is acquired and the sample is then exposed to UV light. Exposure of the caged ATP to UV light results in photolysis of the cage, releasing the ATP. A second single beam spectrum is then acquired and the ratio between the two single beam spectra computed. The resultant spectrum is the spectrum of the difference between the two samples. Using this technique it is possible to detect spectral changes related to photolysis of the caged ATP, hydrolysis of ATP by the Ca^{2+}ATPase, and changes in the protein itself. The changes in protein structure detected by such techniques are small, absorbance changes in the amide I region being less than 2% of the total amide I absorbance.

As an alternative to photolytic release of substrates a methodology utilizing ATR spectroscopy of films of membrane proteins has been used (13). In this methodology, a film of the membrane protein is deposited upon the surface of an IRS crystal. (Note this technique is not suitable for water soluble proteins, which may be denatured when deposited upon the surface of the IRS crystal.) A single beam spectrum of the protein film in its resting state is acquired with a ligand-free buffer flowing over the film. A single beam spectrum of the ligand-bound protein is then recorded from the same film, only this time the necessary ligand is added to the buffer flowing over the film. The ratio between the two single beam spectra is then calculated to produce the difference spectrum. This approach has been successfully used to study structural changes in the nicotinic acetylcholine receptor associated with binding of acetylcholine and a number of antagonists (13). Subtle differences in the structure of this large protein (M ~ 30 kDa) reflecting rearrangements of the polypeptide backbone and changes in the environment and/or the protonation state of amino acid side chains could be detected, demonstrating the sensitivity of this approach.

References

1. Elliot, A. and Ambrose, E. J. (1950). *Nature*, **165**, 921.
2. Jackson, M. and Mantsch, H. H. (1991). *Can. J. Chem.*, **69**, 1639.

3. Jackson, M. and Mantsch, H. H. (1995). *CRC Crit. Rev. Biochem. Mol. Biol.*, **30**, 95.

4. Khachfe, H., Mylrajan, M., and Sage, J. T. (1997). *Cell. Mol. Biochem.*, **44**, 39.

5. Harrick, N. J. (1967). *Internal reflectance spectroscopy*. John Wiley, New York.

6. Jackson, M. and Mantsch, H. H. (1991). *Biochim. Biophys. Acta*, **1078**, 231.

7. Jackson, M. and Mantsch, H. H. (1991). *Biochim. Biophys. Acta*, **1118**, 139.

8. Cameron, D. G. and Moffatt, D. J. (1984). *J. Test. Eval.*, **12**, 78.

9. Zhang, M., Fabian, H., Mantsch, H. H., and Vogel, H. (1994). *Biochemistry*, **33**, 10833.

10. Braiman, M. S. and Rothschild, K. J. (1988). *Annu. Rev. Biophys. Biophys. Chem.*, **17**, 541.

11. Rothschild, K. J. (1992). *J. Bioenerg. Biomembr.*, **24**, 147.

12. Mäntele , W. (1993). *Trends Biochem. Sci.*, **18**, 197.

13. Baenziger, J. E., Miller, K. W., McCarthy, M. P., and Rothschild, K. J. (1992). *Biophys. J.*, **62**, 64.

Chapter 7

Protein–ligand interactions studied by Raman and resonance Raman spectroscopy

Robert Withnall

School of Chemical and Life Sciences, University of Greenwich, Wellington Street, Woolwich, London SE18 6PF, UK.

1 Introduction

The occurrence of the phenomenon of inelastic light scattering (now often called 'Raman light scattering') was first demonstrated by C. V. Raman (1), who was awarded a Nobel prize in Physics in 1930 for his discovery.

The Raman light scattering process involves two photons, one of which is incident on the sample whilst the other is scattered from the sample. The scattered photons travel in all directions over the full solid angle of 4π steradians although, in the Raman experiment, it is common to detect either those which are back-scattered (back-scattering configuration) or those which are scattered at an angle of 90° to the direction of propagation of the incident light (90° scattering configuration). Then, the intensity of the scattered photons is plotted versus their wavenumber displacement from the incident photons resulting in a so-called 'Raman spectrum'. The exciting light induces oscillation of the electrons in the sample, and the induced dipole moment (μ_{ind}) is proportional to the electric field (E), according to *Equation 1*:

$$\mu_{ind} = \alpha.E \qquad [1]$$

where the proportionality constant, α, is the polarizability of the molecule.

The oscillating electric field, E, is generated by the incident laser light and may be expressed as:

$$E = E_o.\sin 2\pi\nu_o t \qquad [2]$$

where ν_o is the frequency of the laser light and E_o is the amplitude of the electric vector.

Also, the polarizability, α, will vary during a vibration according to *Equation 3*:

$$\alpha = \alpha_o + \sum_i \left(\frac{\partial\alpha}{\partial Q_i}\right) Q_i \qquad [3]$$

where α_o is the polarizability of the molecule at rest, Q_i is the normal co-ordinate for the ith vibration, and $\partial\alpha/\partial Q_i$ is the change in molecular polarizability with respect to Q_i. Furthermore, assuming simple harmonic motion, Q_i oscillates according to *Equation 4*:

$$Q_i = Q_o.\sin 2\pi\nu_i t \qquad [4]$$

where ν_i is the natural frequency of oscillation of the ith vibration.

Equation 1 can be recast as *Equation 5* by substituting for E, α, and Q_i from *Equations 2, 3,* and *4* respectively.

$$\mu_{ind} = \left[\alpha_o + \sum_i \left(\frac{\partial\alpha}{\partial Q_i}\right)Q_o \sin 2\pi\nu_i t\right]E_o\sin 2\pi\nu_o t$$

$$= \alpha_o E_o\sin 2\pi\nu_o t + \sum_i \left(\frac{\partial\alpha}{\partial Q_i}\right)\frac{Q_o E_o}{2}[\cos 2\pi(\nu_o - \nu_i)t - \cos 2\pi(\nu_o + \nu_i)t] \qquad [5]$$

The first term in *Equation 5* represents the Rayleigh scattered light which originates from induced dipoles having a frequency ν_o. However, some incident photons from the laser give up some of their energy to the sample in the scattering process, whereas some gain some energy from it. In the former case the scattered photons have frequencies lower than those of the incident photons (given as $\nu_o - \nu_i$ in the second term of *Equation 5*) and the plot of wavenumber displacement (to lower energy of the laser line) versus intensity is known as a Stokes Raman spectrum. Conversely, the analogous plot in the energy region above the laser line is known, by convention, as an anti-Stokes Raman spectrum; in this case the scattered photons have frequencies of $\nu_o + \nu_i$ (see *Equation 5*). The peaks in a Raman spectrum correspond to internal transitions of a molecule; these transitions may be rotational, vibrational, or electronic, although vibrational Raman spectroscopy is most commonly practised.

As is shown in *Figure 1*, the incoming photon promotes a vibrational oscillator from the ground vibrational level (v = 0) to some virtual level, which is a non-stationary state. In a Stokes Raman scattering process, the oscillator returns to the v = 1 level. Alternatively, the oscillator may already be in the v = 1 level due to thermal excitation, and it can relax to the ground (v = 0) level following an anti-Stokes scattering process. Clearly the wavelength of the incident laser light is not important for Raman light scattering to occur, and indeed the laser light may be chosen to have infrared, visible, or ultra-violet energies. From a theoretical viewpoint, it is advantageous to use exciting light of high energy, because the light scattering efficiency is proportional to the frequency of the scattered light to the power of 4. If this were the only consideration, ultra-violet light would be the exciting radiation of choice, but lasers which provide ultra-violet light are generally more costly than those which only operate in the visible region. In addition, high energy ultra-violet photons can damage samples, particularly when fragile biological materials are under investigation. Despite these drawbacks, Raman spectroscopy is sometimes performed in the ultra-violet region, as it can provide a wealth of unique information (see later). Also, there can be experimental reasons for choosing infrared light as the exciting

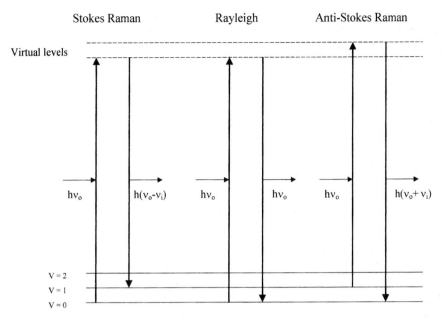

Figure 1 Idealized model of Rayleigh scattering and Stokes and anti-Stokes Raman scattering.

radiation, the primary one being to keep the level of interfering fluorescence to a minimum (see later). Furthermore, since the mid-1970s microscopes have often been coupled to Raman spectrometers in order to facilitate the examination of samples of micrometre dimensions (2), and commercial microscopes can easily be adapted for this purpose when the Raman spectrometer is operating in the visible region of the electromagnetic spectrum.

Although infrared and Raman spectroscopy both give information about molecular vibrations, the selection rules which govern the activities of infrared and Raman vibrations are fundamentally different (3). Whereas the condition for infrared activity is a change in dipole moment during the vibration, Raman activity requires a change in molecular polarizability during the vibration. Indeed, it is only through the interaction of the electronic and nuclear motions (i.e. the breakdown of the Born Oppenheimer approximation) (4) that a vibrational Raman effect is observed at all.

The appropriate quantum mechanical integral is:

$$\frac{\partial \alpha}{\partial Q} = \int \varphi_m \, \mu_{ind} \, \varphi_n \, dQ \qquad [6]$$

where φ_m and φ_n are the ground and excited state vibrational wavefunctions respectively.

A consequence of the different infrared and Raman activities of molecular vibrations is that the two techniques can often provide complementary information and, in the special case of a molecule having a centre of symmetry, the

267

activities for any given vibration are mutually exclusive. The complementary nature of infrared and Raman spectra can be seen for the water molecule; it is a strong infrared absorber but weak Raman scatterer, all three of its fundamental vibrations giving rise to strong infrared bands, but weak Raman ones. This is an advantage when performing Raman spectroscopy of proteins in aqueous solution, as the solute bands are not usually masked by those of the solvent. On the other hand, in the analogous infrared experiment, certain measures are often taken in order to reduce the interference from the water absorptions, for example very small solution cell path lengths are used (see Chapter 6 in this volume). A merit of Raman microscopy is that no preparation of solid samples is required prior to collecting the spectrum. Thus the technique can be conveniently applied to examine samples in both the solid state and in aqueous solutions, and it is a very useful tool for ascertaining whether the structure of a water soluble protein is the same in both. Diffraction techniques can, of course, be employed to determine structures in the solid state, but not in solution. Solution structures can often be obtained by NMR spectroscopic techniques, but this can be very time-consuming (see Chapter 10 in this volume). It is a big asset of the technique that it has the capability of determining structure and structural changes under functional conditions. Furthermore the scattering process is very fast (ca. 10^{-14} sec), and as a result the Raman technique is unrivalled in its ability to detect transient species and provide insight into protein dynamics. Two drawbacks are that fluorescence and sample degradation due to the laser beam can be concerns for some protein samples. However, there are ways of overcoming the fluorescence and sample degradation problems in many instances, as will be mentioned below.

2 Raman experiments

It is possible to perform a Raman spectroscopic experiment in many different ways, depending on the information which is sought. In this section the various types of Raman spectroscopy will be classified, and their experimental set-ups will be considered. Experimental protocols for spectral acquisition will only be given for two experiments (Sections 2.1 and 2.3) because it is felt that the procedures, other than those related to sample handling, depend to a great extent on how the experiment has been configured. The various types of experiment are summarized in Table 1.

2.1 Visible/near infrared Raman spectroscopy

The intensity of Raman scattered light is weak ($< 10^{-6}$) in comparison to that of Rayleigh scattered light, resulting in two experimental challenges. First, it is desirable to detect as much of the Raman scattered light as possible and, secondly, it is necessary to separate the weak Raman scattered light from the much stronger Rayleigh scattered radiation. The former is accomplished by maxi-

Table 1 Raman spectroscopic techniques and the information which they can provide

Region in the electromagnetic spectrum of the exciting radiation		Sample state
Infrared **Visible**	**Ultra-violet**	
FT Raman **Visible/NIR Raman**	**UV Raman**	Solutions or solids
← **Difference Raman spectroscopy**	→	Solutions
Vibrational information about structural changes taking place on ligand binding		
← **Resonance Raman spectroscopy**	→	Solutions or solids
Vibrational and electronic information about structural changes taking place on ligand binding		
← **Time-resolved resonance Raman spectroscopy** →		Solutions
Information on fast conformational changes and short-lived intermediates		
← **Raman optical activity**	→	Solutions
Information on asymmetric centres. Very informative about secondary structure.		
← **SERS and SERRS**	→	Solutions, solids, or sols
Improved limits of detection		
← **PSCARS**	→	Solutions
Information on changes in protein secondary structure on ligand binding		

mizing all of the parameters which determine the light intensity, I_s, registered at the detector, as given by *Equation 7*:

$$I_s = I_o \times \sigma \times n_s \times \nu^4 \times K(\nu) \times A(\nu) \qquad [7]$$

where n_s is the number of Raman scattering molecules in the interrogation volume, σ is the Raman cross-section of a given vibration of the analyte, I_o is the intensity of the incident light, $K(\nu)$ is the spectrometer response, and $A(\nu)$ is the self-absorption of the medium.

As with most techniques, there is a trade-off between spectral resolution and signal intensity. However when working in the condensed phase there is a limit to the required spectral resolution, which is determined by the band-widths, and these in turn depend upon broadening mechanisms resulting from perturbations of the vibrational transitions by the surrounding molecules. Usually the full width at half height (FWHH) of Raman bands of condensed phase molecules is > 1 cm^{-1}, and in the case of biological molecules it is often more than 10 cm^{-1}, so an equivalent spectral resolution is normally sufficient. For a dispersive spectrometer, the experimental spectral resolution ($\Delta\nu$) is given by *Equation 8* (5):

$$\Delta\nu = D_L^{-1} \cdot (\text{S.W.}) \qquad [8]$$

where D_L^{-1} is the reciprocal linear dispersion which depends on the length of the monochromator, the groove density of the grating(s), the order of the dispersion, and the angle of diffraction, and S.W. is the slit width at the entrance to the monochromator.

Clearly, as can be seen from *Equation 8*, the smaller the slit width, the better the spectral resolution with the added benefit that stray light rejection is improved. Obviously, though, it is important that very little of the precious Raman scattered light is also rejected by the slit. Thus it is desirable to bring the Raman scattered light to a tight focus at the entrance slit by means of a collection lens. When using a 90° scattering configuration (see *Figure 2*), a separate lens is required in order to focus the laser light onto the sample. The light gathering power of an optic is determined by its f/ value (i.e. 'f number' value), which depends on the size of the cone of acceptance of the light, according to *Equation 9* (7):

$$f/ \text{ value} = (2NA)^{-1} = \tfrac{1}{2}\,(\mu.\sin\theta)^{-1} \qquad [9]$$

where NA is the numerical aperture, μ is the refractive index of the medium ($\mu = 1$ in air), and θ is the half angle of the cone of acceptance of light.

The f/'s of the collection lens and spectrometer are critical parameters in the Raman experiment. A slow f/ spectrometer has a high resolving power and a low cone half angle of light acceptance, whereas a fast f/ spectrometer has low resolving power but high luminosity. It is the latter type which is favoured for aqueous solution studies of protein–ligand interactions, because most proteins have low solubilities in water giving rise to weak Raman signals (see later).

It is the geometric etendue which characterizes the light gathering (or emitting) ability of an optical system. The etendue is expressed as the product, $A\Omega$, of the area (A) of the detector (or emitting source) and the solid angle (Ω, where $\Omega = 2\pi\,(1 - \cos\theta)$) into which it is accepted (or from which it is propagated). The etendue should be constant throughout the optical train, otherwise losses in light flux will result. It is worth noting in passing that, with Raman microscopy, the microscope is usually the limiting part of the optical train in terms of the etendue (8). Photomultiplier tubes have large areas, so when they are used as detectors it is common to increase the solid angle, Ω, at the sample, as the area of the focused laser spot is small. For example, a 20 μm diameter spot size can be magnified to fill a 160 μm slit width by choosing an f/1 collection lens in conjunction with an f/8 spectrometer. This process is known as f/ matching (9).

The problem of how to reject efficiently the intense Rayleigh scattered light has traditionally been approached by using more than one diffraction grating in double or triple grating spectrometers. However, the relatively recent development of holographic notch and edge filters has provided an alternative approach. These filters have a very high optical density (ca. 8) at the wavelength of the laser light with a steep cut-off slope, but have good transmission characteristics at other wavelengths. Thus a single grating polychromator (i.e. a spectrograph) having a CCD (charge coupled device) multichannel detector can be used in conjunction with a holographic notch or edge filter. This has the advantage that the light throughput is greater than that of a double- or triple-grating spectrometer, since light losses cannot be avoided at reflective diffraction gratings. A disadvantage is that bands at low wavenumber shifts (< 50 cm^{-1}) cannot be observed, since they come below the cut-off of the filter. This can be particularly disadvantageous in single crystal Raman spectroscopic studies, if bands due to lattice vibrations need to be observed, since these can appear in this low wave-number region. The quantum efficiency versus wavelength curve for a CCD detector, with silicon as the photosensitive element, peaks in the red region and extends out to ca. 1000 nm on the long wavelength side; the useful range of this curve determines the laser wavelengths which can be used for excitation. An advantage of using a CCD detector is that it has extremely low dark current (ca. 1 electron per pixel per hour when cooled with liquid nitrogen), and the readout noise is insignificant compared to the shot noise (which originates from the signal itself). This permits the use of long integration times which build up the signal to noise ratio. Also, the silicon CCD has a good quantum efficiency in the 800 nm range, enabling relatively low power laser diodes to be used to provide excitation in the near infrared region (e.g. the GaAlAs diode laser which operates at a wavelength of 783 nm). Far fewer samples fluoresce when excited at 783 nm, compared to 514.5 nm, but there is a sensitivity trade-off due to the ν^4 dependence of scattering cross-sections. Diode lasers are cheap and robust, being solid state devices, and they have acceptably narrow linewidth (< 1 cm^{-1}). How-ever they are limited in power and need excellent temperature control (± 0.01 °C) and power supply stability in order to minimize the frequency drift (10).

A CCD chip is a two-dimensional array of pixels, typically 1024×256 with an overall width of ca. 2.5 cm as each pixel is about 25 μm wide. In the spectro-graph the Raman scattered light is dispersed into its component wavelengths across the width of the array, i.e. the 1024 pixels in any given row. As the Raman signals of proteins and protein–ligand complexes at low concentrations are weak, Dong *et al.* (6) have pointed out that it is advantageous to use the full height of the CCD pixel array. By selecting a slow focusing lens, the laser light can be brought to a fine focus with a long, thin waist (see *Figure 2*). This focused light can be imaged on to the full-length of the CCD detector with the collection optics, when a tall slit (ca. 8 mm in height) is employed; consequently, the full height of the CCD is binned for maximum sensitivity. The scattered light is dispersed by a transmissive diffraction grating (shown in *Figure 2*), which has been recently developed by Kaiser Optical Systems, Inc., rather than a reflective diffraction grating. The former is favoured here due to its excellent efficiency in transmit-ting the Raman scattered light. Since each pixel of a CCD detector is relatively much smaller than the active area of a PMT detector, there is little to be gained from f/ matching with CCD systems. Thus, it is effective to keep the f/ of the collection optics (achromatic lenses L1-L4 in *Figure 2*) equal to that of the fast f/1.4 spectrometer. When using CCD detectors, it is necessary to bring the scattered light into focus in the plane of the detector, making any astigmatism in the spectrograph undesirable as it will degrade signal and resolution.

In addition to providing laser light, gas lasers also give unwanted plasma emissions (although they are often used for frequency calibration). If these are not filtered out, they appear as very sharp lines (originating in the gaseous plasma of the laser tube) in the Raman spectrum and should not be confused with the Raman bands. Tabulations of wavelength locations and intensities of plasma lines of lasers, e.g. the argon ion laser, are given in the literature (11) and can be

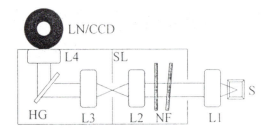

Top view

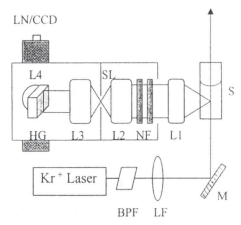

Side view

Figure 2 Diagram of the lay-out of a high throughput Raman spectrometer used for the study of protein–ligand interactions (reprinted from ref. 6).

used for frequency calibration. One way of filtering out plasma lines is to use interference filters, which are generally of the Fabry–Pérot type (5). An interference filter has the opposite transmission profile to a notch filter, being optimized in order to transmit efficiently laser light of a given wavelength whilst rejecting light at other wavelengths. The wavelength of maximum light transmission can be varied to a small degree by angle-tuning of the filter. Pre-monochromators can also be used to 'clean up' (i.e. reject plasma emissions) the laser light. One type of pre-monochromator is the Pellin–Broca prism which diverts the laser beam by 180°, another type is a prism pre-monochromator which does not change the direction of propagation of the beam.

The choice of diffraction grating will affect the wavelength-dependent throughput of the spectrometer. Most diffraction gratings are blazed which optimizes them at a particular wavelength, and the efficiency reduces by ca. 50% at 0.67 λ and 1.8 λ. The throughput efficiency curve of triple grating spectrometers can be quite steep due to the combined effect of all three gratings.

Protocol 1 outlines a procedure for obtaining Raman spectroscopic data on protein–ligand interactions, using an instrument such as that shown in the diagram of *Figure 2*.

Protocol 1

Obtaining Raman spectroscopic data on protein–ligand interactions

Equipment and reagents

- Raman spectrometer
- Centrifuge
- Centrifuge concentrator
- UV/vis spectrometer

- Glass cuvette
- Solution of protein
- Solution of protein–ligand complex
- Methanol

Method

1 Check the purity of the sample by, for example, NMR spectroscopy or SDS–PAGE.

2 Check the enzyme specific activity (12).

3 Centrifuge the sample and remove any precipitated particles. Any particulates present in the solution will degrade the signal to noise quality of the spectrum.

4 Use a centrifuge concentrator to adjust the protein concentration to ca. 1 mM, this can be checked by means of UV/vis spectroscopy.

5 Make up the solution of protein–ligand complex by introducing the ligand to an aliquot of the solution from step 4, above. Then the concentration of protein in the protein–ligand solution will be identical to that of the protein solution containing no ligand. Determine whether the protein–ligand solution contains any free ligand. If so, the difference spectrum will exhibit bands due to free ligand.

6 Switch on the laser and allow time for it to stabilize (ca. 30 min). The laser should be operated in 'light only' mode rather than current mode if this option is available.

7 Check the alignment[a] and cleanliness of laser beam steering optics.

8 Make sure the Rayleigh line is set to a relative wavenumber displacement of zero.

9 Load the protein in buffer solution into a glass cuvette, making sure any grease (e.g. finger grease) has been removed from the outer glass wall by washing with methanol.

10 Adjust the focus and position of the image of the scattered light (which can be viewed by the CCD camera when the laser beam is attenuated with neutral density filters) on the spectrometer entrance slit by a relative movement of the sample and collection lens.

11 Peak up on a Raman band by monitoring it in real time and finely adjusting the on-axis position of focusing and collection lenses.

12 Collect the spectrum of the protein in buffer solution. A laser power of ca.1.0 W is suggested as a starting point, when using exciting light of 752.5 nm wavelength

(krypton ion laser), although the optimum will be found by trial and error. The spectral acquisition time should be sufficient to give a signal to noise ratio of at least 300:1 (6).

13 Repeat steps 10–12 for the protein–ligand complex.[b]

[a] The alignment of the spectrometer optics is normally checked with a back alignment laser (e.g. HeNe or diode laser) which defines the spectrometer axis. No off-axis movement of the image should occur on defocusing the lenses. The exciting beam axis (e.g. for a 90° exciting/scattering geometry) can also be defined during this procedure; it will intersect the spectrometer axis at the focus of the collection lens.

[b] Identical conditions should be used for the protein–ligand complex in buffer as for the protein sample in buffer. When the ligand is a strong scatterer sophisticated difference techniques may not be necessary (13). Instead, in order to maintain the same laser focus and identical collection geometry, the solution cuvette can be kept in the same position (care being taken not to disturb it) while the buffered solution of the protein is swapped for a buffered solution of the protein–ligand complex *in situ*.

Having collected a spectrum, re-run a part of it to make sure there is no sample degradation. In the event of any signs of degradation, such as a decrease in signal with time or bands shifting and broadening with time (which might be the case if the protein denatures), repeat the experiment with a lower laser power. Then perform a frequency calibration of the bands by, say, collecting a standard emission spectrum (e.g. of a neon lamp in the red region). The wavenumber scale does not have an exact linear dependence on the angular displacement of the grating(s) (for a dispersive instrument), which are driven by, e.g. cosecant or sine bar mechanisms. Therefore it is essential to use a nonlinear frequency calibration algorithm (e.g. fitting a cubic function) across the whole wavenumber range.

2.2.1 Trouble-shooting

If there is no signal but a high background due to fluorescence, extra purification steps may be required in order to suppress the fluorescence (when it originates from impurities). Where solids are concerned, e.g. protein powders, the fluorescence signal can sometimes reduce if the laser beam is exposed to the sample for a prolonged period of time. If the fluorescence persists, a different form of Raman spectroscopy may be required, such as one or more of the techniques outlined in the following sections.

The features due to protein–ligand interactions are obtained from the subtraction of (spectrum of the protein–ligand complex in buffer) − x.(spectrum of the protein in buffer), where x is a scaling constant. If the concentration of protein in buffer is the same for both solutions, the scaling factor, x, can be taken to be equal to one (6). The result is a difference spectrum by virtue of this data manipulation. The experiment can be especially designed in order to attempt to optimize the quality of the difference spectrum (see Section 2.3). When the ligand is a strong scatterer and its modes dominate the Raman difference spectrum, the approach given in this section can be used. However, if the ligand is not a strong Raman scatterer, the more sophisticated difference technique of Section 2.3 may be required in order to minimize interference from protein bands.

The experiment described above was recently reported by Carey and co-workers (6) to give a spectrum of aqueous lysozyme (using a concentration of 2.9 mg/ml, i.e. 200 μM, 100 mW of 752 nm laser excitation, and 1 minute accumulation time), which is comparable to that reported in 1970 by Lord and Yu (14) using a concentration of 300 mg/ml, 30 mW of 632.8 nm laser excitation, and several hours accumulation time. In spite of the fact that the laser power in the former case was higher (100 mW of 752 nm light rather than 30 mW of 632.8 nm light), the reduction of concentration by ca. 2 orders of magnitude, along with the reduction in acquisition time by more than two orders of magnitude is illustrative of the enormous improvement in performance of today's technology. In view of the potential and capabilities now offered by the technique, it is perhaps not surprising that Raman spectroscopy has become more popular in recent years.

2.2 FT Raman spectroscopy

The FT Raman spectrometer, like the FT infrared spectrometer, is based on an interferometer (see *Figure 3*) (15). In fact, some commercial instruments are joint FT Raman/FT infrared spectrometers. The exciting light is provided by the fundamental of a neodymium: YAG (yttrium aluminium garnet) laser which comes in the near infrared region at a wavelength of 1064 nm. The scattered light (Rayleigh and Raman) is divided into two by a beam-splitter in the interferometer; half of the light travels a fixed distance, whereas the path length of the other half is varied by reflecting it off a moving mirror. Thus the path difference (or optical retardation) of the divided beams varies with time and, when the beams are recombined at the beamsplitter, the light waves of the two beams interfere either constructively or destructively, depending on their wavelengths. Beyond the interferometer, a notch filter is positioned in the beam path. This blocks the intense Rayleigh scattered light and transmits the Raman scattered light which is detected by a germanium diode or InGaAs diode detector.

There are three fundamental advantages of FT spectroscopy over conventional, dispersive spectroscopy. First, all of the wavelengths are measured over the whole spectral acquisition period all of the time (multiplex or Fellgeth advantage). Secondly, there is no need for a slit in the FT experiment, thus giving a greater throughput (throughput or Jacquinot advantage). Thirdly, there is good wavenumber accuracy and the spectral resolution is constant over the whole spectrum. The last can be determined accurately by measuring the precise path difference of the two optical paths in the interferometer (from the number of fringes of a HeNe reference laser). The major practical advantage of FT Raman

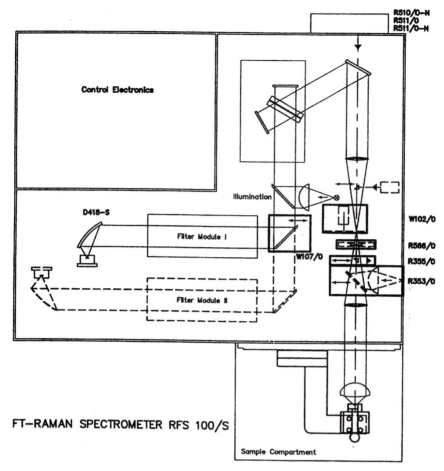

R510/0-N
R511/0
R511/0-N

Control Electronics

Illumination

D418-S

Filter Module I

W102/0

W107/0

R566/0

R355/0

R353/0

Filter Module II

FT—RAMAN SPECTROMETER RFS 100/S

Sample Compartment

Figure 3 Diagram of the lay-out of a Bruker RFS 100/S FT Raman spectrometer.

spectroscopy over dispersive visible Raman spectroscopy, particularly for protein samples, is that fluorescence is frequently much less of a problem.

There are disadvantages of FT Raman spectroscopy over dispersive Raman spectroscopy, however, when examining proteins or protein–ligand complexes. In order to compensate for the loss in scattering efficiency, due to ν^4 the dependence of scattering cross-sections, the laser power in an FT Raman experiment is usually high and this can 'cook' protein samples. Another drawback is that FT Raman spectroscopy is not the technique of choice for solutions of proteins in H_2O, because the Nd: YAG laser fundamental at 1064 nm is absorbed by overtones and combinations of the O–H stretching vibrations of H_2O.

As with Raman spectroscopy using visible excitation, the spectrometer can be configured with a microscope so that FT Raman microscopy can be performed, although the limitation in the etendue imposed by the microscope compromises the throughput (Fellgeth) advantage in FT Raman microscopy. However, the technique opens up the possibility of investigating protein–ligand interactions at the cellular level *in vivo*, e.g. the interaction of a specific retinoid binding protein (retinoic acid receptor, RAR) with retinoids and carotenoids (16).

There can be resonances of the 1064 nm exciting light with low lying electronic transitions. In these instances FT resonance Raman spectroscopy may be carried out, e.g. to probe the primary donor hydrogen bonding interactions in reaction centres of *Rhodobacter sphaeroides* and some of its mutants (17–19). A drawback of FT resonance Raman spectroscopy, at the moment, is that commercial instruments use 1064 nm excitation exclusively so that there is no possibility of obtaining resonance excitation profiles (REP's, see Section 2.4). This could change in the future if FT Raman instruments become modified in order to take advantage of the tunable near infrared laser emission of the titanium sapphire laser. This approach has been demonstrated by Schulte *et al.* (20) who obtained the FT Raman spectrum of retinal using 843.2 nm excitation provided by a Ti: sapphire laser.

2.3 Difference Raman spectroscopy

Raman difference spectroscopy can be used on most protein systems. The approach has been used extensively by Callender and co-workers (21–24) who have studied numerous protein–ligand complexes. In addition, spectra of membrane-bound proteins may also be obtained (25). As previously mentioned, the technique involves subtracting a Raman spectrum of a protein from that of the protein–ligand complex. The difference spectrum exhibits bands not only due to the ligand but also bands which are sensitive to the binding interaction and conformationally sensitive bands of the protein. Thus Raman spectroscopy can represent a powerful means of assessing the strength and origins of the interactions between a protein and its ligand, e.g. an enzyme and its substrate.

The key to the difference technique is having identical experimental conditions, such as the laser power, laser focus, collection geometry, etc., when collecting the spectra of the protein and the protein–ligand complex. Kieffer (26) designed a split rotating cell for difference spectroscopy with a gated electronic amplifier incorporated into the detection system of the experiment. Callender and co-workers (21, 22, 27) have used a specially fabricated split cell (see *Figure 4*), each half being loaded with ca. 30 μl of sample. These solutions are maintained at 4 °C using a bath/ recirculator. The cuvette holder is mounted on a dual translator/ stepping stage which can shuttle each half of the cuvette in and out of the laser beam with a positioning accuracy of within ± 1 μm. This is apparently adequate to ensure signal and wavelength repeatability (21).

The protocol prior to the acquisition of spectra will be similar to *Protocol 1*, steps 1–10. A protocol for the acquisition of spectra is given overleaf.

The features due to protein–ligand interactions are obtained from the subtraction of (spectrum A) − x.(spectrum B), where x is a scaling constant. The concentration of protein in buffer may not be exactly the same for both solutions and there may be small mismatches in the sample alignment, in spite of the precautions which are taken. Consequently, the scaling factor, x, may not be

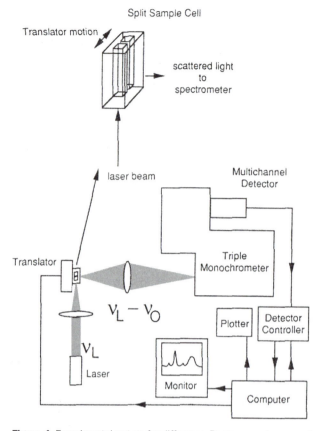

Figure 4 Experimental set-up for difference Raman spectroscopy (reprinted from ref. 21).

exactly equal to one (21). So for obtaining accurate difference spectra, in which background peaks of the protein subtract out, x is determined by trial subtractions.

Protocol 2

Acquisition of spectra

Equipment and reagents

• See *Protocol 1*

Method

1. Load the protein solution into one half of the split cell cuvette.

2. Load the protein–ligand solution into the other half of the split cell cuvette.

3. The spectral region of the Spex Triplemate instrument (shown in *Figure 4*) will be limited if the gratings are not to be repositioned during the experiment.[a] Therefore the spectrograph needs to be centred on the spectral range of interest.

4 Collect the scattered light (spectrum A) from one side of the split cell for 10–60 min so as to give a signal to noise ratio of at least 300:1 and preferably closer to 1000:1.

5 Translate the cell and collect the scattered light (spectrum B) from the second side of the cell for an equivalent period of time as in step 4.

[a] The wavenumber range of the spectral window will depend on the physical width of the multi-channel detector, the wavelength of the excitation, and the groove density of the grating in the spectrograph. It is likely that a better wavelength repeatability can be achieved if the gratings are not moved during the experiment. This is because it is difficult to reproducibly reposition gratings with precision, on account of backlash in the gears of grating drive mechanisms.

A procedure has been developed for checking systematic errors, which may adversely affect an A − B difference spectrum (28, 29). This involves collecting a series of interleaved spectra, ABAB...AB, then adding alternate A spectra and subtracting them from the sum of the rest of the A spectra. The same manipulation is done for the B spectra. These A −A and B −B differences should give featureless spectra with noise levels equivalent to the simple shot noise of individual A and B spectra.

2.4 Resonance Raman spectroscopy

When the wavelength of the exciting radiation is chosen to coincide with an electronic absorption of the sample, there can be an intensification (up to ca. 10^6) of some of the Raman bands due to the resonance effect. These bands are associated with the chromophoric group to which the resonant electronic transition belongs. This selectivity can be very useful when studying proteins and protein–ligand complexes, because the spectrum is simplified, being dominated by these resonantly enhanced bands. For example, it has been used to probe pigment–protein interactions, such as hydrogen bonding in photosynthetic proteins (30, 31).

It should be noted that there is no resonance counterpart in infrared spectroscopy so infrared spectra of proteins, like normal (off-resonance) Raman spectra, can exhibit a very large number of bands.

The resonance Raman effect also provides a means of identifying electronic absorptions of chromophores. The intensity of a Raman band is given by:

$$I_{mn} = \frac{2^3 \pi}{3^3 c^4} I_o (\nu_o - \nu_{mn})^4 \sum_{\rho\sigma} (\alpha_{\rho\sigma})^2_{mn} \qquad [10]$$

where I_o and ν_o are the intensity and frequency of the exciting light, ν_{mn} is the vibrational frequency, ρ and σ are the x, y, and z components of the polarizability tensor, and the $(\alpha_{\rho\sigma})_{mn}$ are the scattering tensor components which are given by the KHD (Kramers-Heisenberg-Dirac) equation:

$$(\alpha_{\rho\sigma})_{mn} = \frac{1}{h} \sum_e \left[\frac{(M_\rho)_{me} (M_\sigma)_{en}}{\nu_e - \nu_o + i\Gamma_e} + \frac{(M_\sigma)_{me} (M_\rho)_{en}}{\nu_e + \nu_o + i\Gamma_e} \right] \qquad [11]$$

where the summation is over all excited states, e, ν_e is the frequency of an electronic transition from the ground state to an excited state, ν_o is the exciting frequency, $(M_\rho)_{me}$ is the transition moment from the ground state to an excited state along the direction ρ, $(M_\sigma)_{en}$ is the transition moment from the excited state to the final state along the direction σ, and Γ_e is a damping constant.

In normal (off-resonance) Raman spectroscopy, the scattering tensors are independent of the exciting frequency as $\nu_o \ll \nu_e$ for all excited states. However, when the exciting light is in resonance with one of the electronic transitions, $\nu_e - \nu_o$ becomes very small. Then, the first term on the right-hand side of the KHD equation can become very large giving selective enhancement of certain Raman bands.

The total wavefunction can be expressed as the product of the electronic and vibrational wavefunctions, if the Born–Oppenheimer approximation is assumed. The result is that the scattering tensor components are approximately the sum of two terms, A and B (often referred to as the Albrecht A and B terms), as given by (32, 33),

$$(\alpha_{\rho\sigma}) \approx A + B \qquad [12]$$

The A term is given by:

$$A \cong M_e^2 \frac{1}{h} \sum_\nu \frac{\langle j | v \rangle \langle v | i \rangle}{\nu_{vi} - \nu_o + i\Gamma_v} \qquad [13]$$

where M_e is the electronic transition moment connecting the ground state to the resonant excited state, e, and $\langle j | v \rangle$ and $\langle v | i \rangle$ are Franck–Condon overlap integrals, ν_{vi} is the transition frequency from the ground vibrational level, i, to the excited vibrational level, v, ν_o is the frequency of the laser beam, and Γ_v is the band width of vibrational level, v.

A term resonance Raman scattering (also called Franck–Condon resonance Raman scattering) requires the electronic transition moment, M_e, to be large. Therefore this type of resonance Raman scattering is observed when there is a resonance with a fully allowed electronic transition (i.e. the molar extinction coefficient, ε, for the electronic transition $\approx 10^4$ M^{-1}cm^{-1} or more). It is only the bands of totally symmetric vibrational modes which are enhanced in A term resonance Raman spectra, and the spectra often exhibit progressions in overtones and combination tones, which are due to non-zero Franck–Condon overlap integrals in *Equation 13*. Furthermore, bands which are most enhanced tend to be those of modes involving motions along vibrational co-ordinates which connect the ground and excited state geometries (Tsuboi's rule) (34). Consequently, although the wavenumber positions of the A term resonance Raman bands are characteristic of a molecule in its ground electronic state, their intensities contain information about the resonant excited state of the molecule. Resonance enhancement profiles (REP's) of bands can be obtained; these are plots showing how the resonance enhancement of a band varies with the exciting laser wavelength.

The B term in *Equation 12* is given by:

$$B \cong M_e M_e' \frac{1}{h} \sum_\nu \frac{\langle j | Q | v \rangle \langle v | i \rangle + \langle j | v \rangle \langle v | Q | i \rangle}{\nu_{vi} - \nu_o + i\Gamma_v} \qquad [14]$$

where Q is the normal co-ordinate of a certain vibration:

$$M'_e = \frac{\mu_s \langle s | \frac{\partial H}{\partial Q} | e \rangle}{\nu_s - \nu_e} \qquad [15]$$

ν_e is the transition frequency of the first excited state, ν_s is the transition frequency of the second excited state, μ_s is the transition dipole moment of the second excited state, and H is the electronic Hamiltonian.

B term (or Herzberg–Teller) resonance Raman enhancement involves two excited states which are coupled by vibrational modes of appropriate symmetry. This mixes the two excited states providing 'additional' intensity to the weaker one, thereby making possible a strong resonance Raman enhancement. Unlike the A term resonance Raman spectra, bands arising from non-totally symmetric modes can be enhanced by this B term mechanism and the spectra do not exhibit overtone progressions.

Self-absorption of the scattered light is a serious problem in the resonance Raman experiment, and there is an optimum solution concentration which minimizes self-absorption (35):

$$A(\text{optimum}) = (2kr)^{-1} \qquad [16]$$

where A(optimum) is the optimum absorbance, r(cm) is the path length of the scattered light in solution, and k is log(e) = 2.303.

Sample degradation can also be a problem due to absorption of the exciting light, so it is common to use rotating solution cells, or recirculating solutions. Care must be taken to prevent wobble of rotating solution cells, because this can lead to movement of the image of the collected light on the entrance aperture of the spectrometer, which produces noise in the spectrum. One design uses a rotating NMR tube in a wobble-free spinner which consists of a compressed air-driven turbine and a collet that centres the tube (36, 37).

Another method of minimizing the possibility of sample degradation by the laser beam is to study the protein sample at low temperature as a frozen solution. The protein solution can be loaded into a miniature cell which is connected to the cryotip of a refrigerator (38). The frozen sample is maintained at a sub-ambient temperature, which may be as low as 4 K if a commercial three stage closed cycle helium refrigerator (e.g. an Air Products Displex) is used. This method is particularly useful for precious protein samples because only one or two drops are required. Another reason for cooling protein–ligand complexes is that it may enable some information to be obtained on their dynamics. Carey's group have cooled solutions containing enzyme–substrate intermediates in order study the dynamics of key groups in the active sites of some dithioacyl papains (13, 39, 40).

Whichever sampling method is used in the resonance Raman experiment, it is always advisable to check that the band intensities have not diminished during spectral acquisition by repeating the measurement of some bands without changing any of the experimental parameters.

The choice of the wavelength of the exciting radiation is determined by the wavelength region in which the chromophoric group absorbs. Therefore the experimental set-up needs to offer the possibility of changing the wavelength of the exciting light over a wide range. Consequently, for resonance Raman spectroscopy, double grating or triple grating spectrometers are often used rather than single stage spectrographs with notch filter/ interference filter combinations, which are only optimized for one wavelength. Double monochromators (e.g. the Spex 1403 spectrometer) with PMT detectors can be used for such studies. Alternatively, triple monochromators with multichannel detectors, such as CCD or intensified photodiode array (IPDA) detectors, may be used. These have the first two monochromators configured in double subtractive mode. In this configuration, a spectral band-pass is selected by passing the dispersed radiation through a wide intermediate slit, positioned between the first and second monochromators. The dispersion effects of the first and second monochromators

Table 2 Emission lines of the most commonly available discrete-wavelength lasers (reproduced from ref. 41)

λ/ nm	Laser	λ/ nm	Laser	λ/ nm	Laser
157	Fluorine	454.5	Argon	543.5	Helium–neon
173.6	Ruby × 4	457.7	Krypton	568.2	Krypton
193	Argon fluoride	457.9	Argon	578.2	Copper
222	Krypton chloride	461.9	Krypton	595.6	Xenon
231.4	Ruby × 3	463.4	Krypton	628	Gold
248	Krypton fluoride	465.8	Argon	632.8	Helium–neon
266	Nd:YAG × 4	468.0	Krypton	647.1	Krypton
308	Xenon chloride	472.7	Argon	657.0	Krypton
325	Helium–cadmium	476.2	Krypton	676.5	Krypton
333.6	Argon	476.5	Argon	687.1	Krypton
337.1	Nitrogen	476.6	Krypton	694.3	Ruby
347.2	Ruby × 2	482.5	Krypton	722	Lead
350.7	Krypton	484.7	Krypton	752.5	Krypton
351	Xenon fluoride	488.0	Argon	799.3	Krypton
351.1	Argon	495.6	Xenon	904	Gallium arsenide
353	Xenon fluoride	496.5	Argon	1047	Nd:YLF
355	Nd:YAG × 3	501.7	Argon	1060	Nd:glass
356.4	Krypton	510.5	Copper	1064	Nd:YAG
363.8	Argon	514.5	Argon	1092.3	Argon
406.7	Krypton	520.8	Krypton	1152.3	Helium–neon
413.1	Krypton	528.7	Argon	1315	Iodine
415.4	Krypton	530.9	Krypton	1319	Nd:YAG
428	Nitrogen	532	Nd:YAG × 2	2940	Er:YAG
437.1	Argon	534	Manganese	3391	Helium–neon
441.6	Helium–cadmium	539.5	Xenon	3508	Helium–xenon

cancel (hence 'double subtractive'), but the combination of the two mono-chromators acts as a filter, both of the Rayleigh line and of stray light, which precedes the spectrograph. The Spex Triplemate 1877 and Dilor XY instruments, for example, are designed so that they may operate in this way.

The wavelengths of exciting laser light, which may be employed in an experiment, will depend on the laser source(s) available in the laboratory. The emission lines of the most commonly available discrete-wavelength lasers are listed in *Table 2* (although it should be noted that some of the ultra-violet lines can only be obtained for the higher power lasers). Also it is necessary to change the front (output coupler) and rear mirrors when changing wavelength regions, as the end mirrors give optimum performance over a specific range. The laser manual will undoubtedly outline this procedure, but it is a useful tip to maintain laser action if possible by changing one mirror at a time.

Dye lasers are very useful for resonance Raman spectroscopy, because they can be tuned to provide laser light of any wavelength across the visible spectrum. Each dye only operates over a certain wavelength range, however. A powerful laser is needed to pump the dye laser, and the pump beam must have a shorter wavelength than the desired wavelength of the dye laser light. Dye lasers have the disadvantage that the dyes have limited stability (the rhodamine dyes have a better stability than most), and consequently the dye laser power can markedly diminish over a period of a few days. If laser light in the 650–1100 nm region is needed, an alternative tunable laser is the titanium sapphire laser. This also requires a powerful pump laser, particularly towards the ends of the afore-mentioned tunability range, but it has the advantage of better output power stability.

Resonance Raman spectroscopy has been used extensively to investigate the binding of exogenous ligands such as O_2 and CO to haem proteins (42, 43). These proteins contain the haem group, iron protoporphyrin IX (see *Figure 5*) as the chromophoric centre, which is well suited for resonance Raman spectroscopic study due to the strong π–π^* absorptions of the porphyrin ring. The visible absorption spectra of haem proteins are characterized by a strong band (the Soret or B band) in the blue region and a weaker band (the Q_o or α band) to

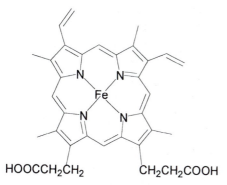

HOOCCH$_2$CH$_2$ CH$_2$CH$_2$COOH

Figure 5 Structure of iron protoporphyrin IX.

longer wavelength, in addition to a vibronic side band (the Q_1 or β band) which comes to higher energy of the Q_o band. Franck–Condon (or Albrecht A term) resonance Raman spectra result from excitation within the Soret band and these show bands due to totally symmetric vibrations (A_{1g} for a porphyrin core of D_{4h} symmetry). On the other hand, excitation within the Q_o and Q_1 bands gives rise to Herzberg Teller (or Albrecht B term) resonance Raman spectra and the spectra show bands due to non-totally symmetric modes (B_{1g}, B_{2g}, and A_{2g} vibrations in D_{4h} symmetry) (44). These resonance Raman spectra of the haem protein–ligand complexes are particularly informative about the spin, oxidation, and ligation states of the iron atom since reliable marker bands are known, which originate from vibrations of the porphyrin core (see Table 3). The identification of these bands in the resonance Raman spectra is aided by a measurement of their depolarization ratios. These measurements are performed by incorporating a polarization analyser into the experimental set-up; it is inserted into the path of the scattered light in between the collection optics and the polarization scrambler (which is placed before the entrance slit of the spectrometer). (Note: The polarization scrambler is required because Raman spectrometers usually have different throughput efficiencies for vertically and horizontally polarized light.) Two spectra are obtained under identical conditions, one with the polarization analyser transmitting light polarized in a direction parallel ($I_{\parallel}$) to the polarization of the incident light and the other with the transmitted light polarized perpendicular to the incident polarization ($I_{\perp}$). The depolarization ratio (ρ) of a band is then given by the ratio of $I_{\perp}$ and $I_{\parallel}$ according to:

$$\rho = \frac{I_{\perp}}{I_{\parallel}} \qquad [17]$$

The porphyrin bands due to totally symmetric A_{1g} vibrations (in D_{4h} symmetry) should be polarized, having ρ values $< 3/4$, whereas bands due to B_{1g} and B_{2g} vibrations should be depolarized, having ρ values of $3/4$. Bands which are due to A_{2g} vibrations are expected to exhibit inverse polarization with ρ values $\sim \infty$ (44).

Table 3 Wavenumber locations (cm^{-1}) of bands which are sensitive to oxidation and spin state for FePPIX haems (taken from ref. 45)

Spin state	Oxidation state	Molecule	Oxidation state markers		Spin state markers		Oxidation and spin state markers	
			A(p)	B(dp)	C(ap)	D(p)	E(p)	F(dp)
ls	Fe(III)	Ferricyt c	1374	1562	1582	1582	1502	1636
ls	Fe(III)	CNMHb	1374	1564	1588	1583	1508	1642
hs	Fe(III)	FMHb	1373	1565	1555	1565	1482	1608
hs	Fe(II)	Deoxy Hb	1358	1546	1552	1565	1473	1607
ls	Fe(II)	Ferrocyt c	1362	1548	1584	1594	1493	1620
ls	?	Oxy Hb	1377	1564	1586	1582	1506	1640

2.5 Ultra-violet Raman and ultra-violet resonance Raman (UVRR) spectroscopy

Ultra-violet resonance Raman spectroscopy permits a drastic reduction in sample concentrations (they can go into the 10 μmol region) due to the intensification of the signals resulting from resonance enhancement. However the total amount of sample which is required is nearly the same as for off resonance Raman spectroscopy, as sample volumes of 1–50 cm^3 are needed if the solution is recirculated or else rotating sample cells are necessary in order to avoid sample decomposition. In some cases, neither of these sampling arrangements is suitable if photo-aggregation and precipitation occurs. In these instances a wire-guided jet sampling system has been successfully used (46, 47).

In the UV resonance Raman experiment the determination of the cross-sections (σ) of the vibrations of an analyte is often important since they provide important structural information (see later). When using *Equation 7* in order to determine σ values, a knowledge of the intensity of the incident laser beam is required as well as the solid angle of collection of the scattered light, and the interrogation volume of the laser. The last two quantities are not easily measured, but fortunately it is not necessary to make absolute measurements of scattering cross-sections. Instead, an internal intensity standard can be used for which the Raman cross-section ($\sigma_s(\nu_o)$), excited at a frequency, ν_o, is already known. Then the Raman cross-section of the analyte ($\sigma_a(\nu_o)$), excited at ν_o, can be calculated relative to that of the standard from:

$$\sigma_a(\nu_o) = \frac{I_a(\nu_o)C_s}{I_s(\nu_o)C_a} \times \sigma_s(\nu_o) \qquad [18]$$

where $I_a(\nu_o)$ and C_a are the Raman intensity and concentration of the analyte respectively, and $I_s(\nu_o)$ and C_s are the Raman intensity and concentration of the standard.

Raman cross-sections values may be found in the literature quoted as integrated values (over the full solid angle, Ω, of 4π steradians), σ, or as differential values $(d\sigma/d\Omega)_{90°}$ evaluated using a 90° scattering geometry. These values can be related via:

$$\left(\frac{d\sigma}{d\Omega}\right)_{90°} = \frac{3}{8\pi}\left(\frac{1 + \rho}{1 + 2\rho}\right)\sigma \qquad [19]$$

where ρ is the depolarization ratio measured using a 90° scattering geometry (48). Internal intensity standards are frequently used in Raman spectroscopy. Their use enables quantitative measurements to be made, somewhat analogous to taking a reference spectrum in absorption spectroscopy. However, care must be taken to correct the relative intensities of bands of an analyte band and an internal standard (I_a/I_s) for the throughput of the spectrometer and the response of the detector, both of which are wavelength dependent. Internal standards should be chosen judiciously for proteins and protein–ligand complexes in aqueous solution. The following criteria should be borne in mind.

Criteria for choosing an internal standard:

(a) Useful at low concentrations.

(b) Does not interact with the protein.

(c) Is not photochemically active.

(d) Does not absorb the excitation or scattered light.

(e) Has few Raman bands to minimize interference with the analyte.

Song and Asher (49) have reported the Raman cross-sections for the 605, 634 cm^{-1} doublet (due to symmetric and anti-symmetric As–C stretching respectively) of cacodylic acid and the 834 cm^{-1} band (due to Se–O symmetric stretching) of SeO_4^{2-} between the wavelengths of 218 and 514.5 nm, and both appear to satisfy the requirements for internal intensity standards given above. In particular, both were found to have absorption cross-sections for UV excitation which are ca. 5-fold greater than those of ClO_4^- and SO_4^{2-}, which have been commonly used in the past as intensity standards. Consequently, cacodylic acid and SeO_4^{2-} can be used at lower concentrations. Furthermore ClO_4^- and SO_4^{2-} were both found to perturb the tertiary structures of aquomethaemoglobin (met-Hb) and its fluoride (met-HbF) and azide (met-HbN$_3$) complexes, and reduce the fluoride ligand affinity. They also hindered the quaternary R → T transition, induced by inositol hexaphosphate, even at concentrations as low as 0.1 M. Conversely, cacodylic acid and SeO_4^{2-} at concentrations of 0.1 M appeared to have little effect on the haemoglobin tertiary or quaternary structures, or indeed upon the R → T transition, induced by inositol hexaphosphate. Cacodylic acid is a particularly useful internal standard, when working with proteins and protein–ligand complexes, because it buffers solutions at physiological pH values.

The band due to the librational mode of water at ca. 500 cm^{-1} can also be a useful internal standard in UVRR spectroscopic studies of aqueous protein solutions (50).

2.6 Raman optical activity (ROA)

ROA is a far more incisive probe of biomolecular conformation than conventional vibrational spectroscopy. It is in fact a form of Raman difference spectroscopy, involving the subtraction of a spectrum originating from the use of left circularly polarized light from that using right circularly polarized light. There is more than one form of ROA since, besides the different possible scattering geometries, four different forms of modulation may be employed. These are, namely, incident circular polarization (ICP), scattered circular polarization (SCP), in-phase dual circular polarization (DCP$_I$), and out-of-phase dual circular polarization (DCP$_{II}$), and the interested reader is referred to ref. 51 (and references therein) for a description of each. Up until 1988, ICP ROA was practised exclusively; it involves the measurement of Raman intensities scattered when using right and left circularly polarized incident light. An electro-optic modulator, switches the incident laser light between left and right circular polarization states alternately at a frequency of typically less than 1 Hz. Hecht and Barron (52) have shown that the back scattering ICP ROA configuration gives an ROA signal-to-noise ratio which is $2\sqrt{2}$ greater than that of polarized 90° scattering, *inter alia pares*. Furthermore a detailed analysis of all possible ROA experiments (53), together with practical considerations, has shown that the backscattered CID (circular intensity difference) measurement is the ultimate strategy, and is necessary for biochemical ROA studies (54). ROA is due to the interaction of electric dipole–magnetic dipole and electric dipole–electric quadrupole transition moments. The ROA signals are weak because they are typically about 10^{-3} to 10^{-4} of the parent Raman scattering intensities and these parent Raman signals are themselves only ca. 10^{-6} times as strong as the Rayleigh scattered light. Thus, in order to observe ROA signals of proteins, high concentrations (~ 150 mg/ml) are required along with a high laser power (hundreds of milli-watts), a high-throughput spectrometer, a sensitive detector, and long collection times (~ 10 hours).

The high concentrations of samples required to obtain ROA spectra from aqueous solutions represent a drawback of ROA. RROA might afford higher sensitivity but it has not yet been observed, even though magnetic ROA was observed (of cytochrome *c*) in the early days of ROA measurements (55). Nafie (56) has reported some theory of RROA which suggests that it is a remarkably simple form of ROA in the single electronic state limit; the RROA bands are all predicted to have the same sign which is opposite that of the pure electronic circular dichroism of the single, resonant electronic transition. On the other hand, non-linear techniques such as Raman gain/loss spectroscopy may well provide the means of measuring optical activity spectra of dilute solutions in the future (57).

It does appear possible to monitor small conformational changes which take place on the binding of a ligand to a protein, for example during complex formation in enzyme catalysis, by means of ROA. A report on the lysozyme protein bound to the trimer of *N*-acetylglucosamine (NAG$_3$) serves as an example (58). A positive ROA band at ca. 1340 cm^{-1} showed an approximately two-fold increase in intensity in the protein bound complex relative to the free protein. This band was attributed to the loop structure (possibly 3_{10}-helix), known to be present from X-ray diffraction studies, and it was suggested that the intensity change points to a transformation between mobile and rigid loop structures upon ligand binding (58, 59).

Although it is clear from the foregoing that ROA spectra of proteins and protein–ligand complexes can be informative about secondary structure, an understanding of the origin of a number of ROA features is by no means complete at the present time. Also, ROA spectroscopy is currently somewhat of an esoteric field, as it is performed by only a handful of groups world-wide. Nevertheless, it is an important technique for stereochemical studies of biomolecules in aqueous solution; one reason is that it has an inherent advantage over VCD in that the ratio of the interaction difference for left and right circularly polarized light to the average interaction (which has been termed the chirality number, q) is between one and two orders of magnitude larger for ROA than for VCD on account of the $1/\lambda$ factor (60).

2.7 Time-resolved resonance Raman (TR^3) spectroscopy

X-ray crystallography and NMR spectroscopy are powerful techniques for studying static protein–ligand complexes which can give much useful information for the interpretation of Raman and resonance Raman spectra of such samples. However, resonance Raman spectroscopy is at the forefront of those techniques which can be applied to the study of the structure and dynamics of transient species. Such information is important because it is needed in order to understand biological function. Kitagawa and Ogura (61) have written a very useful review on the TR^3 spectroscopy of haem proteins. Another article by Kincaid (62) provides a very informative account on TR^3 spectroscopic studies of the structure and dynamics of transient species.

When carrying out TR^3 experiments, a well defined initial point in time is required for the exposure of a ligand to a protein. For a limited number of proteins, this can be achieved by photolysis of a pre-existing protein–ligand complex. For example, the carbon monoxide adduct of haemoglobin, HbCO, can be photolysed to generate a transient form of haemoglobin, Hb*. This may recombine with the CO or undergo bimolecular reactions with other ligands such as O_2, as shown below:

$$HbCO + h\nu \rightarrow Hb^* + CO$$

$$Hb + O_2 \rightarrow HbO_2$$

Hb* is a transient species which has been shown by TR^3 spectroscopic studies, when generated on a 7 nsec time scale, to be different in structure to deoxyhaemoglobin, Hb (63). Slight downward shifts in the core size marker bands on going from deoxyhaemoglobin at equilibrium, Hb, to Hb* were interpreted to indicate that Hb* has a slightly expanded core relative to that of Hb. This would suggest that the out of plane displacement of the iron atom in the 7 nsec photoproduct, Hb*, is less than in Hb. Such an interpretation was supported by the observed wavenumber locations of the iron–histidine stretching frequencies at different time delays of the probe pulses relative to those of the pump (64).

Rapid mixing TR^3 spectroscopic studies have been conducted in order to investigate the reaction of Hb with O_2 (62). In such experiments, the initial point in time is defined by the mixing of the reactants and, clearly, the more rapid the mixing, the better the time resolution. Various devices, which have been designed for rapid mixing RR spectroscopy, are discussed in the literature (65–68). The time resolution of rapid mixing TR^3 spectroscopic experiments is much lower than that of experiments where the reactants are generated by a short photolysis pulse. This is because there is a time delay before the reactants are homogeneously mixed, known as the dead time. Typical dead times are in the millisecond range, and a protocol has been devised for the measurement of this critical parameter for any particular experimental arrangement (66).

Rapid mixing TR^3 spectroscopic experiments, in conjunction with flash photolysis, have been used to study the cytochrome oxidase/oxygen reaction at room temperature (69). Information on the initial oxygen adduct is difficult to obtain by other methods due to the rapid rates of O_2 binding ($\sim 10^8$ M^{-1} s^{-1}) and electron transfer reactions ($\sim 10^5$ to 10^3 s^{-1}) reactions. The CO cytochrome oxidase complex was prepared and photolysed with 532 nm light in order to generate initially the reduced deoxy enzyme. A photolabile intermediate species, which is present after 40 μsec, was shown to have an oxidation state marker band (v_4) at 1378 cm^{-1} and a spin state marker band (v_2) at 1588 cm^{-1} (69). This species was identified as being due to adduct formation of dioxygen with cytochrome a_3.

Following the binding of the dioxygen to the fully reduced enzyme, multistep proton and electron transfer reactions ensue as the enzyme functions to reduce the dioxygen molecule to water. TR^3 spectroscopic studies have provided detailed information on a number of the key intermediates in this reaction sequence (70–76). The interpretation of these spectra has been aided greatly by the use of the $^{18}O_2$ isotopomer in addition to dioxygen containing the oxygen atom isotopes in their natural abundance. In particular, the difference spectra can be expected to exhibit features associated only with the bound dioxygen molecule and its reduction products, since all other features should cancel. Furthermore isotopic shifts and multiplet patterns provide very useful information on the structure and bonding of moieties containing the oxygen atoms.

2.8 Surface-enhanced Raman spectroscopy (SERS) and surface-enhanced resonance Raman spectroscopy (SERRS)

Surface-enhanced Raman spectra are obtained when the analyte is in the presence of a roughened metal surface (silver and gold are among the metals which are known to give strong surface enhancements). Both electromagnetic and chemical (or charge transfer) mechanisms contribute to SERS enhancement (77), and enhancements of four orders of magnitude can result. If there is resonance Raman enhancement as well as surface enhancement (i.e. SERRS), bands can intensify enormously by up to ten orders of magnitude, since the two effects are roughly multiplicative. The metal may take the form of a colloidal suspension where the metal particles have diameters of the order of tens of nanometres, or it may be an electrode surface which is roughened electrochemically. There is a big caveat when studying proteins using SERS, namely that the metal surface will at best perturb the protein secondary, tertiary, and quaternary structure and at worst denature the protein. Clearly, it is undesirable for the metal surface to be directly in contact with the protein, so chemical spacers can be used. Even so, it is advisable to check that the structure has not been altered by the colloid by means of other techniques, e.g. CD spectroscopy.

Silver sols for SERS may be made by reducing aqueous silver nitrate solutions with a chemical reducing agent, e.g. sodium borohydride or sodium citrate, the latter being preferred for studies of proteins because the citrate is complexed to the silver and provides a protective layer between the metal and the protein. A protocol for making a silver sol by sodium citrate reduction of silver nitrate, according to the method of Lee and Meisel (78), is given below (79).

Protocol 3

Preparation of a silver sol by sodium citrate reduction of silver nitrate[a]

Equipment and reagents
- UV/vis spectrometer
- Glassware
- Silver nitrate
- 1% (w/v) sodium citrate

Method

1 Clean all glassware thoroughly before use. Final cleaning is with aqua regia followed by rinsing in distilled water.

2 Dissolve 90 mg of silver nitrate in 500 cm^3 of distilled water and bring to the boil.

3 Add 10 cm^3 of 1% (w/v) sodium citrate in distilled water.

4 Stir the solution vigorously and continuously while maintaining it at its boiling point for approx. 1 h.

5 Take a UV/vis absorption spectrum of the colloidal suspension. The wavelength of maximum absorption, λ_{max}, should be approx. 406 nm.[b]

[a] Gold colloids are an alternative, and these have the advantage over silver colloids of being biochemically less active.

[b] In addition, the full width at half height (FWHH) of the absorption band of the silver colloid, which is a measure of the distribution of particle size, should be less than 60 nm. It was Mie (80), who first recognized that, for gold colloids, the λ_{max} shifts to longer wavelengths with increasing particle size for spherical particles of radii approximately $> \lambda/20$. This dependence can be exploited by adding an aggregating agent, which is a solution of an electrolyte such as potassium chloride or sodium perchlorate, to the aqueous silver hydrosol in order to shift the λ_{max} so as to maximize the SERS signals. This is not always necessary, because some proteins aggregate the colloid themselves (79). As well as the increased detection sensitivity of SERS, the discrimination against fluorescence, in some cases, is another significant advantage; the reason being that energy transfer from the protein to the metal is a competing deactivation pathway.

Beljebbar *et al.* (16) have used aqueous silver hydrosols in order to perform FT-SERS on complexes of a specific retinoid binding protein, RAR-γ, with dimethylcrocetin, DMCRT, (and all-*trans*-retinoic acid) in different ratios. For the 1:1 complex at a concentration of 1×10^{-5} M, an interaction was manifested by an increase in wavenumber of about 4 cm^{-1} for the DMCRT band at 1210 cm^{-1}, as well as a decrease of the I_{1541}/I_{1210} intensity ratio (2.5 for free DMCRT and 1.9 for the DMCRT-RAR-γ complex).

2.9 Coherent anti-Stokes Raman spectroscopy (CARS)

CARS is a four wave mixing experiment for which the state diagram is shown in *Figure 6*. Laser light of two frequencies, one fixed and one tunable, is incident on the sample. The tunable frequency is scanned in the range above the fixed frequency, and a spectrum can be obtained by plotting the intensity of the CARS emission against the difference in frequency of the two incident beams. Care must be taken in order to satisfy the phase matching (or momentum matching) condition in the CARS experiment. This is difficult for liquid samples when scanning a wavenumber range of many reciprocal centimetres, due to the variation with wavelength of both the wave vector of the electromagnetic field of the Stokes beam and the refractive index of the medium. The detector and collection optics need to be rotated synchronously about the point of intersection of the pump and Stokes laser beams as the Stokes frequency is scanned, in order to fulfil the phase matching condition (81).

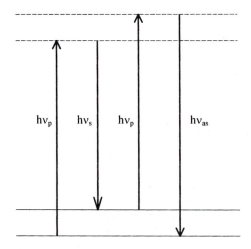

Figure 6 Energy level diagram for the CARS process. The subscripts on the photon energies indicate: p, pump; s, Stokes; as, anti-Stokes.

Although the intensity of the coherent resonant emission is high (ca. four or five orders of magnitude greater than that for spontaneous Raman scattering), there is a high non-resonant background which limits the achievable signal to noise ratio of the vibrational bands in the experiment. Chikishev *et al.* (82) have used polarization sensitive CARS (PSCARS) as a means of discriminating against this strong, coherent, non-resonant background. A very high degree of polarization of the pump and Stokes laser beams is required in such experiments for good non-resonant background suppression. When this is done, however, the PSCARS experiment can enable changes in secondary structure to be characterized upon the binding of a ligand to a protein. This was demonstrated for the binding of anthranilic acid to chymotrypsin (82).

3 Interpretation of protein spectra and band assignments

3.1 Amide group frequencies

A protein consists of amino acids which link through amide groups (see *Figure 7*).

Figure 7 The amide or peptide linkage.

These amide groups have characteristic vibrational frequencies which typically come in particular ranges (see *Table 4*). The amide I, II, and III bands have been used much more than the others for structural interpretations, although some of the others (particularly the amide A band) may be informative. The amide I band is due to the C=O stretching vibration, and it is very strong in the infrared spectrum and moderately strong in the Raman spectrum. The amide II band is primarily due to a mixture of in plane N-H bending and C-N stretching vibrations, and it is strong in the infrared spectrum but weak in the Raman spectrum. However, in contrast to the normal (off-resonance) Raman spectrum, the UV resonance Raman spectrum usually exhibits a strong amide II band when there is a resonance with the amide $\pi-\pi^*$ electronic absorption in the 200 nm region. The amide III band is also due to a mixture of in-plane N-H bending and C-N stretching vibrations, but these co-ordinates have opposite phase in this normal mode compared to that giving rise to the amide II band, and it typically has weak infrared and medium Raman intensities. In order to identify the amide III band unambiguously it is necessary to convert the labile N-H groups into N-D's by exchanging with D_2O. The rate of exchange can vary between groups, because it depends on their accessibility to aqueous solvent so this exchange may need to be monitored as a function of time. Any amide III bands will disappear from the 1200–1350 cm^{-1} region, shifting down to the ca. 950 cm^{-1} region (amide III' region). These bands exhibit a large H/D shift on

Table 4 Wavenumber ranges and vibrational assignments for the amide vibrations (taken from ref. 81)

Name	Wavenumber/ cm^{-1}	Approximate description
Amide A	3250–3300	N–H stretch
Amide I	1630–1700	C=O stretch (C–N, C–N–H)
Amide II	1510–1570	N–H deformation (C–N)
Amide III	1230–1330	N–H/C–H deformation
Amide IV	630–750	O=C–N deformation
Amide V	700–750	N–H out-of-plane deformation
Amide VI	~ 600	C=O out-of-plane deformation

account of the appreciable percentage character of N–H in-plane bending in this mode.

The precise positions of the amide I and III bands are sensitive to the secondary structure of a protein, and correlations exist in the literature between frequencies of amide I and III bands and secondary structural motifs (see *Table 5*). These have been obtained by comparing the positions of the bands for a large number of proteins, which have had their secondary structures characterized by means of X-ray diffraction. In normal (off-resonance) Raman spectra it is difficult to assign features to specific groups of atoms. Thus, it is often hard to interpret the spectra beyond assigning percentages of secondary structural motifs from amide I and III band intensities and profiles. This is where resonance enhancement and difference techniques are advantageous, because the spectra can become simplified to a level which permits a fuller interpretation.

Table 5 Wavenumbers (cm^{-1}) of Raman active backbone and side chain vibrations of proteins (reproduced from ref. 83)

Origin and wavenumber location (cm^{-1})	Assignment	Structural information	Refs.
Backbone			
Skeletal acoustic			
25–30	Mode of large portion of protein; possibly intersubunit mode or side chain torsion	Overall structure; subunit interactions	84–86
75	Torsion mode	α-Helix	86
Skeletal optical			
935–945 (950 D_2O)	C–C stretch (or C^α–C–C stretch)	α-Helix	87–91
900	C–C stretch	These broaden and lose intensity	87
963		with denaturation	
1002 (H_2O)	C^α–C' or C^α–C^β stretch	Suggestive of β-pleated sheet	87
1012 (D_2O)	–	–	
1100–1110	C–N stretch	Conformation change marker broadens and loses intensity with denaturation	90
Amide I			
	Amide C=O stretch	Strong band; hydrogen bonding lowers amide I wavenumbers	87,92,93
1655 ± 5	–	α-Helix (H_2O)	87, 92
1632 (D_2O)	–	α-Helix (D_2O)	87
1670 ± 3	–	Antiparallel β-pleated sheet	87,92,93
1661 ± 3 (D_2O)	–	α-Helix (D_2O)	87, 92
1655 ± 3	–	Disordered structure (solvated)	87, 91
1658 ± 2 (D_2O)	–	–	87, 88
Amide III			
	C–N stretch, N–H in plane bend	Strong hydrogen bonding raises amide III wavenumbers	87, 90

Table 5 *Continued*

Origin and wavenumber location (cm^{-1})	Assignment	Structural information	Refs.
Backbone *continued*			
> 1275	Amide III weak	α-Helix, no structure below 1275 cm^{-1}	87, 90
1235 ± 5 (sharp)	Amide III strong	Antiparallel β-pleated sheet	87, 90
983 ± 3 (D$_2$O)	Charged coil?	Disordered structure	87, 89
1245 ± 4 (broad)	Charged coil?	Disordered structure	87, 89
Amino acid chains			
Tyrosine doublet			
850/830	Fermi resonance between ring fundamental and overtone	State of tyrosine -OH I$_{850}$/I$_{830}$ = 9:10 to 10:3, H bond from acidic proton donor; 10:9 to 3:10, strong H bond to negative proton acceptor	94
Tryptophan			
880/1361	Indole ring	Ring environment; sharp intense line for buried residue; intensity diminished on exposure or environmental change	95, 96
Phenylalanine			
1006	Ring breathing	Conformation-insensitive frequency/intensity reference	97
624	Ring breathing	Ratio with tyrosine 664 to estimate Phe/Tyr	97
Histidine			
1409 (D$_2$O)	*N*-deuteroimidazole	Possible probe of ionization state; metalloprotein structure; proton transfer	97
-S–S-			
510	S–S stretch	Gauche-gauche-gauche; broadening and/or shifts may indicate conformational heterogeneity among disulfides	98–100
525	S–S stretch	Gauche-gauche-trans	98–100
540	S–S stretch	Trans-gauche-trans	98–100
C–S			
630–670	C–S stretch	Gauche	98–100
700–745	C-S stretch	Trans	98–100
S–H			
2560–2580	S–H stretch	Environment, deuteration rate	93, 99
Carboxylic acids			
1415	-COO$^-$ sym stretch	State of ionization	97
1730	-COOH, C=O stretch -COOR C=O stretch	Metal complexation	97

3.2 Group frequencies of side groups

Distinct features in the Raman spectrum which can give very useful information are given in *Table 5*. The notation systems used for the vibrational modes of phenylalanine, tyrosine, and tryptophan and the atomic displacement vectors of some of their vibrational modes are given in an article by Harada and Takeuchi (101).

Tryptophan has an intense band at ca. 1550 cm^{-1} due to an indole ring mode (W3 mode). An empirical relationship has been found between the frequency of W3 and $\chi^{2,1}$, the dihedral angle connecting the indole ring and the Cβ atom of the tryptophan side chain (102).

$$\nu(\text{W3}) = 1542 + 6.7 \times \{\cos(3 \times |\chi^{2,1}|) + 1\}^2 \qquad [20]$$

This relationship has been used in order to identify bands due to specific Trp amino acids in proteins (103, 104).

4 Specific information on proteins and their interactions with ligands

4.1 Secondary structure

A protein changes its conformation upon ligand binding (105). One may assume that a conformational change does not just involve the active site, which is a small part of the protein molecule, but that it involves the whole protein molecule. Thus, it is of considerable interest to determine the secondary structure of a protein, both when free and bound to a ligand.

The amide I region generally consists of a number of inhomogeneously broadened, overlapping bands, due to a mixture of different secondary structures. One method of obtaining the percentages of pure secondary structures is to perform a curve fitting analysis (see *Figure 8*). Commercially available software packages which provide the means for doing this include Grams 32 (supplied by Galactic Industries), Microcal Origin, and Opus/IR. There are five main variables which are as follows: the number of bands, the band positions, the band widths, the band shapes, and the baseline. Of these, the band shapes can be fitted to Gaussian, Lorentzian, or a linear combination of Gaussian and Lorentzian functions, and the baseline can be corrected by subtracting out a spectrum of the pure solvent and any other background signals (e.g. bands due to amino acid side chains). The number of component bands used should be the minimum number required to fit the composite band shape. This can be found from the derivative spectra; the spectra can be differentiated a number of times (106), in order to obtain the number of bands and their positions. These last two parameters can then be constrained and the band widths varied in the fitting routine. The best-fit bands may then be assigned to secondary structure types based on the literature, although in certain cases they are not completely unequivocal. The percentages of secondary structure types can be determined from the integrated intensities of these bands. This procedure has been carried

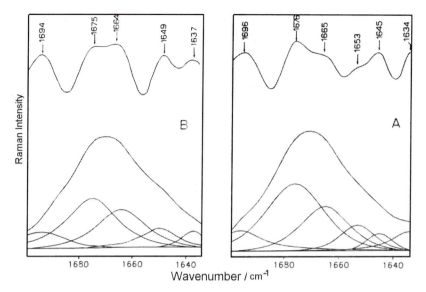

Figure 8 Curve fitting analysis of the amide I Raman band (lower field) and fourth derivative spectra (upper field) of (a) avidin and (b) the avidin–biotin complex (reproduced from ref. 106).

out in order to study the binding of biotin and its analogues by avidin and streptavidin (106, 107).

Another method of obtaining the constituent bands is to perform a Fourier self-deconvolution, for which a protocol is given in Chapter 6 on Infrared Spectroscopy in this volume.

Yet another method of obtaining percentages of secondary structure from amide I bands, which has been used successfully by a number of groups, has been devised by Williams (108–112). This method, which estimates the percentage content of total helix, β-strand, turn, ordered α-helix, disordered helix, and undefined involves analysing the amide I band as a linear combination of the spectra of reference samples for which the structures are known. This relies on the supposition that, for globular proteins, the structures are nearly identical in the solution and solid states. The Raman spectrum of each reference protein is represented as normalized intensity measurements at p different wavenumbers, and for n proteins:

$$\mathbf{A}\,\mathbf{x} = \mathbf{b}$$

where $\mathbf{A}$ is a p by n matrix, $\mathbf{b}$ is a p-dimensional vector that contains the normalized spectrum of the protein being analysed, and $\mathbf{x}$ is an n-dimensional vector that maps the intensities of the n reference proteins at a given wavenumber into the intensity at the same wavenumber of the protein being analysed.

The fraction, f, of each type of secondary structure is given as a weighted sum of the structures of the reference proteins:

$$f = \mathbf{F}\,\mathbf{x}$$

where $\mathbf{F}$ is an $m \times n$ matrix containing m classes of secondary structure and n reference proteins. This type of analysis is extremely valuable because it permits the measurement of the change in secondary structure of, for example, a DNA-binding protein on binding to DNA (111, 112).

Berjot *et al.* (113) have also published a method for the estimation of protein secondary structures from Raman spectra. Their method (the reference intensity profiles method) involves the use of reference spectra, corresponding to α-helix, β-sheet, and undefined structures, which were computed by an iterative fit of the amide I spectra of 17 proteins with known secondary structures.

Recently, an article has appeared on a holistic approach to protein structure characterization using Raman amide I bands (114). This technique combines the superposition of reference spectra for pure secondary structure elements with simultaneous subtraction of bands due to aromatic side chains and background signals due to fluorescence and the solvent. It was demonstrated that this method can be used to estimate secondary structures of proteins in solution, suspension, and dry solid forms. Furthermore allowance was made for different scattering efficiencies for different secondary structures, and more than one peak was allowed per structure (114).

UVRR spectroscopy is a powerful technique for determining the secondary structure of a protein, since it provides electronic as well as vibrational information. Chi *et al.* (115) used an exciting wavelength of 206.5 nm (from the frequency doubling of the 413 nm line of a krypton ion laser), in order to achieve resonance with the amide π–π^* transitions. They reported a quantitative methodology, based upon the observation of the amide I, II, and III bands and the band due to the C_α–H bending vibration, for obtaining the percentage α-helix, β-sheet, and unordered secondary structures. Factor analysis (116) was used to determine that these three structures were the minimum number required to model the spectra. The pure secondary structure Raman spectra (PSSRS) for α-helix, β-sheet, and unordered secondary structures were determined using a least-squares deviation minimization method for a series of 13 proteins with well-known X-ray structures. Then these PSSRS were used to calculate the secondary structures of unknown proteins.

The quantitation of secondary structure by means of UVRR was found to compare very well with that of electronic CD, since essentially identical correlation coefficients were found for α-helical structural predictions but superior correlation coefficients were found for β-sheet and unordered conformations (115). The underlying reason for the inferior structural prediction of α-helix, compared to β-sheet and unordered conformations, in UVRR spectroscopy is due to its smaller contribution to the spectra on account of the hypochromism of the α-helix electronic absorption band. Furthermore the sensitivity of UVRR spectroscopy may well be superior to that of electronic CD spectroscopy, since the UVRR spectra exhibit richer details due to the different secondary structures than do the CD spectra. Chi *et al.* (115) have used typical protein concentrations of ca. 0.5 mg/ml for UVRR spectroscopy, but lower concentrations (by at least an order of magnitude) could apparently be used for precious samples. An additional advantage of UVRR over electronic CD spectroscopy is that side chains and, in the case of glycoproteins, carbohydrate components do not interfere with the secondary structure measurements.

4.2 Solvent exposure of aromatic side chains

The Raman scattering cross-sections of tryptophan, tyrosine, and phenylalanine are enhanced in the ultra-violet region, with their REP's showing the same trends as their electronic absorption spectra. Tryptophan residues have their $B_{a,b}$ transition at 218 nm ($\varepsilon = 34\,000$ M^{-1} cm^{-1}), tyrosine residues (below their pK_a of 10.1) have an L_a transition at 222 nm ($\varepsilon = 7000$ M^{-1} cm^{-1}), and phenyl-alanine residues have an L_a transition at 207 nm ($\varepsilon = 7000$ M^{-1} cm^{-1}) (117). A term (Franck–Condon) UVRR spectra are obtained from tryptophan, when in resonance with the allowed $B_{a,b}$ transition, whereas B term (Herzberg–Teller) UVRR spectra are obtained from tyrosine and phenylalanine, when in resonance with their symmetry-forbidden L_a transitions. The enhancement factors for selected bands of these aromatic amino acids are given in Table 6 for excitation wavelengths of 200, 218, and 240 nm (118). Histidine has a broad absorption with a maximum at ca. 207 nm and a relatively low molar extinction coefficient of 5000 $M^{-1}cm^{-1}$ (119). Unfortunately, the resonance enhancement of its Raman bands via the π–π^* transitions, which have apparently been resolved at 204 and 218 nm from the REP (119), is weak compared to that of tryptophan, tyrosine, and phenylalanine (as can be seen from *Table 6*). Excitation of proteins in the 185–200 nm region would give further enhancement of Tyr, Phe, and His modes, but there would be interference from enhanced bands due to amide modes of the polypeptide backbone.

Information from UVRR spectra about solvent exposure of the Trp and Tyr amino acids is useful because it can be correlated with secondary structure at the Trp and Tyr attachment sites (104). Unfortunately, the Fermi resonance doublet of Tyr in the 830–850 cm^{-1} region of normal (off-resonance) Raman spectra, which is useful as an indicator of H bonding interactions involving the Tyr -OH group (see *Table 5*), is not observed in UVRR spectra. Instead, the Raman cross-sections of the Trp and Tyr amino acids have been shown to be indicators of their local exposure to water (104). The absorptions at ca. 220 nm of both Trp and Tyr shift to higher energy on increased exposure to polar solvents, such as water, due to stabilization of their ground electronic states. When the UVRR spectra are obtained with 229 nm excitation, the effect of increased exposure of

Table 6 Molar enhancement factors ($\times$ 10^{-2}) for selected Raman bands of the aromatic amino acids at excitation wavelengths of 200, 218, and 240 nm (taken from refs. 118 and 119)

	Tryptophan	Tyrosine	Phenylalanine	Histidine
Selected band (cm^{-1})	1016	1617	1000	1575
200 nm	29	202	36	8
218 nm	506	126	96	32
240 nm	53	30	8	4

Trp and Tyr to polar solvents is to shift their absorption maxima (and REP maxima) away from the exciting wavelength (104). Hence the magnitudes of their Raman cross-sections decrease on increased exposure to polar solvents such as water.

The advantage of using 229 nm excitation in order to probe Trp and Tyr residues is that their bands are selectively enhanced with minimal interference from the bands of the amide backbone. In the case of Tyr, however, stronger enhancement is given with 200 nm excitation (as can be seen from Table 6), since this is close to resonance with its allowed $B_{a,b}$ transition. Bands of Trp which show enhancement are located at 759, 857, 879, 1012, 1339, 1359, 1453, 1516, and 1557 cm^{-1}, whereas enhanced Tyr bands occur at 1179, 1207, and 1613 cm^{-1} (104).

4.3 pK$_D$'s of protein–ligand complexes

In a rare instance, it was possible to estimate a pK_D value for a protein–ligand complex from the information on the protein secondary structure obtained by means of Raman spectroscopy (106). The percentages of α-helix and β-sheet secondary structural content of the avidin protein in complexes with biotin and a number of its analogues showed a linear correlation with known pK_D values. Using this correlation and the percentages of α-helix and β-sheet secondary structural content of avidin complexed with biotin methyl ester, a value of 14.7 was calculated for the pK_D of this complex (106).

4.4 pK$_a$'s of specific protein groups

Ligand-binding properties of proteins often show pH profiles. The pK_a's of specific protein carboxyl groups can be measured from Raman titration curves using difference techniques and compared with the pK_a of the pH profile for ligand-binding. A close correspondence of the pK_a value of a specific protein group at or near a binding site with that of ligand-binding can suggest that the ionization state of the group is a key determinant in the interaction. A specific amino acid is tagged by site-directed mutagenesis and the differences between the tagged and native proteins are compared as a function of pH. The pK_a is obtained from a fit of the titration curve to the Henderson–Hasselbach equation (see *Figure 9*) (103):

$$\frac{A_{diff}}{A_{std}} = \frac{C}{1 + K_A/[H^+]}$$

where A_{diff} is the relative amplitude between a positive band (due to the wild-type protein) and its negative counterpart (due to the mutant) in the difference spectrum, A_{std} is the amplitude of a protein band, used as an internal standard, in the wild-type protein spectrum, C is the maximum amplitude, and K_A is the acid dissociation constant. The pK_a value is obtained from a best fit to the Raman titration curve using non-linear least squares techniques (103, 120).

Tonge and Carey (40) measured the spectroscopic pK_a values based on the

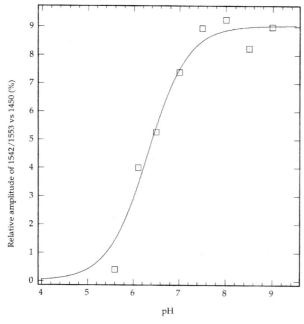

Figure 9 Raman titration curve for pK$_a$ determination. The value found for the pK$_a$ was 6.3 (± 0.1) (taken from ref. 103).

acyl C=O stretching wavenumbers of some acyl–enzyme complexes. For three substrates, these were found to be equivalent to the pK$_a$ values measured from the deacylation kinetics, suggesting that the carbonyl feature reflects activation of the deacylation mechanism and can be used to probe that mechanism (see next section) (121).

4.5 Entropic and enthalpic changes on ligand binding

When flexible molecules, e.g. nucleotides, bind to proteins, their bands may be observed to be sharper in the difference spectra compared to those in the solution spectrum of the free (unbound) ligand. This qualitative difference has been interpreted as being due to a conformationally more constrained molecule (24), and this suggests that its entropy has been reduced.

Although high resolution X-ray diffraction studies can locate all of the atoms in a protein–ligand complex, the precision is often only ca. 0.1–0.3 Å. In comparison, vibrational techniques can provide a much higher precision of ca. ± 0.004 Å using empirical correlations of band wavenumber locations with bond lengths. Furthermore, for a substrate with a C=O group which hydrogen bonds to the enzyme in the enzyme–substrate complex, the location of the Raman band due to the C=O stretching mode can be used in order to estimate the hydrogen bond strength (see *Protocol 4*). In order to obtain quantitative results, calibrating studies need to be carried out for models of the substrate molecule of interest. In the example below these calibrating studies are considered for analogues of the acyl methyl ester which forms a complex with either chymotrypsin or subtilisin (13).

Note that, although the enthalpy of formation of a hydrogen bond, C=O···H, will be related, by the correlation, to the shift of the Raman C=O stretching band, it is FT infrared spectroscopy which is used here for the calibrating studies. This is because the O–H and C=O stretching modes, which are monitored, give strong infrared bands which can be measured with good wavenumber and intensity accuracy. A non-aqueous solvent, e.g. CCl_4, is required which has poor H bond acceptor and donor properties, in order to minimize solvent inter-actions. Note, also, that the O–H stretching region, which is of interest here, would be masked by strong water bands in an aqueous environment.

Protocol 4

Estimation of hydrogen bond strength

Equipment and reagents

- Spectrometer
- H bond donor molecule
- H bond acceptor molecule

Method

1 Make up a mixture of known concentration of H bond donor molecule, e.g. ethanol, in CCl_4 solution, and take the infrared spectrum. The solution should be sufficient-ly dilute that only monomeric H bond donor molecule is present in the solution. This can be gauged from the O–H stretching region; a single band should be present due to the H bond donor molecule.

2 Make up a mixture of H bond donor molecule (e.g. ethanol) and acyl methyl ester (i.e. hydrogen bond acceptor) in CCl_4. The concentration of H bond donor (e.g. ethanol) should be the same as in step 1. Take the infrared spectrum at the same temperature as in step 1.

3 Repeat steps 1 and 2 for more mixtures of different H bond acceptor concentra-tions.

4 From the intensity of the O–H band determine the concentrations of free donor and acceptor and the H-bonded complex.

5 Calculate K from a least squares fit to the equation,

$$K = \frac{[A - H \cdots B]}{[AH][B]}$$

6 Repeat steps 1–5 at a number of different temperatures over the range, say, of 0–50 °C using a heated or cooled solution cell (commercially available from Specac in the UK). This will give values for the equilibrium constant, K, at different tem-peratures.

7   Calculate the enthalpy of formation of the hydrogen bond, ΔH, between the acyl
    methyl ester and the ethanol H bond donor from the van't Hoff equation:

$$\log_e K = -\frac{\Delta H}{RT} + \frac{\Delta S}{R} \qquad [21]$$

This is done by taking the data from step 6 and performing a least squares fit of
$\log_e K$ versus $1/T$ in order to obtain the slope $(= -\Delta H/R)$. Also, measure the wavenumber shift, $\tilde{v}$, for the C=O stretching band when the free acyl methyl ester hydrogen bonds to ethanol. Thus, for a given H bond donor, e.g. ethanol, ΔH and $\tilde{v}$ have
been determined.

8   Repeat steps 1–8 in order to obtain ΔH and $\tilde{v}$ for other H bond donors, e.g. phenol
    and 3,5-dichlorophenol.

9   Obtain a calibration graph by plotting ΔH versus $\tilde{v}$.

Clearly, a value for the entropy change, ΔS, of hydrogen bond formation can also
be obtained from the intercept of the $\log_e K$ versus $1/T$ plot of step 7 in *Protocol 4*.

4.6 Binding rate constants

TR³ spectroscopy is at the forefront of techniques used to study dynamic properties of proteins. When a ligand binds to a protein it can regulate the activity of
the protein via a so-called allosteric effect. This activity regulation is achieved by
means of a conformational change of the protein, which can be followed by
means of TR³ spectroscopy (61).

The Fe–CO stretching wavenumber of the CO adduct of myoglobin (Mb) is
known to be sensitive to the geometry of CO in the haem pocket. It adopts a
bent geometry in the closed structure due to steric hindrance with the distal
histidine (His 64), but an upright geometry in the open structure where it has
been proposed that the His 64 swings out of the haem pocket when its imidazole ring has become protonated. This conformational change opens up a
pathway for the ligand to approach its binding site.

The Fe–CO stretching wavenumbers of neutral and acidic MbCO are 507 and
488 cm⁻¹, respectively, and these had been considered to represent the closed
and open forms.

The TR³ spectra discussed below were reported by Kitagawa's group (61) who
obtained them from MbCO under an atmosphere of CO. The following discussion summarizes their elegant work (see ref. 61 and references therein). Two
sequential 10 nsec laser pulses, having wavelengths of 532 and 416 nm, represented the pump and probe beams, respectively. Photodissociation of MbCO
occurred with high quantum yield on exposure of the sample to the pump
beam, and the recombination of Mb with CO was subsequently monitored with
the probe at various pump-probe time delays. As can be seen from *Figure 10*, at
pH 8.0 the band at 511 cm⁻¹, due to the closed form, disappears for a pump-probe

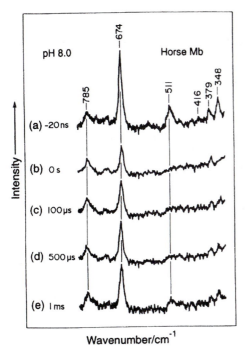

Figure 10 TR^3 spectra in the region of the Fe–CO stretching vibration of horse MbCO at pH 8.0 under 1 atm of CO. Δt_d is given on the left of each spectrum. The wavelengths of the pump and probe laser beams were 532 and 416 nm, respectively. Reproduced from ref. 61.

time delay (Δt_d) of zero nanoseconds but starts to recover for $\Delta t_d = 500$ μsec. *Figure 11* shows the spectra from a similar experiment at pH 4.5, which corresponds to the mid-point pH between the acidic and neutral forms. A band at 492 cm^{-1} grows in at earlier times than a band at 507 cm^{-1}, which is consistent with the former being due to an open form which binds the CO ligand at a faster rate than the closed form (represented by the band at 507 cm^{-1}).

Mutant MbCO's were also examined in order to investigate this possible mechanism. It was found that the wild-type (H64H) and mutants in which the distal His was replaced by Gly (H64G) or Gln (H64Q) gave a band due to the Fe–CO stretching vibration in the 505–510 cm^{-1} region, whereas mutants in which the distal His was replaced by Ala (H64A), Ile (H64I), Val (H64V), and Leu (H64L) gave the band in the 490–495 cm^{-1} region (61).

Figures 12 and *13* show the TR^3 spectra of H64H and H64L, in which the Fe–CO stretching bands grow in at 510 and 491 cm^{-1}, respectively. Plots of the integrated intensities versus Δt_d are shown in *Figure 14* for H64H, H64G, and H64Q. The intensities were normalized against the intensity at $\Delta t_d = -20$ μsec and fitted to a biexponential curve (61):

$$[\text{MbCO}] = A_1[1 - \exp(-k_1 t)] + A_2[1 - \exp(-k_2 t)] \qquad [22]$$

The binding rate constants, k_1 and k_2, and the amplitudes, A_1 and A_2, were derived from these biexponential fits for the recombination of CO both with the

303

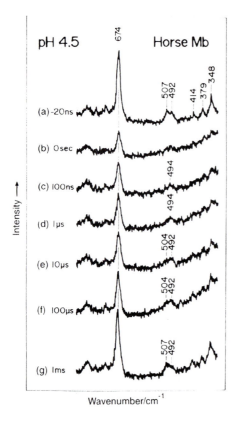

Figure 11 TR^3 spectra in the region of the Fe–CO stretching vibration of horse MbCO at pH 4.5 under 1 atm of CO. Δt_d is given on the left of each spectrum. The wavelengths of the pump and probe laser beams were 532 and 416 nm, respectively. Reproduced from ref. 61.

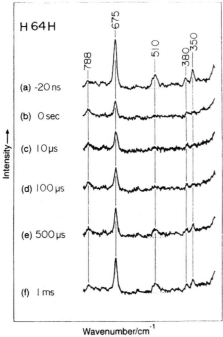

Figure 12 TR^3 spectra in the region of the Fe–CO stretching vibration of the wild-type human MbCO. Δt_d is given on the left of each spectrum (to $\pm$ 2 nsec). Broken lines indicate the assumed base lines. The pump and probe beams had wavelengths of 532 and 416 nm and pulse energies of 4.4 mJ and 100–125 μJ, respectively, and the accumulation time was 4000 sec for each spectrum. Reproduced from ref. 61.

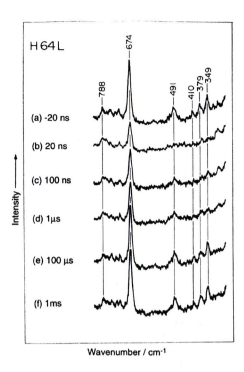

Figure 13 TR^3 spectra in the region of the Fe–CO stretching vibration of the E7-Leu mutant human MbCO. Experimental conditions are the same as those for *Figure 12*. Reproduced from ref. 61.

Table 7 The static wavenumbers and the rebinding rate constants and amplitudes of photodissociated human MbCO and its distal histidine mutants (reproduced from ref. 61)

	v_{Fe-CO}/ cm^{-1}	k_1/ s^{-1}	k_2/ s^{-1}	A_1	A_2
H64H	510	2.0×10^4	6.9×10^2	0.31	0.69
H64G	505	2.5×10^4	3.2×10^3	0.33	0.67
H64Q	508	1.0×10^5	1.1×10^3	0.29	0.71
H64L	490	5.5×10^6	2.5×10^4	0.83	0.17
H64V	493	2.5×10^6	3.1×10^3	0.50	0.50
H64I	492	2.0×10^5	1.2×10^4	0.17	0.83
H64A	495/510	3.0×10^5	3.2×10^3	0.23	0.77

wild-type Mb and its mutants (61). The values derived for these parameters are given in *Table 7*.

As can be seen from Table 7, the rate constant for the fast phase, k_1, is larger when the Fe–CO stretching band occurs in the 490–495 cm^{-1} region than when it appears in the 505–510 cm^{-1} range. This would be consistent with the ca. 490 cm^{-1} and ca. 510 cm^{-1} bands belonging to open and closed forms of MbCO, respectively. However, if the band at ca. 490 cm^{-1} is due to an open form which is a precursor to the closed form with a band at ca. 510 cm^{-1}, the spectra of H64H, H64G, and H64Q which show the band at ca. 510 cm^{-1} should have shown the ca. 490 cm^{-1} band at short Δt_d values (100 nsec to ca. 10 μsec). It was reasoned that the absence of this band in the spectra of H64H, H64G, and H64Q suggests

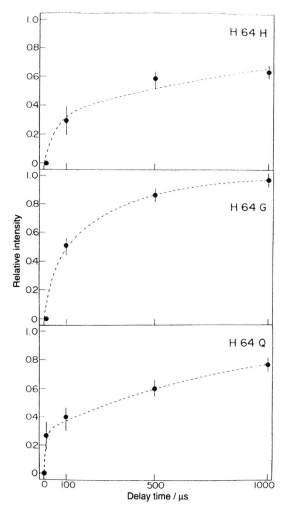

Figure 14 Recovery of the CO-bound form for H64H, H64G, and H64Q. Experimental points represent the relative intensities at each Δt_d with regard to that for $\Delta t_d = -20$ nsec. The broken lines denote the values calculated using *Equation 22* and the parameters listed in Table 7. Reproduced from ref. 61.

that CO entry goes by way of protein rearrangements which are distinct from an open structure intermediate (i.e. the swinging out of the His 64 side chain) (61).

Acknowledgement

We are grateful to the EPSRC for an instrumentation fund (ref: GR/L85176).

References

1. Raman, C. V. and Krishnan, K. S. (1928). *Nature*, **121**, 50.

2. (a) Delhaye, M. and Dhamelincourt, P. (1975). *J. Raman Spectrosc.*, **3**, 37; (b) Rosasco, G. J., Etz, E. S., and Cassutt, W. A. (1975). *Appl. Spectrosc.*, **29**, 396.

3. Wilson, E. B., Jr., Decius, J. C., and Cross, P. C. (1995). *Molecular vibrations*. McGraw-Hill, NY.

4. Born, M. and Oppenheimer, J. R. (1927). *Ann. Physik.*, **84**, 457.

5. Strommen, D. P. and Nakamoto, K. (1984). *Laboratory Raman spectroscopy*. Wiley, USA.

6. Dong, J., Dinakarpandian, D., and Carey, P. R. (1998). *Appl. Spectrosc.*, **52**, 1117.

7. Lerner, J. M. and Thevenon, A. (1998). *The optics of spectroscopy*. Instruments SA, Inc.

8. Treado, P. J. and Morris, M. D. (1993). In *Microscopic and spectroscopic imaging of the chemical state* (ed. M. D. Morris), Chapter 3. Marcel Dekker, NY.

9. Fryling, M., Frank, C. J., and McCreery, R. L. (1993). *Appl. Spectrosc.*, **47**, 1965.

10. Williamson, J. M., Bowling, R. J., and McCreery, R. L. (1989). *Appl. Spectrosc.*, **43**, 372.

11. Loader, J. (1970). *Basic laser Raman spectroscopy*. Heyden/Sadtler, London.

12. For an example, see Austin, J. C., Wharton, C. W., and Hester, R. E. (1989). *Biochemistry*, **28**, 1533.

13. Carey, P. R. (1998). *J. Raman Spectrosc.*, **29**, 7.

14. Lord, R. and Yu, N. T. (1970). *J. Mol. Biol.*, **50**, 509.

15. (a) Hendra, P. J., Jones, C., and Warnes, G. (1991). *Fourier transform Raman spectroscopy*. Ellis Horwood, Chichester, England; (b) Chase, D. B. and Rabolt, J. F. (1994). In *Fourier transform Raman spectroscopy: from concept to experiment* (ed. D. B. Chase and J. F. Rabolt), Chapter 1. Academic Press, San Diego, CA, USA.

16. Beljebbar, A., Morjani, H., Angiboust, J. F., Sockalingum, G. D., Polissiou, M., and Manfait, M. (1997). *J. Raman Spectrosc.*, **28**, 159.

17. Mattioli, T. A., Hoffmann, A., Robert, B., Schrader, B., and Lutz, M. (1991). *Biochemistry*, **30**, 4648.

18. Mattioli, T. A., Hoffmann, A., Sockalingum, D. G., Schrader, B., Robert, B., and Lutz, M. (1993). *Spectrochim. Acta*, **49A**, 785.

19. Mattioli, T. A., Williams, J. C., Allen, J. P., and Robert, B. (1994). *Biochemistry*, **33**, 1636.

20. Schulte, A., Lenk, T. J., Hallmark, V. M., and Rabolt, J. F. (1991). *Appl. Spectrosc.*, **45**, 325.

21. Callender, R. and Deng, H. (1994). *Annu. Rev. Biophys. Biomol. Struct.*, **23**, 215.

22. Callender, R., Deng, H., and Gilmanshin, R. (1998). *J. Raman Spectrosc.*, **29**, 15.

23. Deng, H. and Callender, R. (1998). *J. Am. Chem. Soc.*, **120**, 7730.

24. Deng, H. and Callender, R. (1999). *J. Raman Spectrosc.*, **30**, 685.

25. Spiro, T. G. (ed.) (1987). *Biological applications of Raman spectroscopy*, **1**, 81.

26. Kieffer, W. (1973). *Appl. Spectrosc.*, **27**, 253.

27. Deng, H., Zheng, J., Clarke, A., Holbrook, J. J., Callender, R., and Burgner, J. W. (1994). *Biochemistry*, **33**, 2297.

28. Deng, H., Zheng, J., Sloan, D., Burgner, J., and Callender, R. (1989). *Biochemistry*, **28**, 1525.

29. Chen, D., Yue, K. T., Matin, C., Rhee, K. W., Sloan, D., and Callender, R. (1987). *Biochemistry*, **26**, 4776.

30. Lutz, M. (1984). *Advances in infrared and Raman spectroscopy* (ed. R. J. H. Clark and R. E. Hester), Vol. 11, p. 211. Wiley, Chichester, UK.

31. Lutz, M. and Robert, B. (1988). *Biological applications of Raman spectroscopy* (ed. T. G. Spiro), Vol. 3, p. 347. Wiley, New York.

32. Albrecht, A. C. (1961). *J. Chem. Phys.*, **34**, 1476.

33. Tang, J. and Albrecht, A. C. (1970). *Developments in the theories of vibrational Raman intensities* in *Raman spectroscopy* (ed. H. A. Szymanski), Vol. 2. Plenum Press, NY.

34. Hirakawa, A. Y. and Tsuboi, M. (1975). *Science*, **188**, 359.

35. Strekas, T. C., Adams, D. H., Packer, A., and Spiro, T. G. (1974). *Appl. Spectrosc.*, **28**, 324.

36. Walters, M. A. (1983). *Appl. Spectrosc.*, **37**, 299.

37. Eng, J. F., Czernuszewicz, R. S., and Spiro, T. G. (1985). *J. Raman Spectrosc.*, **16**, 432.

38. Czernuszewicz, R. S. (1986). *Appl. Spectrosc.*, **20**, 571.

39. Carey, P. R. and Tonge, P. J. (1990). *Chem. Soc. Rev.*, **19**, 293.

40. Tonge, P. J. and Carey, P. R. (1989). *Biochemistry*, **28**, 6701.

41. Andrews, D. L. (1990). *Lasers in chemistry*. Springer–Verlag, Heidelberg, Germany.

42. Yu, N. T. (1986). In *Methods in enzymology* (eds. C. H. W. Hirs and S. N. Timasheff), Academic Press, Vol. 130, p. 350.

43. Spiro, T. G. and Czernuszewicz, R. S. (1995). In *Methods in enzymology* (ed. K. Sauer), Academic Press, Vol. 246, p. 416.

44. Ferraro, J. R. and Nakamoto, K. (1994). *Introductory Raman spectroscopy*. Academic Press, London.

45. Spiro, T. G. and Strekas, T. C. (1974). *J. Am. Chem. Soc.*, **96**, 338.

46. Kaminaka, S. and Kitagawa, T. (1992). *J. Am. Chem. Soc.*, **114**, 3256.

47. Cho, N., Song, S., and Asher, S. A. (1994). *Biochemistry*, **33**, 5932.

48. McHale, J. (1999). In *Molecular spectroscopy*, p. 167. Prentice Hall, New Jersey, USA.

49. Song, S. and Asher, S. A. (1991). *Biochemistry*, **30**, 1199.

50. Chi, Z. and Asher, S. A. (1999). *Biochemistry*, **38**, 8196.

51. Nafie, L. A. (1996). *Appl. Spectrosc.*, **50**, 14A.

52. Barron, L. D., Hecht, L., Wen, Z. Q., Ford, S. J., and Bell, A. F. (1992). *SPIE laser study of macroscopic biosystems*, **1922**, 2.

53. Hecht, L. and Barron, L. D. (1990). *Appl. Spectrosc.*, **44**, 483.

54. Barron, L. D., Hecht, L., Bell, A. F., and Wilson, G. (1996). *Appl. Spectrosc.*, **50**, 619.

55. Barron, L. D. (1975). *Nature*, **257**, 372.

56. Nafie, L. A. (1996). *Chem. Phys.*, **205**, 309.

57. Hug, W. (1994). *Chimia*, **48**, 386.

58. Barron, L. D., Hecht, L., Bell, A. F., and Wilson, G. (1996). *J. Raman Spectrosc.*, **50**, 619.

59. Ford, S. J., Cooper, A., Hecht, L., Wilson, G., and Barron, L. (1995). *J. Chem. Soc. Faraday Trans.*, **91**, 2087.

60. Hug, W. (1982). In *Raman spectroscopy: linear and nonlinear* (ed. J. Lascombe and P. V. Huong), p. 3. Wiley, Chichester, England.

61. Kitagawa, T. and Ogura, T. (1993). In *Advances in spectroscopy* (ed. R. J. H. Clark and R. E. Hester), Vol. 21, Chapter 3. Wiley, Chichester, UK.

62. Kincaid, J. R. (1995). In *Methods in enzymology* (ed. K. Sauer), Vol. 246, Chapter 19. Academic Press, San Diego, CA, USA.

63. Dasgupta, S. and Spiro, T. G. (1986). *Biochemistry*, **25**, 5941.

64. Scott, T. W. and Friedman, J. M. (1984). *J. Am. Chem. Soc.*, **106**, 5677.

65. Hester, R. E. (1978). *Advances in infrared and Raman spectroscopy*, **4**, 1 .

66. Sans Cartier, L. R., Storer, A. C., and Carey, P. R. (1988). *J. Raman Spectrosc.*, **19**, 117.

67. Simpson, S. F., Kincaid, J. R., and Holler, F. J. (1993). *Anal. Chem.*, **55**, 1420.

68. Paeng, K. J., Paeng, I. R., and Kincaid, J. R. (1994). *Anal. Sci.*, **10**, 157.

69. Babcock, G. T., Jean, J. M., Johnston, L. N., Palmer, G., and Woodruff, W. H. (1984). *J. Am. Chem. Soc.*, **106**, 8305.

70. Varotsis, C., Zhang, Y., Appelman, E. H., and Babcock, G. T. (1993). *Proc. Natl. Acad. Sci. USA*, **90**, 273.

71. Han, S., Ching, Y. C., and Rousseau, D. L. (1990). *Proc. Natl. Acad. Sci. USA*, **87**, 8408.

72. Ogura, T., Takahashi, S., Hirota, S., Shinzawa-Itoh, K., Yoshikawa, S., Appelman, E. H., *et al.* (1993). *J. Am. Chem. Soc.*, **115**, 8527.

73. Han, S., Ching, Y.-C., and Rousseau, D. L. (1990). *Proc. Natl. Acad. Sci. USA*, **87**, 2491.

74. Han, S., Ching, Y.-C., and Rousseau, D. L. (1990). *Nature (London)*, **348**, 89.

75. Ogura, T., Takahashi, S., Shinzawa-Itoh, K., Yoshikawa, S., and Kitagawa, T. (1991). *Bull. Chem. Soc. Japan*, **64**, 2901.

76. Schelvis, J. P. M., Varotsis, C., Deinum, G., and Babcock, G. T. (1999). *Laser Chem.*, **19**, 223.

77. Campion, A. and Kambhampati, P. (1998). *Chem. Soc. Rev.*, **27**, 241.

78. Lee, P. C. and Meisel, D. (1982). *J. Phys. Chem.*, **86**, 3391.

79. Smith, W. E. (1993). In *Methods in enzymology* (eds. J. S. Riordan and B. L. Vallee), Academic Press, Vol. 226, p. 482.

80. Mie, G. (1908). *Ann. Physik.*, **25**, 377.

81. Diem, M. (1993). *Introduction to modern vibrational spectroscopy*. Wiley, NY.

82. Chikishev, A. Y., Koroteev, N. I., Otto, C., and Greve, J. (1996). *J. Raman Spectrosc.*, **27**, 893.

83. Peticolas, W. L. (1995). In *Methods in enzymology* (ed. K Sauer), Academic Press, Vol. 246, p. 389.

84. Peticolas, W. L. (1979). In *Methods in enzymology* (eds. C. H. W. Hirs and S. N. Timasheff), Academic Press, Vol. 61, p. 425.

85. Brown, K. G., Erfurth, S. C., Small, E. W., and Peticolas, W. L. (1972). *Proc. Natl. Acad. Sci. USA*, **69**, 1467.

86. Genzel, L., Keilmann, F., Martin, T. P., Winterling, G., Yacoby, Y., Frolich, H., *et al.* (1976). *Biopolymers*, **15**, 219.

87. Yu, T.-J., Lippert, J. L., and Peticolas, W. L. (1973). *Biopolymers*, **12**, 2161.

88. Frushour, B. G. and Koenig, J. L. (1974). *Biopolymers*, **13**, 1809.

89. Yu, N.-T., Liu, C. S., and O'Shea, D. C. (1972). *J. Mol. Biol.*, **70**, 117.

90. Chen, M. C., Lord, R. C., and Mendelson, R. (1976). *J. Am. Chem. Soc.*, **96**, 3038; Chen, M. C., Lord, R. C., and Mendelson, R. (1973). *Biochim. Biophys. Acta*, **328**, 252.

91. Kuck, J. F. R., Jr., East, E. J., and Yu, N.-T. (1976). *Exp. Eye Res.*, **22**, 1.

92. Krimm, S. and Bandekar, J. (1986). *Adv. Protein Chem.*, **38**, 181.

93. Yu, N.-T. and East, E. J. (1975). *J. Mol. Biol.*, **250**, 2196.

94. Siamwiza, M. N., Lord, R. C., Chen, M. C., Takamatsu, T., Harada, I., Matsura, H., *et al.* (1975). *Biochemistry*, **14**, 4870.

95. Peticolas, W. L. (1982). In *Raman spectroscopy: linear and nonlinear* (ed. J. Lascombe and P. V. Huong). Wiley, Chichester, England.

96. Miura, T., Takeuchi, H., and Harada, I. (1988). *Biochemistry*, **27**, 88.

97. Spiro, T. G. and Gaber, B. P. (1977). *Annu. Rev. Biochem.*, **46**, 553.

98. Sugeta, H., Go, A., and Miyazawa, T. (1972). *Chem. Phys. Lett.*, 83.

99. Yu, N.-T., DeNagel, D. C., Jui-Yuan Ho, D., and Kuck, J. F. R. (1987). In *Biological applications of Raman spectroscopy* Vol. 1: *Raman spectra and the conformations of biological macromolecules* (ed. T. G. Spiro), p. 47. Wiley, NY.

100. Qian, W. and Krimm, S. (1992). *Biopolymers*, **52**, 1025 and 1503.

101. Harada, I. and Takeuchi, H. (1986). In *Advances in spectroscopy* (ed. R. J. H. Clark and R. E. Hester), Vol. 13, pp. 113–75.

102. Miura, T., Takeuchi, H., and Harada, I. (1989). *J. Raman Spectrosc.*, **20**, 667.

103. Chen, Y.-Q., Kraut, J., and Callender, R. (1997). *Biophys. J.*, **72**, 936.

104. Chi, Z. and Asher, S. A. (1998). *Biochemistry*, **37**, 2865.

105. Fersht, A. (1977). *Enzyme structure and mechanism*. W. H. Freeman, San Francisco, CA.

106. Torreggiani, A. and Fini, G. (1998). *J. Raman Spectrosc.*, **29**, 229.

107. Torreggiani, A. and Fini, G. (1999). *J. Mol. Struct.*, **480–481**, 459.

108. Williams, R. W., Dunker, A. K., and Peticolas, W. L. (1980). *Biophys. J.*, **32**, 232.

109. Williams, R. W. (1983). *J. Mol. Biol.*, **166**, 581.

110. Williams, R. W. (1986). In *Methods in enzymology* (eds. C. H. W. Hirs and S. N. Timasheff), Academic Press, Vol. 130, p. 311.

111. DeGrazia, H., Harman, J. G., Tan, G. S., and Wartell, R. M. (1990). *Biochemistry*, **29**, 3557.

112. Tan, G. S., Kelly, P., Kim, J., and Wartell, R. M. (1991). *Biochemistry*, **30**, 5076.

113. Berjot, M., Marx, J., and Alix, A. J. P. (1987). *J. Raman Spectrosc.*, **18**, 289.

114. Sane, S. U., Cramer, S. M., and Przybycien, T. M. (1999). *Anal. Biochem.*, **269**, 255.

115. Chi, Z., Chen, X. G., Holtz, J. S. W., and Asher, S. A. (1998). *Biochemistry*, **37**, 2854.

116. Malinowski, E. R. (1991). In *Factor analysis in chemistry*, pp 32–82. John Wiley & Sons, New York.

117. Le Tilly, V., Sire, O., Alpert, B., Chinsky, L., and Turpin, P.-Y. (1991). *Biochemistry*, **30**, 7248.

118. Rava, R. P. and Spiro, T. G. (1985). *J. Phys. Chem.*, **89**, 1856.

119. Caswell, D. S. and Spiro, T. G. (1986). *J. Am. Chem. Soc.*, **108**, 6470.

120. Li, H., Hanson, C., Fuchs, J. A., Woodward, C., and Thomas, Jr., G. J. (1993). *Biochemistry*, **32**, 5800.

121. Carey, P. R. and Tonge, P. J. (1995). *Acc. Chem. Res.*, **28**, 8.

Chapter 8

Electrospray ionization mass spectrometry

T. J. Hill, D. Lafitte, and P. J. Derrick

University of Warwick, Department of Chemistry, Coventry CV4 7AL, UK.

1 Introduction

Characterization of protein–ligand interactions involves the determination of stoichiometry, affinity, and conformational changes associated with such interactions. Mass spectrometry is rapidly emerging as an important and powerful new tool in this area to complement the wealth of other techniques described in these volumes. The development of ionization methods, such as electrospray ionization (ESI) and matrix-assisted laser desorption/ionization (MALDI), applicable to biomolecules in solution or as solids has allowed the routine analysis, with picomole (10^{-12} M) to attomole (10^{-18} M) sensitivity, of proteins, peptides, oligonucleotides, carbohydrates, and other classes of biomolecule. Extending beyond simple molecular mass determinations, mass spectrometry is now readily applied to the sequencing of peptides, proteins, and DNA as well as for the study of protein aggregation, folding, and non-covalent interactions in general.

One of the main attributes of mass spectrometry is its ability to separate and detect many components of a mixture. It is, therefore, possible to analyse the interaction at the molecular level without the need for the complex models that are often required when performing such detailed analysis with macroscopic measurements. Characterization of specificity, stoichiometry, and conformational change is now possible by ESI mass spectrometry.

This chapter outlines the potential of mass spectrometry for protein studies, detailing some experimental considerations when choosing an appropriate instrument. Particular emphasis is given to the capabilities of ESI, coupled to either a quadrupole or a Fourier transform ion cyclotron resonance (FTICR), mass spectrometer for the study of protein–ligand interactions, using the calcium-binding protein calmodulin as a case study example.

2 Mass spectrometry

Figure 1 shows the basic components of a mass spectrometer: sample injection, ionization, selection, and detection. Following sample injection, molecule ions

SAMPLE INJECTION   IONISATION   SELECTION   DETECTION

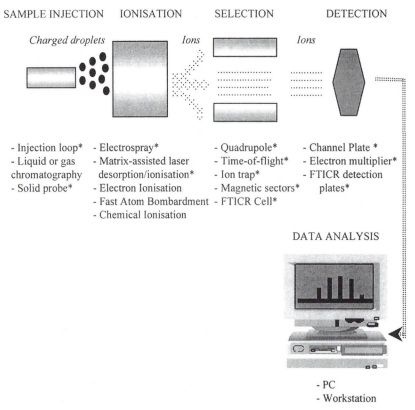

- Injection loop*
- Liquid or gas chromatography
- Solid probe*

- Electrospray*
- Matrix-assisted laser desorption/ionisation*
- Electron Ionisation
- Fast Atom Bombardment
- Chemical Ionisation

- Quadrupole*
- Time-of-flight*
- Ion trap*
- Magnetic sectors*
- FTICR Cell*

- Channel Plate *
- Electron multiplier*
- FTICR detection plates*

DATA ANALYSIS

- PC
- Workstation

Figure 1 Components of a mass spectrometer (* denotes those suitable for the study of non-covalent interactions).

are generated as a result of the loss or gain of a charge (*ionization*) via processes such as protonation or electron ejection. Molecule ions are then separated or distinguished according to their mass-to-charge (m/z) ratios and detected. The resulting signal is transferred to a suitable interface (e.g. PC or workstation) for processing and storage.

There are numerous ionization techniques for the ionization of biomolecules in mass spectrometry, but by far the most commonly used techniques for biomolecules with masses greater than 400 Da are fast atom bombardment, MALDI, and ESI.

Fast atom bombardment (FAB) is an ionization technique which utilizes a non-volatile liquid matrix, such as glycerol, to absorb much of the energy from a high-energy particle beam (e.g. Xe atoms or Cs^+ ions at keV impact energies). The beam bombards the matrix/sample mixture and causes sample ions to be desorbed into the gas phase with little fragmentation. Disadvantages of FAB include a significant drop in sensitivity for higher masses (FAB is most suitable for small peptides, < 3 kDa), confusion in interpretation caused by matrix peaks in the mass spectrum, and the sample solubility requirement which accompanies the use of a liquid matrix.

The principle of MALDI is similar to that of FAB, with a solid matrix and a photon beam replacing the liquid matrix and particle beam. The matrix absorbs the photon energy and the matrix–sample mixture is desorbed. A precise mechanism for the ionization process remains unclear, e.g. whether efficient charge transfer from the matrix to the sample occurs in the gas phase following desorption, or whether it occurs simultaneously with desorption in the condensed phase. The mass range that can be achieved by MALDI significantly exceeds that of any other ionization technique except electrospray ionization and measurement of masses up to 3 000 000 Da has been reported. MALDI is also an extremely sensitive technique that can be applied to a wide range of biomolecules. A further advantage of MALDI is that it has a high tolerance for salt (up to millimolar concentrations), or indeed for many other organic and inorganic contaminants. The energy input with the MALDI technique, however, is often too high to maintain non-covalent associations.

ESI is arguably the ionization technique, of those currently available, that puts the least internal energy into the molecule ions formed. For this reason ESI has evolved as a powerful tool for the analysis of proteins and peptides. The success of ESI, interfaced with mass spectrometry, is primarily due to its unique ability to extract, ionize, and transfer biomolecules *still intact* from solution into the gas phase. The mild electrospray process preserves covalent bonds and under optimal conditions non-covalent interactions, both those intramolecularly involved in the three-dimensional protein structure and those intermolecularly involved in complex formation. There is now a substantial literature to confirm that many non-covalent species existing in solution, e.g. enzyme–substrate, protein–drug, protein–metal, and host–guest complexes, can endure the ESI process (e.g. refs 1–4).

The significance of ESI for the analysis of protein–ligand interactions is clearly indicated. Emphasis is therefore given here to the use of this ionization technique, and the discussion that follows is directed *only* towards ESI.

3 Electrospray ionization

The reader is referred to refs 5–16 for detailed descriptions.

3.1 Principle

The formation of ions by ESI is not understood in complete detail and more than one mechanism has been proposed. Two of the main types of electrospray sources are illustrated in *Figures 2A* and *2B*.

In the source equipped with a capillary (referred to here as 'Source A'; *Figure 2A*), the solution containing the biomolecule of interest emerges from a fine, conductive needle at atmospheric pressure and is attracted towards a heated capillary, which either is made from Pyrex and coated at both ends with platinum or nickel, or is entirely metal. An electrostatic field is maintained by setting appropriate voltages on Vcyl, Vcap, and Vend. Typically, a potential of –3 kV for

A

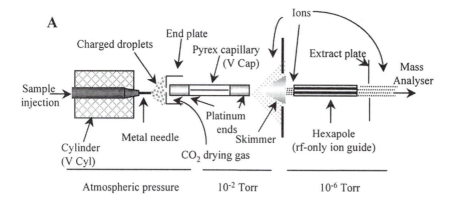

B

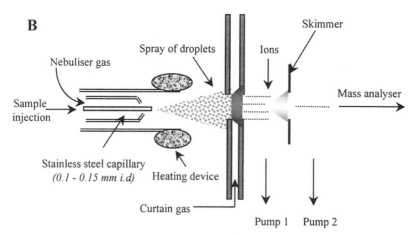

Figure 2 Schematic representations of electrospray sources. (A) ESI source with a capillary (referred to as Source A, e.g. Analytica of Branford, Branford, USA). (B) ESI source with curtain-gas (referred to as Source B, e.g. atmospheric pressure (AP) source used with PE Sciex instruments).

positive ionization and +3 kV for negative ionization is applied at the proximal end of the capillary which causes the sample to be sprayed as a very fine mist of charged microdroplets. A stream of drying gas (N_2 or CO_2, at a temperature of 50–60 °C) applied at the front of the capillary and, opposing the movement of the spray, leads to the shrinkage of the droplet size. Charge density increases and at some stage ion desorption occurs. The skimmer helps to separate the solvent and drying gas from the ions, and the ions are transported through to the mass analyser. The advantage of using this type of ESI source for proteins is the potential for nozzle-skimmer collision-induced dissociation (CID) experiments, which will be discussed later.

In the other type of ESI (Source B; *Figure 2B*) the solution emerges at atmospheric pressure through a heated needle (~70 °C) to which a potential of typically +3 kV (positive ionization) or −3 kV (negative ionization) is applied.

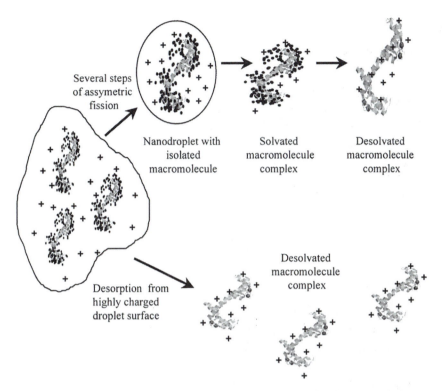

Several steps of assymetric fission

Nanodroplet with isolated macromolecule

Solvated macromolecule complex

Desolvated macromolecule complex

Desorption from highly charged droplet surface

Desolvated macromolecule complex

Figure 3 Models for the formation of macromolecule ions during the electrospray ionization process.

The spray of charged droplets passes through a warm stream of gas where desolvation, induced by the combination of heating and increased vacuum (from atmospheric pressure at the end of the needle to 10^{-6} Torr in the source), eventually leads to ion formation. *Figure 3* depicts the model for the formation of macromolecule ions during the ESI process.

Figure 4 illustrates some of the variations in ESI sources. The axial spray, where the nebuliser is aligned with the orifice of the capillary, is suitable for low flow rates (1–5 μl/min). Pneumatic assistance is required when using high flow rates, e.g. when coupling with a chromatographic interface. One of the main disadvantages of the axial spray is the introduction of a large amount of solvent into the source. Contaminants are, therefore, introduced into the source at the same time as the sample.

The angled spray is pneumatically assisted (with N_2), and the probe is positioned at an angle of perhaps 60° or 90° with respect to the capillary. Charged droplets are attracted by the capillary and the majority of solvent is discarded. The angled spray is a good choice of injection technique when using buffered solutions and those contaminated with salt. It is, however, not suitable for small quantities of sample.

Z-spray™ is a dual orthogonal 'Z' sampling technique developed by Micro-

315

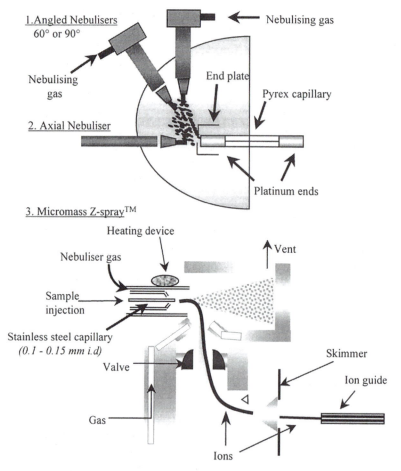

Figure 4 Three variations in ESI source with (1) angled nebuliser, (2) axial nebuliser, and (3) Micromass Z-spray™.

mass to reduce further the level of solvent, salt, and buffer sprayed into the source.

The desolvated gas phase ions generated by the ESI process have often been shown to retain solution phase structural characteristics of the biomolecule. ESI generates multiply charged ions, whereas most other ionization methods generate ions with a single charge. Since the majority of mass spectrometers are limited to analysing ions with m/z ratios of only a few thousand, this feature of ESI extends the analysis from that of smaller compounds (peptides) carrying only one or two charges to that of large biomolecules (and complexes) with masses of several hundred thousands of Daltons carrying many charges.

The number of net charges a protein/peptide carries (its *ionization state*) is determined by the accessibility of ionizable residues. For positive ionization, protonation of the basic residues of the protein (mostly lysine, arginine, histidine, and N-terminus, and possibly asparagine, proline, and glutamine) is involved, whereas negative ionization involves the deprotonation of acidic residues such as aspartic acid, glutamic acid, and the C-terminus. The accessibility of sites for ionization is determined by the conformation of a protein, and as a consequence folded and unfolded proteins demonstrate different charge state distribution patterns in their ESI mass spectra. Proteins electrosprayed from denaturing solutions, such as those containing organic solvents or acid, exhibit a broad pattern of charges with a maximum which usually corresponds to the total number of basic or acidic residues present in the protein, i.e. the majority of sites are exposed. Proteins sprayed from conditions that favour the preservation of their native conformations, e.g. buffered conditions, tend to show a more narrow charge-distribution pattern centred on a much lower charge state. This indicates that fewer basic or acidic sites are accessible for ionization as a result of folding. A protein that is strongly folded in its native state can hide many ionizable sites. For example, consider a case where the folding of, or even ligand binding to, a 20 kDa protein carrying an average of 20 charges causes 15 ionizable residues to be hidden, then the average charge state would shift from 20 to 5 and the m/z ratio of the most abundant peak would increase from 1000 to 4000. The use of a mass analyser with a high m/z range becomes important for the study of non-covalent complexes, as these often have a tendency to exhibit relatively low charge states.

3.2 Sample preparation

There are several important considerations for ESI sample preparation. These include the following.

(a) **Removal of non-volatile impurities**. ESI is particularly intolerant of the presence of salts and detergents, and contaminants such as these should be removed wherever possible. The presence of high levels of impurities results in loss of sensitivity, possibly affecting the formation of the droplets themselves during ESI.

  (i) Protein and peptide samples must be carefully purified and desalted.

  (ii) The sample injection line should be thoroughly cleaned between ESI runs to prevent transfer of material from one sample to the next.

  (iii) The use of glass sample-containers should be avoided as these also introduce unwanted contaminants such as sodium or potassium into the solution. The use of Eppendorf™-brand microcentrifuge tubes is recommended for sample preparation.

  (iv) Samples should be analysed as soon as possible after solution preparation to reduce the possibility of degradation.

(b) **Choice of sample concentration**. High concentrations may lead to precipitation and also increase the concentrations of impurities. The formation of too many ions may also lead to saturation of the detector. Conversely, detection of ions will be difficult if the sample concentration is too low. Typically concentrations for the ESI of proteins are usually in the range of 1–50 μM, depending on the source and the mass spectrometer.

(c) **Choice of solvent**. This is one of the most important considerations. In most cases, the conditions required to maintain protein complexes in solution are not optimal for ESI and vice versa. The use of organic solvents such as methanol and acetonitrile facilitate the formation and desolvation of droplets, but do not mimic physiological conditions, and are not suitable for maintaining non-covalent associations.

(i) When maintaining the integrity of the higher-order structure of a protein is not important, trace amounts of acid (acetic acid with methanol/water mixtures or formic acid with acetonitrile/water mixtures) for positive electrospray or base (ammonia) for negative electrospray also greatly improves the formation of charged species, resulting in better-quality spray and hence spectra. Of course, most protein and peptide samples will precipitate in organic solvents, and so samples are usually solubilized first in water and diluted with an appropriate amount of organic solvent (1:1, v/v ratio is often used).

(ii) The use of organic solvents, acid, or base must be avoided for many non-covalent protein–ligand binding studies, but the higher surface tensions of pure aqueous solutions affect the stability of the electrospray. Low concentrations of volatile buffers such as ammonium acetate (pH 6–7, for positive ESI) or ammonium bicarbonate (pH 7–8, for negative ESI) are often used to improve the spray of charged droplets.

(iii) The analysis of hydrophobic proteins, such as membrane proteins, is often difficult, due to the incompatibility of ESI with the detergents and/or salts required to retain such proteins in solution. However, certain solvent systems have been identified for the ESI study of hydrophobic proteins, including chloroform, acetone, hexafluoroisopropanol, and high percentages of formic acid (70–95%) (17).

3.3 Sample injection

Samples can be introduced into an ESI source in several ways. The most important consideration is that a sample be introduced without significant disturbance of the vacuum. Some of the more common means are listed here.

(a) Rheodyne injector valve coupled with a sample loop (typically 10 μl). Flow rate of 1–5 μl/min through an interface employing a 75 μm i.d. fused silica capillary inlet to the source. This type of injection system requires the use of a carrier solvent (of the same composition as that used in the preparation of the sample).

(b) Direct coupling with chromatographic interfaces, e.g. liquid chromatography and gas chromatography, is achieved with the use of a capillary column. The flow rate is considerably increased and the use of pneumatic assistance is required. Continuous sample-flow introduction can be achieved in this way.

(c) Syringe pump operating at a typical flow rate of 1–5 μl/min.

(d) Direct loading of special needles (glass, coated with gold or palladium) designed for nano-flow electrospray.

Automated sample introduction is now routinely used for the rapid screening of a large number of samples, which is particularly useful for combinatorial chemistry.

Nano-flow electrospray spray is invaluable when dealing with small quantities (< 10 μl) or for maintaining fragile non-covalent interactions. It is also better suited to deal with the problems associated with spraying aqueous solutions. A stable signal can be sustained for at least 30 minutes by nano-flow electrospray with just 2 μl of solution. *Figure 5* shows a schematic of a nano-flow ESI set-up, and the protocol for use is given in *Protocol 1*.

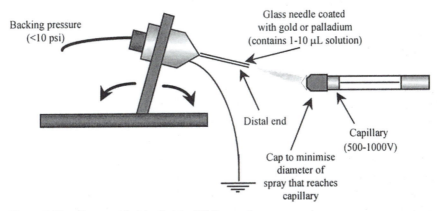

Figure 5 Nano-flow sample injection for ESI Source A.

Protocol 1

Nano-flow with a capillary ESI source

Equipment and reagents

- Nano-flow needles
- Gas supply, N_2 or CO_2
- Capillary ESI source

Method

Care must be taken during sample loading. The distal end of a nano-flow needle is extremely fragile.

1   2 μl or 10 μl of solution is introduced into the needle using gel loader tips.

2   Mount the needle in the holder; ensure that the needle is grounded.

3   Apply a small pressure to the proximal end of the needle with a syringe or by connection to a gas flow (at a pressure less than 10 psi).

4   Drive the needle in front of the capillary, monitoring movements with CCD cameras or a powerful microscope.

5   If the needle is pre-opened align the needle with the orifice of the capillary at a distance of approx. 1 mm. To open a needle, move the needle towards the capillary and gently touch the surface of the capillary. Only a slight touch is required so that the fine tip of the needle is not destroyed.

6   Increase the capillary voltage to between 500–1000 V to obtain a stable current. NB. Prevent arcing by careful control of the current and the distance between the needle and the capillary.

7   The spray can be monitored with the microscope.

8   Heating of the source and the use of drying gas are not required.

3.4 Creating a vacuum

Molecule ions generated in the source must reach the mass analyser and ultimately the detector without colliding with other gaseous molecules. Ion–molecule interactions within a mass spectrometer greatly diminish the sensitivity and resolution of the instrument and are avoided by the use of low pressures, i.e. high vacuum. The pressure that must be achieved and maintained varies with the type of analyser and detector used, but typically falls in the range of 10^{-6} to 10^{-10} Torr. Transferring the sample from atmospheric pressure (760 Torr) into a region of such high vacuum, while still maintaining the pressure in the instrument, therefore poses a great problem and is often accomplished with multiple stages of differential pumping. All instruments possess one or more mechanical pumps, which achieve an initial pressure of (at best) 10^{-3} Torr. A combination of diffusion, turbomolecular, and cryogenic pumping systems can then be used to achieve the lower pressures required for each particular instrument.

3.5 Mass analysis—selection and detection

The common types of mass analyser that can be used to separate ions according to their m/z ratios include magnetic sector, ion trap, quadrupole, time-of-flight, and FTICR. The choice of mass analyser is particularly important since this largely determines the accuracy, range, and sensitivity of the mass spectrometer.

3.5.1 Magnetic-sector analyser

The first mass analysers used a magnetic field to separate ions according to their m/z (mass-to-charge) ratios. An ion accelerated through a magnetic field experiences a force and traverses a path with a radius of curvature that is dependent upon its m/z ratio. Ions are separated according to their radii of curvature and only ions of a given m/z reach the detector for the particular magnetic field applied. Single-sector magnetic mass analysers are limited by their relatively low resolution, but this can be overcome by the addition of an electrostatic sector.

Such double-focusing magnetic-sector instruments can have resolving powers up to 100 000 and can offer m/z ranges in excess of m/z 10 000. The addition of the electrostatic sector compensates for the kinetic energy spread of ions, allowing ions of different kinetic energies to be focused at the same point in front of the detector.

Although sector instruments are characterized by high resolution and mass accuracy, they are generally more expensive and less routine than other mass analysers, such as the quadrupole, and are also more difficult to interface with many inlet systems. These instruments are most commonly used with FAB and EI sources, although coupling of ESI and MALDI sources is possible.

Multi-sector instruments (three- and four-sector) were developed for the study of ion fragmentation. Such instruments (*tandem mass or MS-MS spectrometers*) are particularly useful for biomolecule sequencing. High-energy collision-induced dissociation (CID) with a four-sector instrument produces higher yields of fragment ions than many other MS-MS techniques, and is particularly invaluable for peptide sequencing where comparison of side chain cleavages is essential for distinguishing between amino acids with the same nominal mass, e.g. leucine and isoleucine.

3.5.2 Quadrupole

The quadrupole is one of the commonest types of analyser. An electric field is established by applying a dc voltage and an oscillating ac voltage (at radio frequencies) on four, parallel metal rods (*quadrupoles*). Opposite rods are electrically connected and adjacent rods have opposite polarity. The applied field affects the wave-like trajectory of the ions travelling down the centre between the rods, and only ions of a certain m/z pass through to the detector for any given voltages. The quadrupoles thereby function as a mass filter, and a mass spectrum is constructed by monitoring the ions that pass through as the voltages on the rods are varied.

Quadrupole mass spectrometers are fast, compact, relatively inexpensive, and can analyse ions with m/z ratios of up to at least 5000. Quadrupole analysers can also withstand relatively high pressures, thereby facilitating the coupling of an electrospray ionization source for the efficient analysis of proteins and their non-covalent interactions. Such instruments are also ideal for coupling with chromatographic interfaces.

Triple-quadrupole instruments comprised of two mass analysers separated by a collision cell (the 'third quadrupole') have been developed for ion fragmentation experiments. An ion is selected with the first quadrupole and is transferred into the collision cell (often also a quadrupole). Fragmentation is achieved in the collisional cell by the use of a collision gas such as argon, or xenon, and fragments are then analysed in the third quadrupole.

3.5.3 Ion trap

The ion trap analyser is similar in design to the quadrupole, but uses three electrodes to trap ions in a small volume. The mass analyser consists of a ring electrode separating two hemispherical electrodes. A mass spectrum is obtained by changing the electrode voltages to eject the ions from the trap. Ion trap mass spectrometers are very compact in size, and increase the signal-to-noise ratio of a measurement by trapping and accumulating ions.

The experimental set-up for ion fragmentation with ion traps is different to those of sectors and quadrupoles. The ion population is trapped in a physical location and so the steps of tandem mass spectrometry (MS-MS) experiments must be separated in time, rather than in space as with sectors and quadrupoles. The high CID efficiencies that can be achieved with ion trap MS-MS experiments allow sequential CID reactions to be possible ($(MS)^n$ experiments).

3.5.4 Time-of-flight

The time-of-flight (TOF) analyser normally separates ions according to their velocity. The velocities of ions that are accelerated with the same energy through a voltage is representative of their mass-to-charge ratios. In simple terms, smaller ions reach a detector before larger ions as a result of a greater velocity, and hence the m/z of an ion can be determined from its time of arrival at the detector. An advantage of the TOF analyser is that ions with a very high m/z can be detected ($> 500\,000$) (18) whereas the downside is that the resolving power of the instrument is relatively low (typically ~ 1000). The resolution achieved by TOF mass analysis can be improved to approximately 1500 with the use of an ion mirror (*reflectron*) which compensates for the kinetic energy distributions of ions that reach the detector, and delayed-extraction methods with MALDI give still higher resolution.

The TOF analyser is commonly used with MALDI, but has also been successfully coupled with ESI in many commercial instruments allowing the detection of large ions, including non-covalently associated species, with a m/z ratio beyond the range of most mass analysers.

The recent development of mass spectrometers that couple a quadrupole (or a triple quadrupole) analyser with an orthogonal time-of-flight has extended the sensitivity and resolution boundaries for these instruments. High-quality data achieved for MS-MS experiments, with low-femtomole and attomole sensitivity and mass accuracies better than 0.1 Da offers great advances for protein sequencing by mass spectrometry.

3.5.5 Fourier transform ion cyclotron resonance

A powerful advantage of FTICR over other mass analysers is accurate mass and high-resolution capability. A radiofrequency signal is used to excite ions as they orbit in a magnetic field, and as a result the ions produce a detectable time-dependent image current which is Fourier transformed to give component frequencies proportional to the m/z ratios of the ions present. Ion sources were originally placed within the magnetic field of FTICR instruments, but with the development of transfer ion optics and the advent of multiple stages of differential pumping, ion sources could be moved outside of the magnetic fields for convenience. Location outside the magnet is particularly beneficial for the coupling of for example ESI, which operates at high pressures and therefore must be separated from the mass analyser. FTICR has been successfully coupled to virtually every type of ion source.

FTICR is very useful for multiple collision experiments, i.e. tandem mass spectrometry (MS^n, typically n $\sim$ 4). ESI does not routinely produce fragment ions, but rather maintain the structural integrity of biomolecules such as proteins. It has therefore been necessary to develop fragmentation techniques such as tandem mass spectrometry. FTICR can be used to select ions of a particular m/z (by ejecting all but those required from the cell) and fragmentation can be induced by the introduction of a collision gas. In terms of protein–ligand interactions, such a technique would potentially be useful as a means of determining the site of attachment of a ligand to a protein.

The performance of the FTICR mass spectrometer improves as the magnetic field is increased, and the use of superconducting and resistive magnets to achieve very high field strengths has permitted accurate mass analysis of proteins as large as 67 kDa (porcine albumin) to be practicable (19). Unit-mass resolution has been achieved for two 112 kDa chondroitinase enzymes, whose masses were determined within a 3 Da accuracy (20). FTICR is the most advanced of all mass spectrometers in terms of resolving power. The highest molecular mass reported for a heterogeneous non-covalent complex detected by ESI-FTICR is currently that of a chaperone–ligand complex of 127 kDa (21).

The coupling of ESI with FTICR mass spectrometry has greatly extended the study of proteins. The resolving power attainable with FTICR allows molecule ions to be easily distinguished from adducts which are close in mass, e.g. sodium adducts, making this technique perfect for protein–ligand studies.

3.5.6 Ion detectors

The FTICR analyser also functions as the detector, but other mass analysers described in this section must be accompanied by an ion detector. The most commonly employed methods for the detection of ions involve the use of electron multipliers or scintillation counters which rely on the ions striking a secondary emitting material (*a dynode*), resulting in the emission of electrons. In the case of the electron multiplier further dynode plates are arranged in series at increasing potentials, which results in significant signal amplification by a cascade process. The scintillation counter works in much the same way except that the electrons, that are emitted from the first dynode, strike a phosphorus screen to trigger the release of photons, which are amplified and detected by a photomultiplier that amplifies according to the same cascading principle as the electron multiplier.

Table 1 summarizes the merits of the mass analysers discussed in this section.

3.6 Data analysis

3.6.1 Mass calibration

Two types of calibration are commonly used. The mass spectrometer can be calibrated prior to the sample injection (*external calibration*) with a reference sample,

Table 1 General comparison of mass analysers

Instrument	Sensitivity	m/z range	Resolution	Cost	Features
Quadrupole	Femtomole	< 3000	2000–3000	Low	Compact; easy to handle; fragmentation of small complexes
Quadrupole with orthogonal time-of-flight	Attomole	< 5000	5000–6000	Medium	Compact; easy to handle; fragmentation of small complexes
Ion trap	Femtomole	< 5000	2000	Low	Compact; easy to handle; multiple stages of fragmentation; reactions in cell
Time-of-flight	Attomole	No theoretical limitation	1000–1500 (with reflectron) 5000–10 000 (delayed extraction)	Low	Compact; easy to handle; salt tolerance
Fourier transform ion cyclotron resonance	Attomole	< 10 000	Dependent on magnet, e.g. 200 000 at 9.4 Tesla (> 1 000 000 in high-resolution mode)	High	Massive; less routine; reaction in cell; multiple stages of fragmentation
Sector	Femtomole	< 10 000	> 10 000	High	Massive; less routine; high-energy fragmentation

e.g. horse heart myoglobin, or polyethylene glycols for a wider mass range. Alternatively the reference sample can be injected at the same time as the sample (*internal calibration*) which theoretically gives a better accuracy. In both cases it is recommended that a calibrant with a wider mass range than the sample be used.

3.6.2 Calculating mass from chemical formulae

There are several ways to calculate the mass of a sample from its molecular formula:

(a) The **average mass** is calculated using the average atomic weight for each element. The average atomic weight is an average of the isotopic masses of each element present, weighted by the relative abundances of the isotopes. This is used when the individual isotopes of a molecule ion cannot be distinguished.

(b) The **monoisotopic mass** is calculated using the exact mass of the most abundant isotope of each element (e.g. C = 12.00000 and H = 1.007825), and can only be calculated when the isotopes are resolved.

(c) The **nominal mass** is calculated using the integer mass of the most abundant isotope for each element.

3.6.3 Calculating mass from ESI spectra

An ESI spectrum is composed of a set of peaks that should characterize the m/z ratios of the different multiply charged ions present in solution. The shape of a protein ESI spectrum is usually Gaussian-like and is centred on the most abundant ionization state of the protein.

Software is generally provided by the manufacturer for the automated analysis of ESI spectra, but an explanation of how to calculate the molecular mass of a biomolecule (M) from an ESI spectrum which is comprised of $[M + zH]^{z+}$ species is given here:

(a) Select two adjacent ionization peaks, m_1 and m_2 (where $m_2 > m_1$) such that:

$$m_1 = (M + z_1)/z_1 \tag{1}$$

$$m_2 = (M + (z_1 - 1))/(z_1 - 1) \tag{2}$$

(b) Substitute m_1 and m_2 with the m/z values of the adjacent peaks and solve the equations for the two unknowns M and z_1.

Multiple molecular mass determinations can be made for the same species using the range of charge states that are often present in a given spectrum. Deconvolution software is available for most mass spectrometers and combines all of the charge states to provide a single peak on a mass scale. *Figure 6* shows the raw ESI spectrum obtained for myoglobin in water/acetonitrile (50:50) containing 1% formic acid. The inset shows the deconvoluted spectrum for this data, obtained using maximum entropy software, e.g. Micromass MaxEnt™.

The charge-state of a given species can be deduced directly from the mass spectrum if the peaks are isotopically resolved. The spacing between the isotopes corresponds to the inverse of the number of charges, e.g. isotopes are separated by 1 mass unit for singly charged species, and by ½ for doubly charged species. The resolving power of the instrument often becomes a limiting factor when determining the isotopic spacing for species with a higher number of charges.

3.6.4 Mass accuracy

Mass accuracy is defined as the difference between the theoretically calculated mass and the experimentally determined mass divided by the theoretical mass, and is expressed in parts per million (ppm). Mass accuracy is strongly dependent on the resolution of the instrument and is high when ions are completely resolved. When coalescence of peaks occurs, the experimentally determined mass is an average of the masses of the coalescent ions, which will be very different to the theoretically determined mass, and hence mass accuracy will be poor.

3.6.5 Resolution

The ability of a mass spectrometer to separate ions of different masses (M_1, M_2) is defined as the resolution R, where:

$$R = M/\Delta M \ (\Delta M = M_1 - M_2 \text{ and } M = M_1 \approx M_2). \tag{3}$$

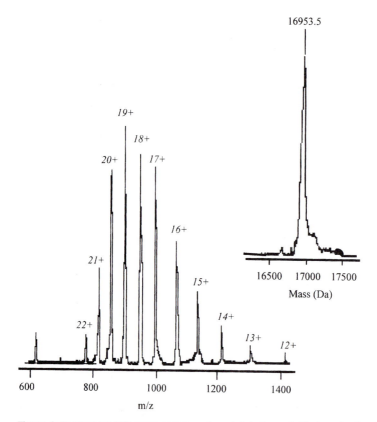

Figure 6 Positive ion ESI mass spectrum (raw data) of myoglobin in water/acetonitrile (50:50) containing 1% formic acid, obtained with a quadrupole (VG Trio-2000, Fisons/VG, Manchester, UK). Inset shows the deconvoluted ESI mass spectrum produced by MaxEnt processing of the raw data.

Resolution can be determined from an ESI mass spectrum by reference to one or two peaks:

(a) **Peak width definition**. ΔM is defined as the full width of a single peak at half maximum, (50%), peak intensity (FWHM). Resolution is then calculated using $R = M/\Delta M$. ΔM can also be defined at other fractions of the peak height, e.g. 5%, as long as this is clearly specified when quoting the resolution.

(b) **Valley definition**. Two peaks of equal height are considered to be resolved if separated by a valley which is more than 10% of their height.

The resolving power of a mass spectrometer is one of its crucial characteristics and generally decreases with increasing m/z for most instruments. Noticeable exceptions are magnetic-sector instruments where variation of the magnetic field maintains the resolution throughout the m/z range.

Resolution is of particular importance for the analysis of complex mixtures. Characterization of small protein–ligand complexes, monitoring of competitive binding of ligands that are close in mass, or distinguishing between covalent and non-covalent associations requires high mass-resolution.

For example, if a 20 kDa protein binds one sodium, the increase in mass is 22 Da (M(Na) − M(H)). The theoretical resolving power required to separate the free protein from the sodium-bound protein is $20\,000/22 = 909$. Such a resolution can easily be achieved at low m/z ratios with most mass spectrometers, but is very difficult above $m/z = 3000$, except with magnetic sector and FTICR mass spectrometers. When the resolving power is not sufficient to resolve two species, coalescence occurs resulting in a broad peak that corresponds to a mixture of components.

3.7 Data storage

Regardless of instrument, the quantity of data that can be readily generated by mass spectrometry, coupled with the fact that individual data files are often large, necessitates the use of high-capacity data-storage media.

For workstations, the storage of data on DAT tapes or optical disks is recommended. DAT tapes are robust and inexpensive but accessing the data is slow. Accessing data from optical disks is much more rapid and convenient, but optical disks are considerably more expensive and are often less reliable.

For a PC, storage on floppy disks is impossible due to the size of the data files and the use of high-capacity removable media, such as writeable CDs, ZIP™, or Jazz™ disks, is recommended.

4 Protein–ligand interactions of calmodulin

Calmodulin (*Figure 7*) is a calcium-binding protein present in all eukaryotic cells, which is involved in many cellular functions including exo- and endocytosis, cellular motility, and calcium transport (22). Calmodulin exhibits a dumbbell shape, composed of two globular domains linked by a long flexible helix. Each globular domain contains two helix-loop-helix motifs (EF-hands) that are the main calcium binding sites. Previous studies have shown that additional sites of weaker affinity are also present (1, 23). The structure/function relationship of calmodulin has been extensively studied by an exhaustive panel of biophysical techniques (24–29). It could be argued, however, that a true understanding of calmodulin function requires analysis at the microscopic level. In this case study calmodulin is used as an example to show the potential of ESI mass spectrometry for the study of protein–ligand interactions. Conformational changes, self-association, and multiple-ligand binding have been investigated.

4.1 Preparation of calmodulin

DNA encoded calmodulin was produced in *E. coli* and purified using fast liquid chromatography on phenyl Sepharose column followed by ion exchange chromatography. The protein was further purified by gel filtration, using a Sephadex column G-25 with 50 mM ammonium bicarbonate pH 7.5 as the eluent buffer, and then freeze-dried.

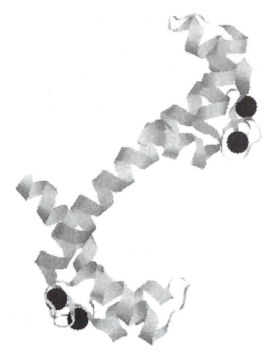

Figure 7 Three-dimensional X-ray structure of calcium-bound calmodulin. Calcium ions are shown in black.

NB. Purification procedures for proteins often use non-volatile buffers such as MOPS (3-[N-morpholino]propanesulfonic acid), Tris (Tris(hydroxymethyl)amino-methane), and Hepes (N-[2-hydroxyethyl]piperazine-N'-[2-ethanesulfonic acid]). ESI is not tolerant of high concentrations of non-volatile material and a careful desalting of the sample is required. Commercial sample batches may also contain significant levels of contaminants.

4.2 ESI mass spectrometry of calmodulin

4.2.1 Conformational studies

Ligand binding typically induces a conformational change in a protein and these can be probed by ESI mass spectrometry. For example, the three-dimensional structure of calmodulin significantly changes when it binds calcium. The principle methodologies that have been pursued so far for the monitoring of protein conformation by ESI-mass spectrometry include the comparison of charge-state distribution patterns and the probing of such chemical reactivities as hydrogen/deuterium-exchange reactions.

i. Charge state distribution

The accessibility of sites for ionization is determined by the conformation of the protein. Folded and unfolded proteins demonstrate different charge-state distribution patterns in their ESI mass spectra. Taking advantage of the fact that the relative amplitudes of the m/z peaks shift with the relative abundances of different conformational states it is possible, in favourable circumstances, to monitor the dynamics of protein folding and unfolding in the presence of ligands.

The first step, however, is characterization of the protein system in the absence of ligands. Charge-state distribution patterns are strongly influenced by the solution environment prior to spraying, and a strict control of this environment is required. It is worth remembering that many other factors, including protein concentration, ionization efficiencies, and declustering processes also contribute to the shape of the charge-state distribution pattern. For example, harsher ESI source parameters can lead to an increase in the collisional energy of a protein and result in charge stripping, which would be manifest as a shift in the charge-state distribution.

Protocols 2 and 3 describe the first steps towards obtaining an ESI spectrum for calmodulin in two different environments.

Protocol 2

ESI mass spectrometry of calmodulin under denaturing conditions

Equipment and reagents

- Electrospray source
- Hamilton syringe
- Rheodyne loop coupled to a fused silica capillary

- 0.1–0.5 mg calmodulin
- 50:50 acetonitrile/water containing 0.1–1% formic acid
- 0.1–1% ammonia

Method

1 Dissolve 0.1–0.5 mg (6–30 pmol) of calmodulin in 1 ml of 50:50 acetonitrile/water containing 0.1–1% formic acid (positive ESI) or 0.1–1% ammonia (negative ESI) to give a calmodulin concentration of 6–30 μM.

2 Load 100 μl of the solution into a Hamilton syringe and pump into the source at a flow rate of 1 μl/min. Or introduce 10 μl of the solution at a flow rate of 2–5 μl/min into the source using a Rheodyne loop coupled to a fused silica capillary (75 μm i.d.).

3 Apply a drying gas (CO_2 or N_2, at a temperature of 50–60 °C) at the capillary entrance (ESI Source A) or a warm counter flow of nitrogen (60–70 °C) (ESI Source B).

4 For positive ESI, apply a typical voltage of –3500 V to the capillary (Source A) or 3500 V to the needle (Source B). Use opposite values for negative ESI.

5 Voltages within the source are very dependent on the design of the source and must be applied according to the manufacturers' recommendation.

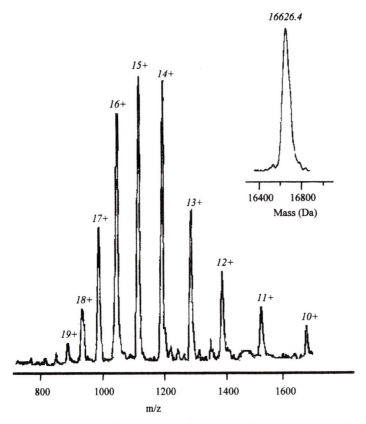

Figure 8 Positive ion ESI mass spectrum of calmodulin in water/acetonitrile (50:50) containing 1% formic acid, obtained with a quadrupole.

Figure 8 shows a typical ESI spectrum obtained in positive mode with a quadrupole (VG Trio-2000, Fisons/VG, Manchester UK) for calmodulin according to *Protocol 2*.

The spectrum exhibits a broad, Gaussian-like shape centred on a maximum of 15 attached protons (15^+ charge state). The maximum number of charges observed (at least $[M + 20H]^{20+}$), exceeds the total number of basic amino acids on the protein, which is 16, on the basis that the N-terminus and lysine, arginine, and histidine residues are basic. The higher number of attached protons observed could be due to protonation of secondary amines, such as those present on the asparagine and glutamine residues. The use of a denaturing solvent results in at least a partial unfolding of calmodulin and the exposure of a significant number of basic amino acid residues. The inset shows the deconvoluted spectrum indicating a relative molecular mass of (16626.4 ± 3.8) for calmodulin.

Figure 9 shows an ESI spectrum obtained under the same conditions using a 9.4 Tesla FTICR mass spectrometer (Bruker Daltonics, Billerica, MA, USA). As can be seen, the shape of the spectrum is not dissimilar to that obtained with the quadrupole, but two significant differences are evident. First, the FTICR spectrum

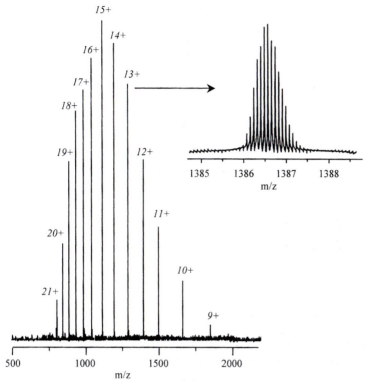

Figure 9 Positive ion ESI mass spectrum of calmodulin in water/acetonitrile (50:50) containing 1% formic acid, obtained with a 9.4T FTICR (Bruker Daltonics, Billerica, MA, USA).

is considerably 'cleaner', i.e. the signal-noise ratio is much higher. Secondly, as revealed by the inset, each charge state is isotopically resolved. The high resolution of a high-field FTICR mass spectrometer allows the mass of calmodulin to be determined with far higher precision. The experimentally determined mass of the most abundant isotope, averaged over several of the observed charge states was (16 626.85 ± 0.02) Da. The isotope pattern can be simulated using the chemical formulae for calmodulin and the software that accompanies the Bruker instrument (Xmass™), and reveals that the most abundant isotope ion contains ten ^{13}C atoms, so that the isotopically pure (^{12}C, ^1H, ^{14}N, ^{16}O) calmodulin would therefore have a mass of 16 616.83 Da. This is in agreement with the theoretical mass of the isotopically pure protein to within 0.01 Da.

Calmodulin can also be studied by negative ion ESI (30–32). The charge-state distribution patterns reported for calmodulin are slightly different in each case, but on the whole reflect the choice of solvent and the ESI parameters used. For example, Hu and co-workers report (31) a pattern centred on the 12^- charge state, with a maximum of 16^- charges observed in water, whereas Veenstra and co-workers (32) show a distribution centred on 15^- and a maximum of 19^- in buffer containing 15% methanol. However, the spectrum given by Nemirovskiy and co-workers (30) in neutral solution (pH 7) shows a higher maximum at 26^-, with the pattern centred on the 19⁻ charge state. Calmodulin contains 39 acidic residues, 13 of which are buried within the protein structure, which implies that in the experiments of Nemirovskiy and co-workers all of the remaining 26 are accessible for ionization at a physiological pH.

The use of denaturing solvents, such as methanol or acetonitrile, must be avoided if the tertiary structure of a protein and any non-covalent protein–ligand interactions are to be preserved. Volatile buffered solutions allow the pH to be carefully controlled to suit the requirements of the protein, and better represent physiological conditions. However, the lack of significant ionic strength of ESI solutions must be noted when assessing the physiological relevance of the interactions observed.

Protocol 3

ESI-MS of calmodulin under buffered conditions

Equipment and reagents

- Electrospray source
- Sephadex PD10 column
- Hamilton syringe
- Rheodyne loop
- Calmodulin
- 10 mM ammonium acetate pH 6 or 7
- 10 mM ammonium bicarbonate pH 7–8
- Bradford reagent

Method

1 Dissolve 2 mg (120 nmol) of calmodulin in 2.5 ml ammonium acetate (10 mM, pH 6 or 7) for positive ESI or ammonium bicarbonate (10 mM, pH 7–8) for negative ESI to give a 48 μM solution.

2 Desalt using a Sephadex PD10 column, collecting approx. 10 × 1 ml fractions.

3 Use Bradford reagent to detect which fraction contains the most protein—fraction 4 is usually the best fraction to use. The final protein concentration is approx. 0.4–0.6 mg/ml (24–36 μM) (determined by UV absorption, using a molar extinction coefficient of $\varepsilon_{280nm} = 1560$ M⁻¹cm⁻¹) (33).

4 For ESI Source A: load 100 μl of the solution into a Hamilton syringe and pump into the source at a flow rate of 1 μl/min. Or for ESI Source B: load 10 μl of the solution into the Rheodyne loop with ammonium acetate or ammonium bicarbonate as the carrier solvent. Flow rate is typically 2–5 μl/min.

5 Apply a drying gas (CO_2 or N_2, *at a lower temperature of 30–40 °C*) at the capillary entrance (ESI Source A) or a warm counter flow of nitrogen (*30–50 °C*) (ESI Source B).

6 For positive ESI, apply a typical voltage of –3500 V to the capillary (Source A) or 3500 V to the needle (Source B). Use opposite values for negative ESI.

7 Positioning the needle slightly off-centre and closer to the capillary is often required for a more stable, buffer electrospray with Source A.

8 The voltages in the source must be kept low (especially the capillary and skimmer voltage) to preserve non-covalent interactions.

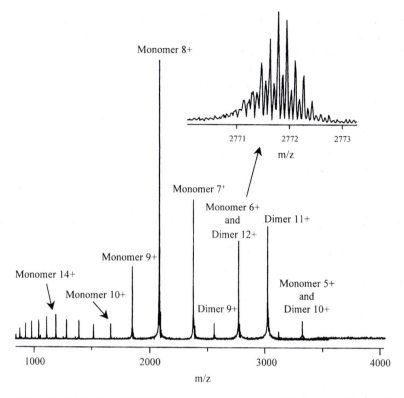

Figure 10 Positive ion ESI FTICR mass spectrum of calmodulin in ammonium acetate buffer (10 mM, pH 6).

Figure 10 shows the positive ESI-FTICR mass spectrum obtained for calmodulin (25 μM) in ammonium acetate (10 mM, pH 6). As can be seen, the resulting spectrum is clearly bimodal. The first envelope is broad and centred on the 14^+ ionization state, similar to that detected with denaturing conditions. The second distribution is centred on the 8^+ charge state and is significantly narrower.

The second envelope, observed only in aqueous buffer solution, is attributed to the presence of a more compact calmodulin conformation that has fewer residues accessible for protonation due to an increased level of folding. It is expected that some of the basic sites would be involved in the formation of secondary structures in the folded conformation.

In addition to monomeric calmodulin species, *Figure 10* also reveals the presence of a species with a mass exactly twice that of the monomer, i.e. a noncovalent calmodulin dimer (34). The inset shows the expansion of the signal at m/z 2770, which reveals the overlapping of two species, corresponding to monomer and dimer. The clear distinction of signals for dimer and monomer species within the same charge state illustrates the resolving power of high-field FTICR mass spectrometry. Furthermore, the mass obtained for the dimer corresponds to exactly twice that of the calmodulin monomer so that high-field FTICR allows the unambiguous assignment of the involvement of non-covalent, as oppose to covalent, associations.

Comparison of charge-state distribution patterns in the absence and presence of ligands may give a qualitative idea of whether ligand binding induces a conformational change in the protein, and whether such a conformational change results in a more folded or unfolded structure. An example of a charge-state distribution shift that accompanies the calcium-induced conformational change in calmodulin is given in Section 4.2.2. The binding of a ligand to a protein may also involve salt links with some of the ionizable sites, so that care must be taken when interpreting pattern-shifts.

ii. Hydrogen/deuterium exchange

Deuterium studies are routinely used with NMR for the characterization of differences between protein conformations. H/D exchange can also be combined with mass spectrometry to explore the conformational relationship between proteins in solution and in the gas phase, and is easily monitored by the mass increase of the protein. The exchangeable hydrogens in solution are those amide protons on the peptide backbone (NH) and on some of the amino acid side chains, e.g. RCO_2H, ROH, RSH, and RNH_2. Those in structured regions will exchange more slowly than those in unstructured regions, and the measured rate of H/D exchange varies with the conformational state. A faster exchange indicates a more open structure and a slower rate suggests a more tightly folded conformation. The total number of protons exchanged is also an important parameter but one which will vary according to the H/D method used. H/D exchange can be carried out in solution prior to mass spectrometric analysis (*Protocol 4*). H/D exchange can also be effected *in situ* in the gas phase (*Protocol 5*), although much less is known about H/D exchange within gaseous ions.

Protocol 4

Solution phase H/D exchange

If purification steps are necessary then it is often more convenient to do these first and freeze-dry the protein before the exchange experiments. Choice and control of temperature for exchange experiments is crucially important.

It may be necessary to reduce disulfide bonds to achieve significant H/D exchange. This is achieved by the addition of a 50-fold molar excess of dithiothreitol to a solution of protein in 0.05 M NH_4HCO_3 pH 8, followed by incubation at 37 °C for 2 h. The protein should be freeze-dried prior to exchange.

There are principally two ways to follow H/D exchange in solution. The incorporation of deuterium can be monitored, or back-exchange can be initiated by the addition of a non-deuterated solvent and monitored.

Equipment and reagents

- See *Protocol 3*
- D_2O or deuterated buffer
- D_2O containing a small percentage of deuterated acetonitrile or deuterated methanol (CH_3OD or CD_3OD)

A H/D exchange

1 Dissolve 0.4–0.6 mg/ml in deuterated solvent. To maintain better the folded conformation of the protein use, for example, D_2O or deuterated buffer prepared with D_2O. To study a more unfolded conformation then use D_2O containing a small percentage of deuterated acetonitrile or deuterated methanol (CH_3OD or CD_3OD). NB. Methanol may also *increase* the α-helicity of a protein.

2 Follow *Protocol 3*, steps 4–6.

3 Acquire several spectra over a period of time to monitor any changes in H/D exchange, e.g. H/D exchange may be slow enough to time-resolve; back-exchange may, however, be significant.

B D/H exchange (35)

1 Deuterate at all exchangeable sites by incubating protein in D_2O. The time required for the exchange of all labile hydrogens will be dependent on the protein and the conditions used, in particular pH and temperature. Some proteins cannot withstand high temperatures or low pH without denaturation occurring and therefore must be incubated for longer periods of time at low temperatures and closer to neutral pH. Complete exchange may be achieved more quickly at elevated temperatures and at a lower pH for more robust proteins.

2 The exchange reaction is initiated by a dilution (typically 100-fold) into H_2O at room temperature.

3 Quenching of exchange can be achieved by lowering the temperature to 0–4 °C prior to injection into the ESI source. The nebuliser gas must also be cooled. When using a rheodyne loop for sample injection, the carrier solvent must also be cooled.

Notes:

(a) To prevent back-exchange, sample vials should be tightly sealed and stored in a desiccator. It may useful to introduce a heavy gas, e.g. argon, above the sample surface. Active deuterated reagents may undergo significant back-exchange in a few minutes. If significant back-exchange is observed, it may be necessary to enclose the electrospray area and flush with dry air.

(b) Protein folding can also be monitored by determining the change in H/D exchange as a denaturant is diluted out of the solution.

(c) Incorporation of deuterium can also be monitored in short regions of a protein by first exchanging in D_2O at pD = 7.0, then digesting with a protease at a lower pH (e.g. pH 2.4 for pepsin, 0 °C), and finally monitoring the deuterium content of each fragment. A correction must be made for back-exchange.

Protocol 5

Gas phase H/D exchange in FTICR

Equipment and reagents

- See *Protocols* 2 and 3
- Deuterated reagent (e.g. D_2O, CD_3OD, CD_3COOD, or ND_3)

Method

1 Follow *Protocol* 2, steps 1–4 or *Protocol* 3, steps 1–6, depending on solution conditions required.

2 Introduce a deuterated reagent, e.g. D_2O, CD_3OD, CD_3COOD, or ND_3 (ordered according to their proton affinity, with ND_3 having the highest proton affinity and therefore being most efficient) into the cell/reaction capillary of the system either via a pulse valve or a leak valve at a pressure of 10^{-6} to 10^{-7} Torr.

3 Leave a reaction delay for the gas to react with the compound.

4 Pump down the pressure to 10^{-9} Torr.

5 Acquire a spectrum.

6 Repeat for several reaction delay times.

7 A reagent gas with a similar proton affinity to the analyte molecule must be used.

8 The uncertainties in pressure measurements are high ($\sim 20\%$) and are often the major source of error.

Figure 11 shows an example of gas phase H/D exchange for calmodulin in buffer. The figure shows an expansion of the 8^+ charge state at several delay times after the introduction of ND_3 into the cell. It is clear that the mass of this species increases, as a result of H/D exchange, with the increase in reaction delay time (36).

Comparison of exchange rates and total number of exchanges for a protein in a range of environments and in the absence and presence of ligands may yield important information about the accessibility of binding sites and also about conformational changes that ligand-binding may induce in a protein structure.

Deuterium exchange rate, in solution, is itself affected by pH, temperature, and the addition of organic solvents, so it is imperative to distinguish between a variation that is truly due to a protein conformational change and one which arises as a result of an environmental change. This is especially important when comparing rates for native and denatured proteins in different solution conditions.

4.2.2 Stoichiometry

The number of ligands that form a complex with a protein is a characteristic that is important to determine and can, in principle, be easily obtained from the

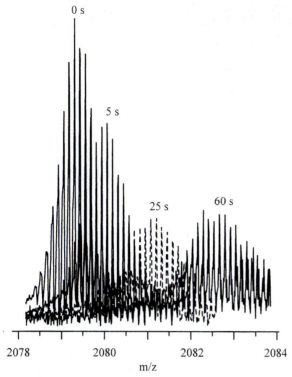

Figure 11 Section of positive ESI-FTICR mass spectrum obtained for calmodulin in ammonium acetate buffer (10 mM, pH 6) showing the 8^+ charge state at various times after the injection of ND_3 into the cell (0, 5, 25, and 60 sec). The inset shows a plot of the number of deuterium exchanged with time for this charge state.

mass spectrum, because the mass of the complex is directly measured. The accurate determination of stoichiometry is arguably one of the strongest features of mass spectrometry for protein–ligand interactions. Care must be taken, however, to select a suitable environment and suitable protein:ligand concentration ratios as these may significantly affect the stoichiometry observed, especially in the case of ion binding.

The fine-tuning of calmodulin function involves the action of many different ions. Calcium is the main regulator of calmodulin but magnesium, potassium, sodium, and zinc are also known to be involved. Calmodulin is known to have four specific calcium-binding sites as well as a number of auxiliary sites. *Figures 12* and *13* show the ESI spectra obtained for calmodulin in buffer in the presence of calcium (*Protocol 6*) using a quadrupole and an FTICR (37) respectively.

As can be seen from both spectra, calcium binding to calmodulin results in a shift in the most abundant charge state from 8^+ (*Figure 10*) to 7^+. This is consistent with the calcium-induced tightening of the calmodulin structure, which has previously been shown by a range of other biophysical techniques. Such a shift is *not* observed when calcium is substituted by magnesium (37), also in agreement with literature reports that magnesium does not significantly alter

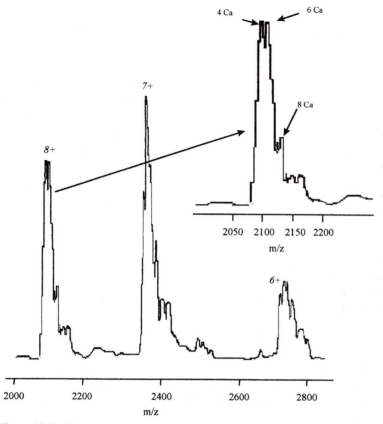

Figure 12 Positive ion ESI mass spectrum of calmodulin in the presence of calcium in ammonium acetate (5 mM, pH 6). Inset shows the deconvoluted spectrum of calcium binding to calmodulin.

the structure of calmodulin upon binding. Again, the difference in the resolving power between the quadrupole and the FTICR is immediately obvious from the quality of the ESI mass spectra obtained.

Protocol 6

Ion binding to calmodulin

Equipment and reagents

- See *Protocol 3*
- 1 M HCl

- 100 mM solution of salt, e.g. calcium chloride, calcium acetate, or magnesium chloride, in 10 mM ammonium acetate buffer

Method

1   Plasticware must be rinsed with 1 M HCl and washed with ultra-pure water.

Protocol 6 continued

2 Desalt and prepare a calmodulin solution in buffer as previously described (*Protocol 3*, steps 1–3).

3 Prepare a 100 mM solution of salt, e.g. calcium chloride, calcium acetate, or magnesium chloride, in 10 mM ammonium acetate buffer. Using a high concentration of stock salt will mean that only small volumes need to be added to the protein stock to give the optimal range of salt concentration, thereby reducing dilution effects.

4 Using a fixed protein concentration add a range of small volumes of salt stock depending on the final concentration of ion required. The maximum salt concentration that can generally be tolerated by ESI is around 2 mM.

5 Extensively wash sample lines and capillary with 10 mM ammonium acetate between each run.

6 Introduce the sample as described earlier (*Protocol 3*, steps 4–6).

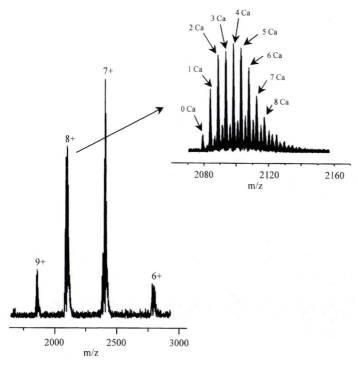

Figure 13 Positive ion ESI-FTICR mass spectrum of calmodulin in the presence of calcium in ammonium acetate (10 mM, pH 6). Inset shows an expansion of the 8$^+$ charge state.

Determination of the number of ligand-binding sites, and distinction of high-affinity sites from low-affinity sites, requires the variation of the ligand concentration. High-affinity, *specific* sites are identified using lower ligand:protein concentration ratios, for example the four main calcium-binding sites of calmodulin become occupied before the low-affinity auxiliary sites.

The presence of low-affinity binding sites can often be misleading when determining stoichiometry, especially for ion binding. At high ligand:protein molar ratios, auxiliary sites become occupied and may dominate a spectrum. The maximum stoichiometry observed then becomes dependent on the concentration of ligand in the solution. In the case of calmodulin at least ten calcium sites have been identified and the occupation of these varies with the calcium/calmodulin ratio (1, 37).

Determination of the stoichiometry of protein–ligand binding involving larger ligands, e.g. peptides, is usually more straightforward because there are often fewer sites involved (38). However, hydrophobic forces can often be more dominant in protein–peptide binding, as opposed to metal ion binding which is primarily an electrostatic interaction, and these complexes are therefore sometimes more difficult to maintain and observe in the gas phase.

4.2.3 Affinity

Absolute ligand-binding affinities are, at present, difficult, if not impossible, to determine accurately by mass spectrometry. This is primarily due to the fact that most compounds have different ionization efficiencies and so it is difficult to correlate quantitatively signal intensities. Ion intensities may be better correlated within the same protein system or when two species, close in mass, have very similar ionization efficiencies. Competition reactions may also provide useful information about *relative* ligand-binding affinities. Relative binding strengths can be estimated by increasing the collisional energy put into a protein–ligand complex during the ESI process, e.g. by altering the capillary/skimmer voltages, and monitoring for the point of dissociation of the complex. While there is no absolute correlation between the voltage required for dissociation and the strength of a given complex, a qualitative indication can be obtained within a particular protein system, e.g. a range of mutants of the same protein can be screened.

4.2.4 Location

The fragmentation of a protein is commonly used to gain information about its sequence. Furthermore, comparison of the fragmentation patterns of a protein alone and that complexed with a ligand may yield precious information about the site of binding. The binding of a ligand may hide some of the sites previously susceptible to fragmentation. Mass spectrometry can be used for protein sequence determination in two ways. Complexes can be fragmented by chemical or enzymatic means and the resulting fragments can be analysed by mass spectrometry, or fragments can be generated during the mass spectrometry experiments, e.g. nozzle/skimmer or tandem mass spectrometry.

i. Nomenclature for protein fragmentation

Figure 14 summarizes the nomenclature used for assigning the fragments that arise from the fragmentation of the protein backbone. Fragment ions can be separated into two categories:

(a) Those that arise as a result of cleavage from the C-terminus and retain the charge on the N-terminus.

(b) Those fragmented from the N-terminus which retain the charge on the C-terminus.

In both cases fragmentation can occur at three positions and fragments are designated as a_n, b_n, and c_n (N-terminus ions) and x_n, y_n, and z_n (C-terminus ions). Fragmentation can also occur within the side chains (R-groups).

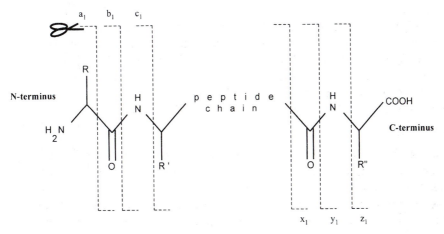

Figure 14 Nomenclature for describing the fragmentation of proteins.

ii. Chemical/ enzymatic digestion

Proteins can be enzymatically cleaved with one or two selected endoproteases or can be chemically digested (*Protocol 7*). The resulting fragments are fractionated by reverse-phase high-performance liquid chromatography (HPLC) or separated by gel filtration or electrophoresis, prior to sequencing. Before the application of mass spectrometry to protein sequencing, peptides were primarily sequenced using automated Edman methods.

Protocol 7

Chemical and enzymatic fragmentation of proteins

Equipment and reagents

- See *Protocol 2*
- See *Table 2*
- 2 mg cyanogen bromide in 100 μl of 70% trifluoroacetic acid

- Reverse-phase HPLC equipment
- 100 pmol of protein in 5 μl of 70% trifluoroacetic acid

A Chemical cleavage

1 Cyanogen bromide cleaves at the C-terminal side of methionine residues.

2 Cyanogen bromide is extremely toxic and should be handled carefully in a well-ventilated fume hood.

Protocol 7 continued

3 Dissolve 2 mg cyanogen bromide in 100 μl of 70% trifluoroacetic acid.

4 Dissolve 100 pmol of protein in 5 μl of 70% trifluoroacetic acid.

5 Mix protein solution with cyanogen bromide solution and leave to digest for 2 h at 25 °C.

6 Dry sample on a vacuum centrifuge equipped with a dry-ice trap.

7 Redissolve sample in 100 μl of 70% trifluoroacetic acid.

8 Separate fragments by reverse-phase HPLC.

9 Analyse fractions by ESI mass spectroscopy, e.g. *Protocol 2*.

10 *N*-chlorosuccinimide can also be used to cleave at the C-terminal side of tryptophan residues.

B Enzymatic digestion

1 Many proteins are protease resistant and often require reduction and *S*-alkylation prior to digestion.

2 Cystine and cysteine residues are often destroyed during sequencing and should be modified, e.g. by reduction with β-mercaptoethanol followed by carboxymethylation of the resulting thiol groups with iodoacetate.

3 Each enzyme has specific sites of action and results in partial hydrolysis of the protein.

4 At least two proteolytic enzymes are usually required to determine a complete sequence. It is important to use enzymes which hydrolyse different sets of peptide bonds, so that an overlap of fragments is obtained. *Table 2* gives the site of action for some proteolytic enzymes.

5 *Protocols* vary depending on the enzyme used.

iii. Nozzle/skimmer sequencing

Ions formed during the ESI process are subject to collisions upon entering the mass analyser and may undergo fragmentation, charge stripping, and declustering. The energy with which ions enter the mass analyser can be adjusted by controlling the electrospray potentials, e.g. the capillary or skimmer voltages. It is thus possible to manipulate the amount of fragmentation that takes place and, in this way, obtain sequence information for a protein system. Fragmentation of protein–ligand complexes can allow the site of ligand binding to be determined, or at least bracketed, provided the non-covalent protein–ligand associations are preserved during fragmentation.

Figure 15 shows an example of ESI-FTICR nozzle/skimmer fragmentation of calmodulin in the presence of magnesium (37). Under harsh source conditions and at this particular magnesium concentration the intact 8^+ charge-state of calmodulin has at least six magnesiums bound. Analysis of the peaks at the lower m/z range reveals that significant fragmentation of calmodulin has occurred.

Table 2 Sites of action of some proteolytic enzymes[a]

Enzyme	Major sites of action	Other sites
Carboxypeptidase A	C-terminal bond of Tyr, Trp, Phe	Does not act at Arg, Lys, or Pro
Carboxypeptidase B	C-terminal of Arg, Lys	None
Carboxypeptidase Y	C-terminal bond	All residues
Chymotrypsin	Carboxyl side of Trp, Phe, Tyr	Leu, Met, Asn, His
Clostripain	Carboxyl side of Arg	
Elastase	Carboxyl side of neutral aliphatic residues	
Leucine aminopeptidase	N-terminal bond	Not for Pro
Papain	Carboxyl side of Arg, Gly, Lys	No action at acidic residues
Pepsin	Carboxyl side of Trp, Phe, Tyr, Met, Leu	Various sites
Staphylococcal protease	Carboxyl side of Glu	Some Asp
Subtilisin	Aromatic and aliphatic residues	Various
Thermolysin	N-terminal side of aliphatic residues	Also Phe
Trypsin	Carboxyl side of Arg, Lys	Not at Lys–Pro or Arg–Pro

[a] Abbreviations: Arg (arginine), Asn (asparagine), Asp (aspartic acid), Gly (glycine), His (histidine), Leu (leucine), Lys (lysine), Met (methionine), Phe (phenylalanine), Pro (proline), Trp (tryptophan) and Tyr (tyrosine).

These fragments are not present when gentler source conditions are used, for an example see *Figure 10*. Closer inspection of the mass spectrum shows the presence of peaks corresponding to the binding of just one or two magnesiums to particular calmodulin fragments.

iv. Tandem mass spectrometry

Tandem mass spectrometry is often a more convenient method for protein sequencing than nozzle/skimmer methods, since it allows the isolation of an ion of interest. The limitation of CID, however, is that is only efficient for small proteins and peptides (< 3 kDa). The resulting fragment ions can be mass analysed without the complications of many other ions. Fragmentation is often achieved by colliding the ion of interest with neutral Ar, Xe, or He molecules (collision-induced dissociation, CID). CID can also be used to investigate the relative stability of a non-covalent complex (*affinity*). The use of ion traps or FTICR instruments allows several stages of ion selection and fragmentation (MS^n).

4.2.5 Relevance to solution phase behaviour

The major forces that govern protein structure and their non-covalent interactions in solution are hydrophobic, ionic, hydrogen bonding, and van der Waals. In contrast, electrostatic forces often dominate gas phase protein interactions, and hydrophobic forces are lacking in the absence of solvent.

Relevance of gas phase mass spectrometric measurements to solution phase

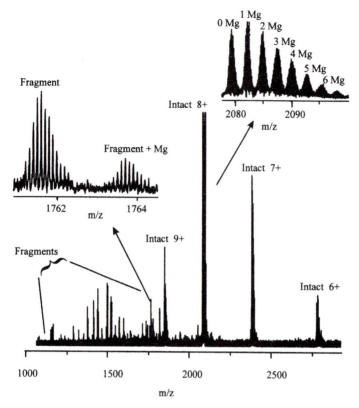

Figure 15 Positive ion nozzle/skimmer CID ESI-FTICR mass spectrum obtained for calmodulin in the presence of magnesium. The insets show magnesium binding to the intact calmodulin and to a calmodulin fragment.

behaviour is a much debated concern. There have certainly been many studies reporting the detection of non-covalent protein–ligand complexes in the gas phase, but questions about whether these complexes result from non-specific coulombic interactions during the electrospray process or whether they arise truly as a result of successful transfer from solution must always be addressed.

Examples of both where ESI results were in agreement with solution phase data (39, 40) *and* where ESI results did not correlate to known solution phase behaviour (41, 42) have been reported in the literature. Control experiments are therefore critical to ensure that non-specific interactions are ruled out.

Mass spectrometry, as with many other methods, is most effective when used as a complement to other techniques.

References

1. Lafitte, D., Capony, J. P., Grassy, G., Haiech, J., and Calas, B. (1995). *Biochemistry*, **34**, 13825.
2. Feng, R., Castelhano, A. L., Billedeau, R., and Yuan, Z. (1995). *J. Am. Soc. Mass Spectrom.*, **6**, 1105.

3. Gale, D. C., Goodlett, D. R., Light-Wahl, K. J., and Smith, R. D. (1994). *J. Am. Chem. Soc.*, **116**, 6027.

4. Ganem, B., Li, Y. T., and Henion, J. D. (1991). *J. Am. Chem. Soc.*, **113**, 7818.

5. Smith, R. D., Loo, J. A., Edmonds, C. G., Barinaga, C. J., and Udseth, H. R. (1990). *Anal. Chem.*, **62**, 882.

6. Siuzdak, G. (1996). *Mass spectrometry for biotechnology*. Academic Press, Inc., California, USA.

7. Przbylski, M. and Glocker, M. O. (1996). *Angew. Chem. Int. Ed. Engl.*, **35**, 806.

8. Smith, R. D., Bruce, J. E., Wu, Q., and Lei, Q. P. (1997). *Chem. Soc. Rev.*, **26**, 191.

9. Gaskell, S. J. (1997). *J. Mass Spectrom.* **32**, 677.

10. Edmonds, C. G. and Smith, R. D. (1990). In *Methods in enzymology* Academic Press, Inc., NY, USA (ed. J. A. McCloskey), Vol. 193, p. 412.

11. Chapman, J. R. (1995). *Practical organic mass spectrometry*, 2nd edn. J. Wiley & Sons, England.

12. Green, M. K. and Lebrilla, C. B. (1997). *Mass Spectrom. Rev.*, **16**, 53.

13. Loo, J. A. (1997). *Mass Spectrom. Rev.*, **16**, 1.

14. Lee, T. D. and Shively, J. E. (1990). In *Methods in enzymology* Academic Press, Inc., NY, USA (ed. J. A. McCloskey), Vol. 193, p. 361.

15. Biemann, K. (1990). In *Methods in enzymology* Academic Press, Inc., NY, USA (ed. J. A. McCloskey), Vol. 193, p. 455.

16. Winston, R. L. and Fitzgerald, M. C. (1997). *Mass Spectrom. Rev.*, **16**, 165.

17. Schindler, P. A., Van Dorsselaer, A., and Falick, A. M. (1993). *Anal. Biochem.*, **213**, 256.

18. Chan, T.-W. D., Colburn, A. W., and Derrick, P. J. (1992). *Org. Mass Spectrom.*, **27**, 53.

19. Senko, M. W., Hendrickson, C. L., Pasa-Tolic, L., Marto, J. A., White, F. M., Guan, S., *et al.* (1996). *Rapid Commun. Mass Spectrom.*, **10**, 1824.

20. Kelleher, N. L., Senko, M. W., Siegel, M. M., and McLafferty, F. W. (1997). *J. Am. Soc. Mass Spectrom.*, **8**, 380.

21. Bruce, J. E., Smith, V. F., Liu, C., Randall, L. L., and Smith, R. D. (1998). *Protein Sci.*, **7**, 1180.

22. Kawasaki, H. and Kretsinger, R. H. (1994). *Protein Sci.*, **1**, 343.

23. Milos, M., Schaer, J. J., Comte, M., and Cox, J. A. (1986). *J. Inorg. Biochem.*, **36**, 11.

24. Babu, Y. S., Bugg, C. E., and Cook, W. J. (1988). *J. Mol. Biol.*, **204**, 191.

25. Ikura, M., Clore, M., Gronenborn, A. M., Zhu, G., Klee, C. B., and Bax, A. (1992). *Science*, **256**, 632.

26. Haiech, J., Klee, C. B., and Demaille, J. G. (1981). *Biochemistry*, **20**, 3890.

27. Bayley, P. M. and Martin, S. R. (1992). *Biochim. Biophys. Acta*, **1160**, 16.

28. Kilhoffer, M. C., Kubina, M., Travers, F., and Haiech, J. (1992). *Biochemistry*, **31**, 8098.

29. Pedigo, S. and Shea, M. A. (1995). *Biochemistry*, **34**, 10676.

30. Nemirovskiy, O. V., Ramanathan, R., and Gross, M. L. (1997). *J. Am. Soc. Mass Spectrom.*, **8**, 809.

31. Hu, P., Ye, Q.-Z., and Loo, J. A. (1994). *Anal. Chem.*, **66**, 4190.

32. Veenstra, T. D., Johnson, K. L., Tomlinson, A. J., Naylor, S., and Kumar, R. (1997). *Eur. Mass Spectrom.*, **3**, 453.

33. Gilli, R., Lafitte, D., Lopez, C., Kilhoffer, M., Makarov, A., Briand, C., *et al.* (1998). *Biochemistry*, **37**, 5450.

34. Lafitte, D., Heck, A. J. R., Hill, T. J., Jumel, K., Harding, S. E., and Derrick, P. J. (1999). *Eur. J. Biochem.*, **261**, 337.

35. Robinson, C. V., Grob, M., Eyles, S. J., Ewbank, J. J., Mayhew, M., Hartl, F. U., *et al.* (1994). *Nature*, **372**, 646.

36. Wallace, J. I., Ph.D. Thesis, University of Warwick, UK (1999).

37. Lafitte, D. L., Hill, T. J., and Derrick, P. J. (2000) Biochemistry, to be submitted.

38. Hill, T. J., Lafitte, D., Wallace, J. I., Cooper, H. J., Tsvetkov, P. O. and Derrick, P. J. (2000). *Biochemistry*, 39, 7284.

39. Light-Wahl, K. J., Winger, B. E., and Smith, D. J. (1993). *J. Am. Chem. Soc.*, **115**, 5869.

40. Tang, X. J., Brewer, C. F., Saha, S., Chernushevich, I., Ens, W., and Standing, K. G. (1994). *Rapid. Commun. Mass Spectrom.*, **8**, 750.

41. Robinson, C. V., Chung, E. W., Kragelund, B. B., Knudsen, J., Aplin, R. T., Poulsen, F. M., *et al.* (1996). *J. Am. Chem. Soc.*, **118**, 8646.

42. Wu, Q., Gao, J., Joseph-McCarthy, D., Sigal, G. B., Bruce, J. E., Whitesides, G. M., *et al.* (1997). *J. Am. Chem. Soc.*, **119**, 1157.

Chapter 9
Nitroxide spin labels as paramagnetic probes

David A. Middleton

University of Oxford, Biomembrane Structure Unit, Department of Biochemistry, South Parks Road, Oxford OX1 3QU, UK.

1 Introduction

This chapter describes how spin label electron paramagnetic resonance (EPR) spectroscopy can be used to examine aspects of molecular structure and dynamics, and changes therein, associated with the binding of ligands to receptors at their sites of action. The scope of the chapter is restricted to studies of proteins that function within cell membranes, where this approach is particularly relevant, although some of the experimental methods can be adapted to accommodate studies on soluble proteins.

Membrane proteins hold key roles in important cellular events such as signal transduction, neurotransmission, and solute transport, and their interaction with ligands is central to the activation, regulation, or inhibition of activities occurring within the cell. Our understanding of how ligands, substrates, and agonists regulate membrane proteins at the molecular level has been hampered by the difficulties in producing crystals for X-ray structure determination, but much has been deduced from spectroscopic methods that observe changes in specific features of proteins accompanying ligand binding.

Spin label EPR spectroscopy has long been a method of choice for the study of lipids and proteins in membranes for a number of reasons. The spectra are anisotropic and give information on molecular order, electronic environment, structure, and dynamics. Moreover, spectra do not suffer from background signal interference and spin labels can be introduced at specific positions within a protein or ligand, helping to make assignment unambiguous. These features make EPR spectroscopy an attractive technique with which to investigate the interactions of ligands with membrane proteins. The rotational mobility of proteins in the membrane can be measured from the EPR spectrum, with the potential to visualize changes in the size of protein assemblies during ligand binding (*Figure 1*). This approach may be useful for examining, for instance, single-transmembrane receptors that require ligand-induced receptor oligomerization for activation (1). EPR is also sensitive to local molecular motions and may be suit-

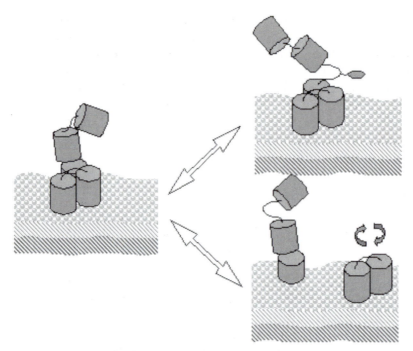

Figure 1 Two possible effects of ligand binding on membrane protein structure. The EPR spectrum of a spin label attached to the protein may be sensitive to changes in protein conformation (*top*) changing the motional characteristics of the probe. The EPR spectrum may also reflect changes in protein–protein associations (*bottom*) that effect the rate of rotational diffusion.

able for reporting ligand-induced conformational changes of proteins (*Figure 1*). G-protein coupled receptors possess hinge residues, and agonist binding might induce conformational changes that influence dissociation and association of G-proteins to the receptors and which are visible to EPR spectroscopy (2).

This chapter demonstrates some of the capabilities of EPR combined with spin labelling, and highlights the strategies that are can be employed for visualizing physical characteristics of the protein that might be influenced by the binding of small molecules. Some general guidance on EPR instrumentation is given, but it is assumed that the reader will have consulted the documentation provided by the spectrometer manufacturer before performing experiments.

2 Origin of the EPR spectrum

2.1 Magnetic resonance

Unpaired electrons, like some atomic nuclei, possess the property of spin, which is a source of angular momentum that confers a magnetic dipole moment μ upon the electron and is fundamental to all magnetic resonance spectroscopy. The spin quantum number, I, of an electron equals ½ and the spins can adopt

$2I + 1 = 2$ quantum mechanical states (S), conveniently defined as $+\frac{1}{2}$ and $-\frac{1}{2}$. Under normal conditions the two spin-states are degenerate, but in the presence of an applied magnetic field the dipole moments associated with the two spin-states adopt orientations in which they are aligned with or against the direction of the magnetic field. The parallel orientation possesses a lower energy than the antiparallel orientation, and the energy separating the two spin-states, ΔE, is proportional to the strength of the applied magnetic field B_0 according to:

$$\Delta E = h\nu = g\beta B_0 \qquad [1]$$

where h is the Planck constant (6.626×10^{-34} J s) and β is the Bohr magneton (9.3×10^{-24} JT^{-1}). The g value, also known as the gyromagnetic ratio or Landé splitting factor, is equal to 2.00232 for a free electron in an isolated atom in a vacuum, but varies according to molecular orientation and electronic environment (see Section 2.5). In terms of classical mechanics, ν is also the frequency at which the electron spins undergo precession in the magnetic field.

The Boltzmann equation dictates that the occupancy of the two spin-states is proportional to the energy separating them, and there is a slight excess of spins in the lower energy spin-state. Magnetic resonance occurs when the electron experiences an applied field of electromagnetic radiation at the (GHz) frequency of the spin as it precesses in the magnetic field. At resonance, electromagnetic radiation is absorbed and transitions between the spin-states occur, giving rise to the EPR spectrum. This phenomenon is exactly analogous to that of pulsed nuclear magnetic resonance (NMR), in which a broad frequency band of applied electromagnetic radiation in the MHz range induces simultaneous transitions between *nuclear* spin states in different magnetic environments. The technique of NMR is covered in the following chapter.

2.2 Instrumentation

Equation 1 implies that magnetic resonance can be induced in two ways. One way is to maintain a constant static magnetic field (thereby fixing the energy separation between the two spin-states) and to vary the frequency of the applied electromagnetic radiation until the absorption frequency is reached. The other is to fix the frequency of the electromagnetic radiation and to vary the energy separating the two spin-states by sweeping the magnetic field. In practice, most EPR spectrometers operate according to the second principle. A klystron provides a source of monochromatic microwave radiation at constant (GHz) frequency and an electromagnet is employed to sweep the magnetic field, typically around a centre value of about 0.3 T (3000 Gauss).

The central feature of an EPR spectrometer is the resonant cavity, which holds the sample, usually contained in a capillary tube (*Figure 2*). Microwave radiation is transmitted into the cavity via a non-reciprocal device called a circulator, which also directs the radiation reflected from the cavity to the detector. When the incident microwave frequency equals the fundamental resonant frequency of the cavity, a standing wave is established and the cavity stores microwave

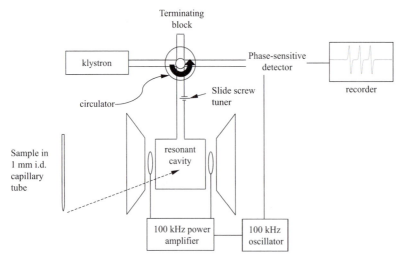

Figure 2 Schematic diagram of a continuous wave EPR spectrometer.

energy. The cavity is said to be critically coupled and will reflect no incident microwaves back to the detector. The sample volume is kept to a minimum in a sealed glass capillary and the largest signal is obtained when the sample is positioned in the centre of cavity where the magnetic field component of the standing wave is highest. The amount of radiation that is coupled to, and reflected from, the cavity is controlled by an iris, which is made to change in size by turning an adjustable iris screw (*Figure 2*). The impedance of the cavity increases as the spins come into resonance, causing small increases in the reflected power that is detected as signal.

The frequency of the incident microwave is constrained by practical limitations and most modern spectrometers operate at a microwave frequency of around 9.5 GHz (X-band).

2.3 Phase-sensitive detection and the first derivative spectrum

The method for detecting the EPR output signal outlined above suffers from the presence of low-frequency noise components, which compromise the sensitivity and detection limits of the spectrometer. Phase-sensitive detection, combined with small-amplitude field modulation (normally at a frequency of 100 kHz) of the output signal, is used to reject the noise components outside a narrow band centred at the modulation frequency. Phase-sensitive detection is achieved by comparing the amplitude-modulated signal with the output of the 100 kHz oscillator; if the signals are opposite in phase the output signal is a minimum, whereas the maximum signal is obtained when the signals are fully in-phase. Hence, the EPR spectrum is obtained as the first derivative of the pure absorption line shape (*Figure 3*; left). A linear relationship exists between the double

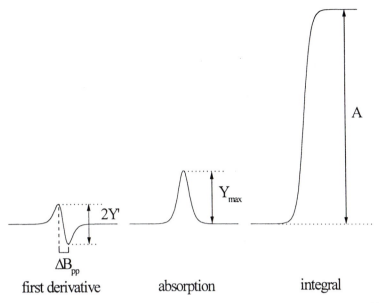

first derivative   absorption   integral

Figure 3 The first derivative EPR spectrum showing the peak-to-peak line width ΔB_{pp} and amplitude 2Y'. The area (A) under the absorption spectrum is obtained by double integration of the first derivative spectrum.

integral of the first derivative spectrum, that is, the area under the absorption spectrum, and the number of spins giving rise to the spectrum.

2.4 Obtaining a spectrum

Once the sample is in place in a critically coupled cavity reflecting little or no microwave power there are a number of tasks to be carried out to optimize spectrometer sensitivity and minimize signal distortions. These can be established using a 1 mM aqueous solution of a nitroxide spin label such as TEMPO (see Section 2.5). The standard sample should ideally be of similar volume and dielectric properties to the protein sample of interest.

It is convenient to start with a set of conditions that will allow a recognizable conventional (first-harmonic, in-phase) spectrum to be obtained, which can then be adjusted systematically to improve the quality of the spectrum. *Protocol 1* below forms a convenient starting point for recording spectra.

Protocol 1

Establishing conditions for running EPR experiments

Equipment and reagents
- EPR spectrometer
- 1 mM aqueous solution of TEMPO

Method

1 This protocol can be carried out using a 1 mM aqueous solution of TEMPO after tuning the klystron, microwave bridge, and resonant cavity according to the instrument user manual.

2 Configure the spectrometer according to the following conditions:

   (a) Centre magnetic field: 3480 G[a]

   (b) Sweep width: 200 G

   (c) Modulation frequency: 100 kHz

   (d) Modulation amplitude: 1 G

   (e) Number of scans: 1

   (f) Harmonic: 1

   (g) Conversion time: 5.12 ms

   (h) Time constant: 1.28 ms

   (i) Number of points: 1024

   (j) Receiver gain: 2×10^{-3}

3 Record a spectrum.

4 If the receiver gain is too high the detector will be saturated and the top of the spectrum will be clipped. If the spectrum is clipped, halve the receiver gain and record another spectrum. If the spectrum is not clipped, double the receiver gain and record another spectrum.

5 Repeat step 4 to obtain the maximum signal amplitude without clipping the spectrum.

6 The sweep width of the applied magnetic field is deliberately broad to ensure that all resonance conditions are met for the spin label so that the spectrum appears within the field of view. After steps 1–5 the centre field value can be changed to correspond to the exact centre of the spectrum.

7 The sweep width can then be cut down to 100 G (1 mT), which is ample for all nitroxide spin labels.

[a] This value is related through *Equation 1* to a frequency of the incident microwave radiation (and the resonant frequency of the cavity) of 9.8 GHz.

The microwave power, modulation frequency, and amplitude between them affect the signal-to-noise ratio, line shape, and resolution of the EPR spectrum. In conventional EPR spectroscopy, the microwave power should be adjusted to provide the maximum signal without saturating the spins, which reduces signal amplitude and causes line shape distortions. Calibration of the microwave power is covered in Section 4, and for conventional spectroscopy the incident power should correspond to the top of the linear phase of a graph (like the one shown later in *Figure 15*) and should be no greater than the power giving maximum signal intensity. Increasing the modulation amplitude will initially have a positive effect on signal intensity but is countered by an increase in the

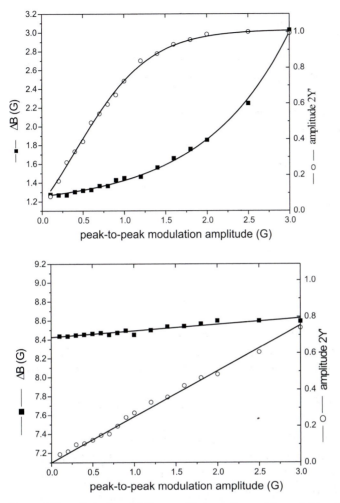

Figure 4 The effect of modulation amplitude on the peak-to-peak line widths (ΔB) and amplitudes (2Y') of a narrow line isotropic spectrum from isotropically tumbling spin label (*left*) and a broad spectrum from spin labelled membrane protein (*right*).

line widths (*Figure 4*). Optimum modulation amplitude is chosen to give good signal without compromising spectral resolution; generally a value is used which is less than the line width observed at low modulation amplitudes. In practice, the broad line widths from nitroxides attached to membrane proteins permit the use of higher modulation amplitudes than for nitroxides in solution without (*Figure 4*).

2.5 EPR spectroscopy of nitroxide spin labels

2.5.1 Spin labels

Unlike the more widely used technique of nuclear magnetic resonance (covered in Chapter 10) EPR spectroscopy rarely enjoys the benefit of intrinsic magnetic

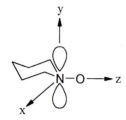

Figure 5 Chemical structure of the nitroxide moiety showing the molecular axis system used by convention. The z axis is defined along the nitrogen–oxygen bond and the y axis is parallel with the nitrogen p-π orbital.

properties of the ligand–protein complex. Chemical labelling methodologies akin to those employed for fluorescent or radioisotope labelling of ligands and proteins often must be applied to make the protein or ligand visible to EPR spectroscopy.

Spin labels are stable free-radical molecules containing one or more unpaired electrons and a reactive functional group able to covalently modify protein side chains and other reactive groups under necessarily mild (low temperature, neutral pH) conditions. A wide variety of spin labels are commercially available from the major fine chemical companies, in which the distinguishing feature is reactivity toward different protein functional groups or residues (3). The common labels are based around the TEMPO (2,2,6,6-tetramethyl-1-piperidinyloxy) and PROXYL (2,2,5,5-tetramethyl-1-pyrrolidinyloxy) structures, which contain the paramagnetic nitroxide moiety (*Figure 5*).

2.5.2 Anisotropy

In the case of an unpaired electron in the p orbital of a nitroxide molecule, the g value in *Equation 1* is anisotropic and can be defined as a second-rank tensor. Deviations of g from the value of 2.00232 (Section 2.1) will therefore depend not only on the electronic environment experienced by the nitroxide, but also on the orientation of the applied magnetic field with respect to the molecular axis system (*Figure 6*). In solutions of low viscosity, rapid isotropic tumbling of the nitroxide reduces the g value anisotropy to a single value averaged over all possible molecular orientations.

From *Equation 1* and *Figure 6* it is clear that the resonance frequency of an electron spin in a magnetic field, and hence the position of a single EPR absorption line, is dependent on the orientation of the g tensor in the magnetic field. Therefore, the position of the resonance will vary according to the orientation of the nitroxide spin label in the magnetic field.

2.5.3 The hyperfine interaction

Although the anisotropic g value carries some limited information about molecular orientation and electronic environment, its effect upon the EPR spectrum is quite small and has been somewhat exaggerated in *Figure 6*. The most important feature of the nitroxide EPR spectrum arises from the hyperfine coupling between the electron and ^{14}N nuclear spin which gives rise to a multiplet structure in the spectral lines.

The ^{14}N nucleus of the nitroxide group has a spin quantum number of 1 and

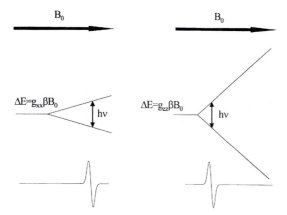

Figure 6 Illustration of the effect of the g value and sample orientation on the position of spectral lines from a hypothetical single crystal. When g is anisotropic (e.g. $g_{xx} g_{zz}$), the separation of energy levels in the increasing magnetic field varies according to the orientation of the tensor elements with respect to the direction of the magnetic field. The illustration shows that aligning g_{xx} and g_{zz} with the magnetic field causes microwave energy to be absorbed at different magnetic field values.

the nuclear magnetic moment adopts three spin-states in an applied magnetic field which can be considered as being aligned with the field ($I = 1$), against the field ($I = -1$) and with no net alignment ($I = 0$). The two energy levels ($S = +\frac{1}{2}$ and $-\frac{1}{2}$) of the electron spin in the applied magnetic field are each split further into three by the local fields of the three orientations of the neighbouring nuclear dipoles. Transitions are allowed between three pairs of energy levels and, hence, the EPR spectrum of a single nitroxide exhibits three lines (*Figure 7*).

The electron density in the nitroxide p-π orbital is non-spherical and the hyperfine interaction with the nucleus is anisotropic, represented by a second rank tensor **A** with principal elements defined according to the nitroxide molecular axis system (*Figure 5*). When the nitroxide spin label undergoes rapid isotropic reorientation in low viscosity solutions, the hyperfine anisotropy is reduced to zero and the spectrum exhibits three lines of equal intensity (e.g. *Figure 7*) separated by the isotropic average of the hyperfine coupling constant a_0.

The appearance of the nitroxide EPR spectrum is extremely sensitive to restrictions of the mobility of the spin label, such as in a viscous solution of glycerol or when attached to anisotropically rotating membrane lipids and proteins. In such cases, the hyperfine and g value anisotropy are *not* fully reduced to zero and spectral line shape analysis can provide important information on the rates and amplitudes of molecular motion.

2.5.4 Nitroxide groups attached to membrane proteins

Biological membranes are complex arrays of structurally diverse macromolecules including lipids, sterols, peptides, and proteins. The effective molecular weight of even the simplest model membranes is on the order of 10^6 to 10^8 Daltons and the correlation time for rotational diffusion of such large structures is on the

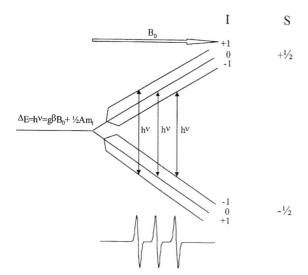

Figure 7 The effect of the nitroxide ^{15}N-electron hyperfine interaction on the electron spin energy levels and the EPR spectrum.

order of milliseconds. Such a slow rate of tumbling is insufficient to reduce the measured hyperfine and g value anisotropy of a nitroxide spin label when restrained at its site of attachment to a membrane protein. In such cases the appearance of the EPR spectrum is close to that observed for a rigid solid (the top spectrum in *Figure 8*) or spin label in a highly viscous solution at the slow motional limit of the EPR time scale.

Often the nitroxide experiences degrees of local freedom when attached to protein side chains, the rates of which are encoded in the EPR line shape through anisotropic averaging of the **A** and **g** tensors (the middle spectrum in *Figure 8*). Labels attached to solvent-exposed regions of a protein, or with long spacer groups between the nitroxide and the site of covalent linkage, are more likely to undergo high amplitudes of local molecular motion than those buried or involved in sites of tertiary contact.

2.5.5 Analysis of spectra

Line shape simulations based on a stochastic Liouville treatment constitute the most rigorous method of analysis of spectra of nitroxides with local rotational correlation times (τ_R) longer than 10^{-9} s. A limited amount of information on spin label dynamics and environment can, however, be deduced from simple inspection of the spectra without recourse to sophisticated line shape simulations (4). Spectra like the example given at the top of *Figure 8*, from nitroxides whose motion is approaching the rigid limit of the EPR time scale, are characterized by an increasing separation ($2A_{zz}$) and line widths (ΔB) of the outer hyperfine extrema. The outer extrema of these spectra correspond to the population of nitroxides having their z axes (*Figure 5*) aligned with the magnetic field direction. Maximum separation of the outer hyperfine extrema ($2A_{max}$) and

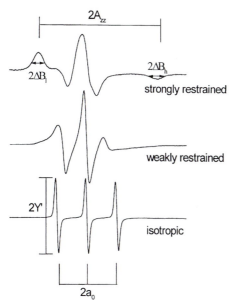

Figure 8 The effect of molecular motion on the EPR spectrum of a nitroxide spin label. Rapid reorientation of the nitroxide in solution gives rise to three narrow lines of equal intensity (*bottom*) from which the isotropic average of the hyperfine constant a_0 can be measured. Weakly restrained nitroxides undergoing motions with a correlation time $\tau > 10^9$ s, when attached to flexible regions of a protein, for example, are characterized by a broadening of the spectral lines and a slight shift in their positions in the spectrum (*middle*). Strongly restrained nitroxides undergoing motions with $\tau > 10^7$ s, have reached, or are approaching, the slow motional limit of the conventional EPR time scale. The width (ΔB) and separation of the lines continue to increase until the anisotropic hyperfine constant A_{zz} reaches the upper limit value, A_{max} (*top*).

maximum line widths (ΔB_{max}) are attained at the rigid limit, when the correlation time for local spin label rotation exceeds 10^{-7} s. The correlation time (τ_R) for rotation of the spin label about its site of attachment can be estimated from the line widths in the low-field (ΔB_l) or high-field (ΔB_h) extrema according to:

$$\tau_R \approx (h/2\pi g\beta\Delta B_{max})\,(\Delta B/\Delta B_{max} - 1)^{-1} \qquad [2]$$

or from the outer hyperfine splitting (A_{zz}) according to:

$$\tau_R \approx (h/2\pi g\beta A_{max})\,(1 - A_{zz}/A_{max})^{-1} \qquad [3]$$

where A_{max} is the maximum attainable (rigid limit) value of the outer hyperfine splitting and all other terms have their usual meanings (Section 2.1). Hubbell *et al.* (5) use two measurements from the EPR spectrum of site-directed spin labelled proteins to relate the extent of nitroxide mobility to the local topography and fold of the protein. The inverse line width of the central resonance, ΔB_0^{-1}, is measured directly from the spectrum (*Figure 8*) while the inverse second moment of the spectrum, $<B_2>^{-1}$, is computed according to the expression:

$$<B_2> = \int_{-\infty}^{\infty} Y(B - B')dB \qquad [4]$$

where B is an arbitrary value of the magnetic field, B′ is the field at the centre of the absorption curve, and Y is the intensity of the absorption spectrum at B (*Figure 3*). Nitroxide groups attached to solvent-exposed loops exhibit high values of both measurements while buried nitroxides generally give the lowest values. Intermediate values suggest sites of tertiary contact or helix surfaces, although there is inevitably some overlap in these areas of measurement. The ability to use the nitroxide spectrum to report on molecular structure hinges on being able to unambiguously identify the modified residue.

The interaction of ligands with membrane proteins may be accompanied by changes in the local mobility of attached spin labels (6). This might occur if the nitroxide moves from a motionally hindered environment to a more exposed region, or the converse, as a result of ligand-induced changes in the protein's secondary and tertiary structure. The effect of ligands on nitroxide mobility may be inferred from changes in single-component spectral line shapes or in the relative contributions of individual superimposed spectra to multiple-component line shapes. Such changes may occur instantaneously after the addition of ligand or may be slow enough to be followed in a time-resolved EPR experiment.

3 Conventional EPR spectroscopy

3.1 Sample requirements

3.1.1 Spin label reactivity

A list of some nitroxide spin labels, and their applications in published studies of membrane proteins, is shown in Table 1. No single label is ideal for modifying all membrane proteins, and it cannot be assumed *a priori* that a particular spin label is suitable for experiments on a protein of interest. In assessing the suitability of a label one might take into account characteristics such as:

- side chain specificity

- size

- flexibility

It is desirable to spin label the protein as selectively as possible, ideally placing labels only at positions that are sensitive to the effects of ligand binding but do not cause loss of protein function or compromise ligand binding affinity. Lysine and cysteine are favoured amino acids for spin labelling because of the reactivity of their side groups. Labels containing the isothiocyanato- substituent preferentially react with the exposed primary amine groups of lysine, but lysines are abundant and often functionally essential residues and the spin label may modify the protein at a number of side groups, or inactivate the protein.

Cysteine occurs somewhat less frequently than lysine in membrane proteins (*Figure 9*) and targeting spin labels at this residue may help to increase the selectivity of protein labelling. Maleimido- and iodoacetamido-labels tend to undergo side-reactions with lysine and, to a lesser extent, methionine side groups, but

Table 1 Nitroxide spin labels and their reactivity toward functional groups in protein side chains

Group modified	R-substituent[a]	Protein/ligand	Reference
Cysteine	Maleimido	Rhodopsin/light	7
		FepA/colicin	8
		Lactose permease/TDG	9
		Aspartate receptor/aspartate	10
		Maltose binding protein/maltose	11
		Aminotransferase	12
	Fluorosulfonylphenyl	Thrombin/hirudin	13
	Iodoacetamido	Ca^{2+}-ATPase/CrATP	14
	Methanethiosulfonato	Bacteriorhodopsin/light	15
		Rhodopsin/light	16
Lysine	Isothiocyanato	H^+/K^+-ATPase/imidazopyridines	6
		5-aminolevulinate dehydratase	17
Methionine	Iodoacetamido	PC transfer protein	18

[a] Introducing aliphatic spacers between the nitroxide group and the reactive functionality can vary the flexibility of the spin label attached to the protein.

spin labels containing functionalities like methanethiosulfonate are much more highly specific for cysteine thiol groups.

The three-dimensional structure and topology of a protein in the membrane may limit further the number of reactive side groups that are available for modification by the spin label. Lysine and cysteine have hydrophilic characteristics and it is common to find both in extramembranous environments of proteins, but cysteine residues have a higher probability than lysine to be found within membrane spanning regions of proteins (*Figure 9*) where they are less available for reaction with the spin label.

In short, when dealing with a wild-type protein prepared from tissue sources there is little guarantee that the protein can be spin labelled selectively and at the desired position in order to maximize the sensitivity of the EPR spectrum to ligand binding. The ability to selectively label protein obtained from an expression system is rather more tractable, however. Hubbell and co-workers (19) pioneered the technique of site-directed spin labelling as a means of introducing nitroxides into specific residues engineered into expressed protein by mutagenic substitution. The technique usually involves replacing all cysteine residues in the protein with an amino acid that does not react with the spin

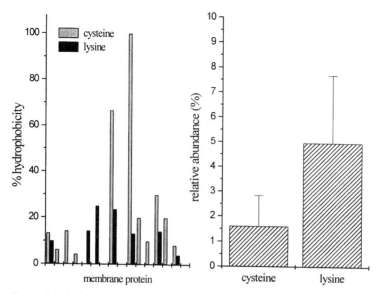

Figure 9 Relative abundance and hydrophobicity of cysteine and lysine residues taken from the primary sequence and hydropathy analysis of representatives of 16 families of membrane protein.

label. Individual cysteines are introduced back into the protein at desired sites to provide the spin label with a unique side group to modify. The advantages of the site-directed approach are clear, but the technique is wholly dependent on access to expression technology capable of producing protein that can be purified in up to nanomole quantities.

3.1.2 Nitroxide-containing ligands

The high selectivity with which many endogenous and synthetic ligands interact with membrane proteins make them attractive targets for incorporating spectroscopic probes such as nitroxide spin labels. By exploiting the site-specificity of a paramagnetic ligand it is possible to circumvent the difficulties of selectivity encountered in side group modifications. In favourable cases the nitroxide ligand binds exclusively to the same recognition site at which the parent ligand induces its biochemical or pharmacological response; hence, the spin labelled ligand acts as a probe of its own binding site. The chemistry used to introduce nitroxides into small molecule ligands too varied to be reviewed in this article, but *Table 2* gives some examples from the literature that are representative of this strategy.

3.2 Modification of protein side groups

3.2.1 Methodology

The structural and functional integrity of membrane proteins is often lost if the protein is removed from its native membrane and not replaced into a similar environment. Spin labelling studies are generally carried out on stable protein

Table 2 Paramagnetic ligands of membrane proteins

Ligand/substrate	Protein	Aim of experiment	Reference
Ouabain-SL	Na$^+$/K$^+$-ATPase	Rotational correlation time	20
Acetylcholine-SL	Acetylcholine receptor	Anaesthetic action	21
Stilbenedisulphonate-SL	Band 3	Probing the binding site	22
XAC-SL	A-1 adenosine receptor	Probing the binding site	23
Phencyclidine	Acetylcholine receptor	Probing the binding site	24
Cholesterol	Cytochrome P450	Binding affinities	25

preparations such as suspensions of purified native membranes or protein reconstituted into well-defined phospholipid bilayers. The protein of interest should constitute the major, or only, protein component of the membrane to avoid excessive spin labelling of contaminants.

Most membrane proteins can be spin labelled rapidly and homogeneously using water soluble agents provided the appropriate protein side groups are exposed to the aqueous medium and available for reaction. In general, the protein is suspended in a buffer solution at a suitable pH for the reaction and incubated with spin label for periods of up to 12 hours. If the protein is temperature-sensitive, overnight incubation on ice or at 4 °C in a cold-room is advisable but most reactions run to completion after 2–3 hours of incubation at room temperature. Removing aliquots at appropriate time points allows the progress of the reaction to be followed by comparing the relative intensities or amplitudes of the lines in the EPR spectrum. It is also a good idea to remove enough material at these time points to carry out a functional assay to monitor the effect that the reaction has upon protein activity. If this is done, a control experiment should also be carried out under the same conditions but in the absence of spin label. After incubation the reaction is quenched and the membranes are washed to remove any remaining spin label that has not reacted with the protein. Finally, the membrane sample is prepared as a small pellet and transferred to a capillary tube ready for EPR analysis.

Protocol 2

Spin labelling membrane proteins with 4-maleimido-TEMPO

Equipment and reagents

- Homogenizer
- Centrifuge
- Beckman 45 Ti polycarbonate centrifuge tube
- Capillary tube
- Nitrogen or argon

- Buffer solution: 5 mM Hepes, 100 mM NaCl, and 1 mM EDTA pH 7.5
- Membranes
- Spin label: 4-maleimido-TEMPO
- Ethanol
- 2-mercaptoethanol

Method

1 De-oxygenate a buffer solution of 5 mM Hepes, 100 mM NaCl, and 1 mM EDTA pH 7.5, by saturating with either nitrogen or argon.

2 Suspend the membranes in 2 ml of the buffer solution to a final concentration of 1 mg protein/ ml using a glass Wheaton homogenizer.

3 Add the spin label in a small volume of ethanol (10 μl) to a final concentration representing a 5-fold excess over the total concentration of protein thiol groups, if known.

4 Stir the suspension on ice or in a cold-room for 18 h (overnight) or for 2 h at room temperature if the protein is robust at higher temperatures. Longer reactions should be carried out under an atmosphere of argon or nitrogen, in a glove box, for example, to avoid oxidation of the spin label.

5 Remove 50 μl aliquots from the suspension at the start of the reaction and at several intervals during the course of the reaction. Add 1 mM 2-mercaptoethanol to the aliquots and keep on ice.

6 Terminate the reaction by adding 2-mercaptoethanol to a final concentration of 1 mM.

7 Dilute the suspension to 60 ml in de-oxygenated buffer and transfer to a Beckman 45 Ti polycarbonate centrifuge tube (or an equivalent). Centrifuge at 100 000 g for 30 min.

8 Repeat step 7 to remove any residual spin label from the suspension.

9 Decant the supernatant, wiping around the interior of the centrifuge tube to drain excess buffer, and transfer the pellet to a capillary tube for EPR analysis.

Cysteine side groups of *Squalus acanthias* Na^+/K^+-ATPase have been modified straightforwardly using the spin label maleimido-TEMPO using a general procedure such as the one described above (26).

The side groups of lysine can be spin labelled using isothiocyanato-TEMPO following a similar procedure to that above.

Protocol 3

Spin labelling membrane proteins with 4-isothiocyanato-TEMPO

Equipment and reagents

- See *Protocol 2*

- Buffer solution: 20 mM Tris–HCl, 100 mM NaCl, and 1 mM EDTA pH 8.3

- Spin label: 4-isothiocyanato-TEMPO

- Buffer containing 1 mM lysine

Method

1 De-oxygenate a buffer solution of 20 mM Tris–HCl, 100 mM NaCl, and 1 mM EDTA pH 8.3, by saturating with argon or nitrogen.

2 Follow *Protocol 2*, steps 2–4. The concentration of spin label should be 5-fold greater than the total number of primary amine groups in the protein.

3 Remove 50 μl aliquots at intervals during incubation, dilute each of them to 1 ml with buffer containing 1 mM lysine, and store at −20 °C.

4 Terminate the reaction by adding lysine to a final concentration of 1 mM.

5 Dilute the suspension to 60 ml in de-oxygenated buffer containing 1 mM lysine and transfer to a Beckman 45 Ti polycarbonate centrifuge tube (or an equivalent). Centrifuge at 100 000 g for 30 min.

6 Continue with *Protocol 2*, steps 8 and 9.

3.2.2 Preparing samples for EPR spectroscopy

It is important to confine the labelled membrane pellet to as small a volume as possible in the capillary tube to concentrate the nitroxide label in the centre of the EPR microwave cavity where the maximum signal can be detected. The sample volume will, of course, depend on the hydrated density of the membranes, but with care it is possible to transfer the sample obtained from the final centrifugation step to the capillary tube as a tight membrane pellet without diluting with buffer. The simple procedure illustrated in *Figure 10* is effective for this purpose.

A   **B**   **C**

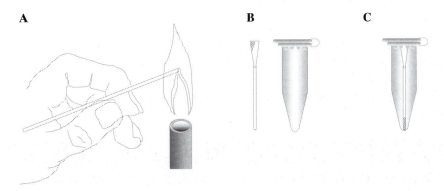

Figure 10 Illustration of a method for transferring a membrane pellet into a glass capillary. (A) The capillary is flame-sealed at one end. (B) A trimmed yellow plastic Gilson pipette serves as a funnel, which sits in the open end of a flame-sealed glass capillary tube. The membrane pellet, which should be obtained using the highest centrifugal force possible, is scraped from the centrifuge tube into the funnel using a spatula and the tube and funnel is placed inside a 1 ml Eppendorf tube. (C) Centrifugation in a bench-top microcentrifuge will transfer the pellet from the funnel into the tube and should occupy no more than 1 cm of the tube length.

Protocol 4

Obtaining a spectrum of spin labelled protein

Equipment and reagents

- 1 mm internal diameter capillary tube
- Flame-drawn pipette
- 5 mm diameter quartz NMR tube
- EPR spectrometer
- Spin labelled membrane protein

Method

1 Transfer the spin labelled membrane pellet to a 1 mm internal diameter capillary tube using a flame-drawn pipette. The sample should occupy about 5 mm of the tube length.

2 Place the capillary tube inside a 5 mm diameter quartz NMR tube in a nitrogen atmosphere and insert into the resonant cavity of the EPR spectrometer so that the sample is in the centre of the cavity.

3 Tune the cavity and record a spectrum at room temperature under the same conditions as those established according to Section 2.4. Ensure the receiver gain is set to avoid saturation of the detector.

4 Several scans should be obtained to accumulate a spectrum with a high signal-to-noise ratio. A sample containing 1 mg of spin labelled protein will typically require 10–50 scans, depending on the number of labels in the protein.

3.2.3 Removing excess label

It is desirable to wash out all or most of the non-reacted spin label from the membrane sample before carrying out the EPR experiments. Free nitroxide representing as little as 5% of the total amount in the sample can dominate the EPR spectrum because the line widths from the free nitroxide are narrow compared with those from the nitroxide attached to protein. For example, two first-derivative Lorentzian or Gaussian lines having the same intensity (as determined by double integration) but different peak-to-peak line widths $(\Delta B_{pp})_1$ and $(\Delta B_{pp})_2$ will have comparative peak-to-peak amplitudes Y'_1 and Y'_2 according to:

$$2Y'_1/2Y'_2 = (\Delta B_{pp})_2^2/(\Delta B_{pp})_1^2 \qquad [5]$$

Hence, free spin label giving rise to first-derivative line widths that are 5 times more narrow than those from the bound label will give rise to peak amplitudes of about 25 times greater than those from the bound label (e.g. *Figure 11*). The presence of such a dominant spectral component can cause sensitivity problems associated with the range of the detector, as well as difficulties with analysis of line shapes, particularly in saturation transfer EPR experiments. The receiver gain can be reduced to prevent the high-amplitude free label signal from overloading the detector, but this makes the detector less sensitive to the broad, low amplitude components of the spectrum.

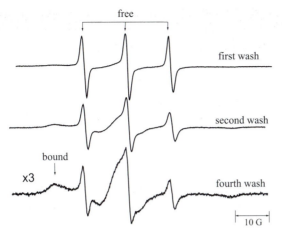

Figure 11 EPR spectra showing the removal of free nitroxide spin label after incubation with a membrane protein preparation. After the first wash by centrifugation and resuspension (*top*) the spectrum is dominated by the free label. Additional washes remove the majority of the free label revealing the outer broad spectral component from the spin labelled protein, although some free label remains visible even after four washes.

Normally the 60-fold dilution of the suspension followed by high-speed centrifugation (as described in *Protocols 2* and *3*) is sufficient to remove over 90% of the free spin label. It will probably be necessary to perform a number of additional washing steps to reduce the small amount of residual spin label to an acceptable level for the EPR experiments. This is usually best accomplished by a number of additional centrifugation-suspension steps as described in the protocols above, but hydrophobic spin labels can partition into the membrane and the remaining few per cent of free label can be extremely difficult to remove. If EPR spectrum does exhibit a component from residual spin label, the pellet can be retrieved from the capillary tube and washed again. Bench-top centrifuging the capillary tube inverted inside a 1 ml Eppendorf tube is the quickest and simplest way of retrieving the sample.

Protocol 5

Dialysis of membranes

Equipment and reagents
- Dialysis membrane tubing (e.g. 50 cm of 2.5 mm diameter Spectra/Por®, molecular weight cut-off 12–14 kDa)
- Centrifuge
- 1 mM EDTA
- 5 mM Hepes pH 7.5 containing 100 mM NaCl, 1 mM EDTA, and 0.05% sodium azide

Method
1　Boil a length of dialysis membrane tubing in distilled water containing 1 mM EDTA. Allow it to cool.

Protocol 5 continued

2 Clip the tube at one end and add up to 2 ml of the spin label/protein reaction suspension (e.g. *Protocol 2*, step 6).

3 Add the tubing to 2 litres deoxygenated 5 mM Hepes pH 7.5 containing 100 mM NaCl, 1 mM EDTA, and 0.05% sodium azide.

4 Stir the dialysis buffer at 4 °C for 24 h, replacing the buffer with a fresh solution at least three times during the course of dialysis.

5 Remove the suspension from the tubing, dilute to 60 ml with buffer, and centrifuge at 100 000 g for 30 min.

6 Transfer the pellet to a centrifuge tube for EPR analysis.

An alternative to centrifugation for removing unwanted spin label is to dialyse the membrane suspension against a large volume of nitroxide-free buffer solution (*Protocol 5*). This approach is less labour-intensive than repeated centrifugation, but is a much slower process and protein losses can be incurred through absorption to the dialysis tubing.

3.2.4 Monitoring the spin labelling reaction

Aliquots taken at regular intervals during the spin labelling procedure can be used to report on the progress of the reaction if very small quantity of membranes can be transferred to capillary tubes with the minimum of sample loss. Samples containing 50–100 µg of protein should first be treated to remove excess unreacted spin label using a scaled-down version of one of the procedures outlined in the last section. It is not crucial, in this case, to remove all traces of free label; dilution of a 50 µl aliquot to 1 ml followed by centrifugation in a bench-top microcentrifuge should remove over 95% of the unreacted agent after 975 µl of the supernatant is decanted. The remaining 25 µl of buffer in the Eppendorf tube can be used to resuspend the tiny pellet at the bottom by agitating with a pipette tip. The suspension will be fluid enough for it to be transferred to a capillary tube using a flame-drawn Pasteur pipette.

The EPR analysis of the aliquots does not require the entire spectrum of the spin labelled protein and only a section of the outer extrema from the strongly immobilized nitroxide (which, in most cases, does not overlap with the lines of the free spin label) need be recorded (*Figure 11*). This enables the receiver gain to be increased without danger of the high amplitude signal from free spin label saturating the detector, thereby improving sensitivity to the low spin label concentration. An increase in the height of the outer extrema with time indicates the progress of the reaction.

Figure 12 illustrates an example of measurements made on a suspension of gastric H^+/K^+-ATPase membranes during a reaction with isothiocyanato-TEMPO. Aliquots containing 100 µg protein were removed at several time points during incubation and divided up to give samples for following the reaction by EPR analysis and by an assay for K^+-stimulated ATP hydrolysis. The progress of

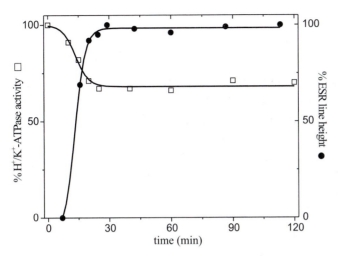

Figure 12 The time course of the reaction between isothiocyanato-TEMPO and porcine gastric H^+/K^+-ATPase as followed by EPR spectroscopy (●) and by the inactivation of ATP hydrolytic activity (□).

the reaction is reflected in the profiles associated with the two independent measurements: after 30 minutes the reaction appears to have run to completion as there is little or no change in either the EPR line heights or residual ATPase activity. The TEMPO group only slightly perturbs the function of the H^+/K^+-ATPase because the spin labelled protein retains 70% of its initial activity.

3.2.5 Further characterization of the labelled protein

While the procedures for preparing and recording spectra of spin labelled proteins are rapid and straightforward, the process of identifying the labelled residues in the protein can be challenging and time-consuming in the absence of site-directed technology. A combination of biochemical techniques is necessary for this purpose and even sophisticated analytical instrumentation, like mass spectrometry and high-performance liquid chromatography, is often unsatisfactory for pinpointing individual sites of modification. The amount of effort applied to this end is left entirely to the reader's discretion.

Protocol 6

Determining the spin label to protein ratio

Equipment and reagents

- EPR spectrometer
- SDS–Lowry reagents
- Bovine serum albumin
- Buffer
- Free spin label in aqueous solution

Protocol 6 continued

Method

1 After the EPR spectrum of the labelled protein has been obtained, remove the sample from the capillary tube and suspend in 1 ml buffer. If *Protocol 2* has been followed, the protein concentration will be approximately 1–2 mg/ml.

2 Determine the protein concentration using the SDS–Lowry method using known concentrations of bovine serum albumin as the protein standard. Express the protein concentration as the total number of moles in the capillary tube (this assumes the protein of interest is 100% pure).

3 The theoretical maximum number of nitroxides per protein molecule can be estimated as n × (protein) (in moles), where n is the total number of reactive side groups in the primary sequence.

4 Make a linear series (six or so) of dilutions of the free spin label in aqueous solution. The maximum spin label concentration should be chosen such that 5 μl of the solution in a capillary tube equals or exceeds n × (protein) in moles.

5 Record spectra of 5 μl of each of the standards and calculate the double integrals. Calculate the double integral of the protein spectrum.[a]

6 Plot the double integrals of the standard solutions against spin label concentration and interpolate the protein integral to find the concentration of protein nitroxides in the capillary tube. The stoichiometry is estimated to be the ratio of spin label concentration to protein concentration in the capillary tube.

[a] Note: Before integration, the spectra should be baseline-corrected by computer. This eliminates the offset of the baseline from zero amplitude and removes any baseline curvature, which becomes more pronounced when signal-to-noise is poor. Baseline correction can be achieved by subtracting a reference spectrum of the cavity without sample, or by fitting a polynomial or sigmoidal curve to the baseline.

We believe it *is* worthwhile, however, to determine the stoichiometry of the reaction between the label and the protein side groups, as this is diagnostic of the selectivity (or lack thereof) of the spin label. The number of labels per protein molecule can be estimated by comparing the doubly integrated protein spin label spectrum with a calibration curve made of double integrals from known concentrations of the spin label in solution, and related to the protein concentration as determined by standard methods.

The accuracy in determining the number of spin labels attached to the protein of interest will clearly diminish when other protein contaminants are present in the membrane.

3.3 Analysis of multiple component spectra

Spin labels do not always undergo uniform rates of motion at their sites of attachment to membrane proteins and the EPR spectrum may contain several overlapping components from nitroxides with varying mobility (*Figure 13*). It is

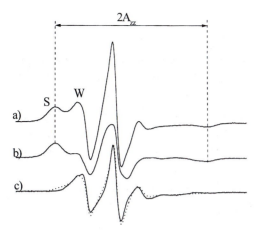

Figure 13 Illustration of the method described in *Protocol 7* for deconvoluting an EPR spectrum containing components from strongly (S) and weakly (W) immobilized spin labels. A suitable single-component spectrum (b) is subtracted from the dual component spectrum (a) to give the difference spectrum (solid line in c). The difference spectrum can be compared with an appropriate spectrum of a weakly immobilized label (dotted line in c) to validate the subtraction.

probable that motional heterogeneity will prevail if several side groups have been modified, particularly if they reside in both buried and exposed regions of the protein, although site-directed spin labelling studies have found that motionally distinct populations of labels can exist even at a single residue (16).

The superposition of two or more spectra with different line shapes can, at first sight, hinder the analysis of the spectrum and limit the amount of available information. It has been found, however, that the relative proportions of the individual spectral components can provide a useful marker for protein conformational changes (16) resulting, for example, from ligand binding (see Section 5). Exposed nitroxides may move to a more restrained environment, and buried nitroxides may become more flexible, thereby changing the contributions to the spectrum made by the individual components. To quantify such a change, one can simply measure the amplitude of the composite spectrum in a sensitive region before and after addition of ligand, but a more rigorous approach is to deconvolute the spectrum into its individual constituents.

Protocol 7

Deconvolution of a two-component spectrum

Equipment and reagents

- EPR spectrometer
- Spin labelled protein (e.g. haemoglobin)

Method

1   Measure the hyperfine splitting $2A_{zz}$ from the separation of the extrema of the outer (strongly immobilized) component of the baseline-corrected spectrum (*Figure 13*).

2 Obtain a spectrum from a strongly immobilized spin label with good signal-to-noise, that closely resembles the outer component of the composite spectrum. To do this, record a series of spectra of spin labelled haemoglobin (for example) at different temperatures until the value of $2A_{zz}$ is identical to that measured in step 1. If $2A_{zz}$ from spin labelled haemoglobin is higher than the desired value, increase the temperature; if $2A_{zz}$ is too small, decrease the temperature.

3 Double-integrate the two-component spectrum.

4 Subtract the haemoglobin spectrum from the two-component spectrum to leave only the component from weakly immobilized spin label. If the difference spectrum is poor, it may be necessary to repeat step 2 to obtain better results.

5 Double-integrate the difference spectrum.

6 The ratio (R) of the integrals from steps 3 and 5 is related directly to the total contribution of the weakly immobilized spin labels to the two-component spectrum.

The simplest case of a multiple component spectrum is that of a protein containing a single weakly immobilized nitroxide and a single strongly immobilized nitroxide as shown in *Figure 13*. Two-component spectra of this type are commonly observed and Protocol 7 is an example of how deconvolution can be achieved in such a case.

4 Saturation transfer EPR

4.1 Aims

Conventional EPR spectroscopy is very sensitive to the dynamics of spin labels occurring in the nanosecond time range; that is, the EPR spectrum reflects rotations around its molecular bonds at frequencies ranging from 10^{12} times per second down to about 10^7 times per second. If, however, the spin label is fully restrained at its site of attachment and is unable to rotate about its molecular bonds, then the mobility of the nitroxide is subject to the overall rate of rotational diffusion of the protein in the membrane.

The rotational diffusion coefficient $D_{R\parallel}$ for the uniaxial rotation of a cylindrical particle, approximating the shape of the membrane-spanning section of a protein, is related to the effective radius (a) of a protein fragment of height h spanning the lipid bilayer, according to (27):

$$D_{R\parallel} = k_B T/(4\pi a^2 h \eta) \qquad [6]$$

where k_B is the Boltzmann constant. Hence, the rate of rotational diffusion of the protein is inversely proportional to its size and to the viscosity (η) of the membrane environment at a given temperature T. The correlation time $\tau_{R\parallel}$ for protein rotation is defined as $1/6D_{R\parallel}$. Determination of $\tau_{R\parallel}$ directly from the EPR spectrum of a locally restrained spin label would provide a very useful measure

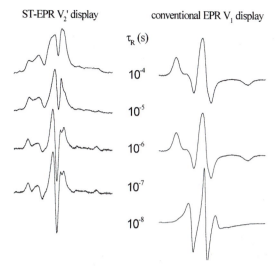

ST-EPR V_2' display   conventional EPR V_1 display

τ_R (s)

10^{-4}

10^{-5}

10^{-6}

10^{-7}

10^{-8}

Figure 14 Line shapes of conventional and ST-EPR spectra of a nitroxide with different rotational correlation times (τ_R) when attached to a membrane protein. At the slow motional limit of the conventional EPR time scale ($\tau_R \sim 10^{-7}$ s) the line shape ceases to be sensitive to further increases in correlation time, while the ST-EPR line shape is sensitive to rotational correlation times of up to 10^{-4} s.

of the size of a molecular assembly and how it responds to ligand binding. Unfortunately, rotational correlation times for membrane proteins are generally on the order of microseconds to milliseconds and outside the time range accessible to conventional EPR, as illustrated in *Figure 14*. Marked changes in the EPR line shape can be seen as the rate of spin label rotation is reduced from 10^{10} per second to 10^7 per second. As the correlation time reaches the slow motional limit of the conventional EPR time scale (10^{-7} s), however, further reduction in the rate of molecular motion has no effect on the EPR line shape (*Figure 14*).

The effective range of EPR can be extended into the time regime of protein rotation using saturation transfer EPR (ST-EPR) spectroscopy, a technique devised by Thomas, Hyde, and co-workers in the mid-1970s. Unlike conventional EPR, the shape of the ST-EPR spectrum is sensitive, in part, to very slow molecular motions with correlation times from 10^{-7} to 10^{-3} s and can provide reasonable estimates of protein rotational diffusion rates (28). The theoretical description of ST-EPR is complicated and the size restraints of this chapter permit only a very cursory description of the technique to be given. The reader is referred to the article by M. A. Hemminga and P. A. De Jager (29) for an excellent introduction to the technique.

4.2 Theory

In a conventional EPR experiment the incident microwave power induces the electron spins to flip from low-energy state to the high-energy state as they come into resonance with the external magnetic field. The spins re-attain their equilibrium (Boltzmann) distribution at a rate T_1^{-1}, the spin-lattice relaxation rate, if the microwave power is sufficiently low that spin-lattice relaxation occurs much faster than the rate at which electrons are flipped into the excited state. The microwave energy absorbed by the spins increases linearly with power until, when the absorption rate is approximately equal to $1/T_1$, the spins are unable to dissipate the absorbed energy fast enough to re-establish the Boltzmann distribution and are said to be *saturated*. If a nitroxide spin label rotates under the condition of saturation, the magnetic anisotropy causes its resonance frequency to change in an orientationally dependent manner. Recalling that EPR spectroscopy employs field modulation and phase-sensitive detection, if the field modulation frequency is such that the electron spin remains in resonance as it rotates in the magnetic field then T_1 relaxation is enhanced and the spins can reabsorb energy. Under such a saturating microwave field the ST-EPR experiment displays maximum sensitivity to the motion of spin labels when the modulation frequency, the spin-lattice relaxation rate and rotational correlation time all have similar values. Since T_1 for nitroxide electrons is about 10^{-5} s the ST-EPR spectrum can be made sensitive to motions with correlation times ranging from 10^{-3} s to 10^{-7} s by selecting a modulation frequency at 50–100 kHz.

The EPR signal can be separated into in-phase and quadrature components oscillating at cos $\omega_r t$ and sin $\omega_r t$, respectively, at a reference frequency of ω_r. Under the non-saturating microwave fields used in conventional EPR, the quadrature component vanishes and only the in-phase signal is detected; under saturating fields, however, the quadrature signal becomes non-zero and it becomes observable. ST-EPR spectra are therefore recorded with the reference frequency of the phase-sensitive detector phase-shifted by 90° relative to the output of the oscillator to eliminate the conventional EPR signal and detect only the quadrature component. In practice, the second-harmonic, quadrature ST-EPR spectra recorded in absorption mode have proved most sensitive to protein rotational rates, and are referred to as the V_2' display. Conventional first harmonic, in-phase EPR spectra are referred to as the V_1 display.

4.3 Recording a spectrum

Hemminga and co-workers (30) proposed a set of conditions for recording ST-EPR of samples contained in glass capillaries, that enable direct comparison of experimental spectra from different sources and also pays attention to the effects of microwave field distribution. These conditions are summarized in *Table 3*.

The recommended saturating microwave field (B_1) of 25 μT refers to the field at the centre of the sample. The mean square value of B_1 can be expressed as:

$$<B_1^2>_s = \mu_0 h Q_L P_i / 2 V_s \omega_0 \qquad [7]$$

Hence, B_1 is dependent not only on the incident microwave power P_i, but also on the quality factor Q_L of the loaded cavity and on the sample volume V_s, which will vary from sample to sample depending on the characteristics of the

Table 3 Recommended standard conditions for recording second-harmonic, out-of-phase ST-EPR spectra (30)

Sample conditions		Microwave				Modulation		Detection	
Spin label concentration (μM)	Sample length (mm)	ν_0 (GHz)	B_1 (mT)	Phase		ν_m (kHz)	B_m (mT)	ν_1 (kHz)	Phase
> 30	5	9.5	25	Absorption		50	0.5	100	Quadrature
< 30	> 20	9.5	25	Absorption		50	1.0	100	Quadrature

material and the cavity. The magnitude of the microwave field and the modulation amplitude B_m can have a profound effect on the ST-EPR line shape and it is important to quantify B_1 to relate the spectral features to molecular dynamics. A number of calibration procedures have been suggested for determining B_1 at a given microwave power using reference compounds with known dielectric and relaxation properties. *Protocol 8* describes a commonly used method, which involves measuring the amplitude of the conventional EPR spectrum of PADS (peroxylamine disulfonate or Fremy's salt) as a function of the square root of the incident microwave power.

Protocol 8

Calibration of the microwave field

Equipment and reagents

- 1 mm i.d. capillary
- 5 mm quartz NMR tube
- EPR spectrometer

- 0.9 mM PADS (peroxylamine disulfonate or Fremy's salt) in 50 mM potassium carbonate

Method

1 Prepare a degassed aqueous sample of 0.9 mM PADS in 50 mM potassium carbonate solution.

2 Dispense 20 μl PADS solution (or enough to give a sample height equivalent to the membrane sample of interest) into a 1 mm i.d. capillary and place in the microwave cavity within a 5 mm quartz NMR tube in a nitrogen atmosphere.

3 Record a series of spectra under the same conditions as described in *Protocol 1*, each time varying the incident microwave power from low to saturating.

4 Below saturation the signal amplitude increases linearly with increasing microwave power (*Figure 15*). Saturation occurs at higher power levels, at which the signal amplitude reaches a maximum and reduces to zero. Sufficient spectra should be collected at different incident microwave powers (P) to follow the signal amplitude until it reaches half the value of the extrapolated linear region (*Figure 15*). $P^{1/2}$ is the value of P that satisfies this condition.

Protocol 8 continued

5   The sample-averaged rf field microwave field (B_1) in μT at incident power P is given
by:

$$B_1 = 10.62(P/P_{\frac{1}{2}})$$

Hence, the recommended microwave field of 25 μT (Table 3) would correspond to
an incident microwave power P = 2.35$P_{\frac{1}{2}}$ as determined using the PADS sample.

The modulation amplitude B_m can be determined by following its effect on
the spectral line widths of a suitable reference material, a small crystal of tetra-
cyanoquinodimethane, for example, which has a very small natural line width
(13 mT). When the sample is placed in the centre of the cavity, the peak-to-peak
distance of the over-modulated V_1 spectrum equals B_m when the modulation
frequency is around 100 kHz (29), as seen in *Figure 4*. The calibration is valid for
all further experiments because the value of B_m is not sensitive to other factors.

To ensure that only the quadrature signal is detected, so that the conventional spectrum is eliminated and only the effects of saturation are observed,
the phase setting of the phase-sensitive detector (PSD) must be determined. A
widely employed procedure is the 'self-null' method, so-called because the sample
is used to calibrate the detector before it is observed by ST-EPR. This procedure
involves recording a series of first harmonic spectra under non-saturating
conditions to find the precise setting of the PSD that is 90° out-of-phase with
respect to the reference signal. This procedure is made simple by the fact that,
below saturation, no 90° out-of-phase signal is detected by the PSD, and one
simply finds the setting of detector at which the conventional EPR spectrum is
nulled.

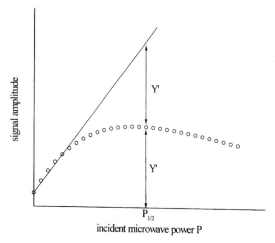

Figure 15 Saturation plot showing the relationship between incident microwave power and
signal amplitude. The value $P_{\frac{1}{2}}$ is the incident power at which Y' is half the value of the
extrapolation from the linear region of the plot.

Protocol 9

Recording a ST-EPR spectrum of a spin labelled membrane protein

Equipment and reagents

• See *Protocol 4*

Method

1 Record a conventional (V_1) EPR spectrum of the sample as described in *Protocol 4*. The spectrum should contain very little or no contribution from weakly im-mobilized (or isotropic) spin label.

2 Collect a series of conventional spectra, each time incrementing the setting of the PSD by 10°. The signal amplitude will decrease systematically as the phase setting of the PSD approaches 90°. The size of the increments can be reduced as the signal approaches the 'null point', to determine the precise setting ($\pm$ 2°) at which no signal remains.

3 Gradually increase the incident microwave power P to the level corresponding to the desired B_1 field (e.g. 25 μT) and re-tune the cavity.

4 Change the remaining parameters to their desired values (refer to Table 3 for the appropriate parameters and recommended values). Ensure that the temperature of the cavity is constant.

5 The spectrometer should now be configured to obtain a second-harmonic, 90° out-of-phase ST-EPR (V_2') spectrum of the membrane sample. The signal amplitude will be somewhat less than that of the conventional spectrum and more scans will be necessary to achieve good signal-to-noise.

6 Record the spectrum.

4.4 Interpretation of ST-EPR spectra

The aim of the ST-EPR experiment here is to use the information contained within the V_2' spectrum to estimate the rate at which a protein rotates in the lipid membrane. Because of the complexity of the theory relating the saturation transfer phenomenon to molecular dynamics, users of this technique have historically tended to rely on semi-empirical, qualitative methods of analysis rather than rigorous computer simulation of spectral line shapes.

Ratios of the line heights in the high-field (H″/H), central (C′/C) and low-field (L″/L) regions of the ST-EPR (V_2') absorption spectrum (*Figure 16*) have been found to be diagnostic of the correlation time (τ_R) defining protein rotational diffusion. The ratios H″/H and L″/L are rather more reliable for determining τ_R than is C′/C, which is more sensitive to the positioning and dielectric properties of the sample. Calibrated diagnostic line height ratios provide good estimates of protein rotational correlation times when the spectrum is free of components from

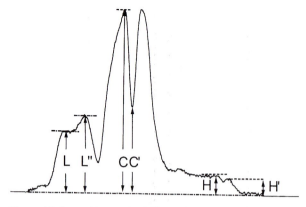

Figure 16 Second-harmonic, 90° out-of-phase (V_2' display) ST-EPR spectrum of a spin labelled membrane protein (200 000 molecular weight), showing the diagnostic line heights (L, L", C, C', H, H") used for determination of protein rotational diffusion rates.

weakly immobilized or isotropic (free) spin label. Correlation times can be obtained more accurately in some cases using the quantity I_{ST}, the normalized ST intensity, defined as the ratio of the integral of the ST-EPR (V_2') spectrum, and the double-integral of the conventional (V_1) spectrum (28).

In practice, the relationship between the line height ratios (or I_{ST}) and τ_R is obtained by plotting calibration curves of line height measurements from a number of standard spin labelled protein solutions of variable viscosity under identical spectrometer conditions to those employed for the sample of interest. Recalling the modified Stokes–Einstein relation in *Equation 6*, the rate of protein rotational diffusion is a function of two variables, molecular radius and the viscosity of the medium. By using a soluble protein of known dimensions such as haemoglobin (r = 29 Å) in water–glycerol solutions of known viscosity one can relate the line heights directly to the calculated *isotropic* rotational correlation times τ_R of the protein in each solution (*Figure 17*). The effective rotational correlation time τ_R^{eff} of the spin labelled membrane protein is then obtained by numerical interpolation. It is advisable that the calibrations are carried out at least once on the same instrument as used for analysis of the membrane protein, but one can also refer to calibration curves in the literature provided the experimental conditions are the same in both cases (31).

Calibration curves like those in *Figure 17* provide a relationship between the shape of the ST-EPR spectrum and the rate of isotropic reorientation of a soluble protein such as haemoglobin. A membrane protein, on the other hand, is restricted in its degrees of rotational freedom by the lipid bilayer and can be considered to rotate about a single axis parallel with the membrane normal. Hence, correlation times for haemoglobin and a membrane protein like Na^+/K^+-ATPase are describing essentially different modes of molecular motion. The value of τ_R^{eff} for the membrane protein, obtained by interpolation, can be corrected to give the anisotropic rotational anisotropic correlation time $\tau_{R\parallel}$ of

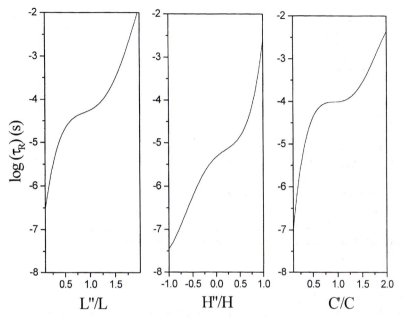

Figure 17 Calibration curves relating ST-EPR line height ratios in the diagnostic regions of spectra from spin labelled haemoglobin to the protein isotropic rotational correlation time τ_R. The curves represent the best fitting polynomials to the experimental data (28).

the membrane-spanning section of a membrane protein by taking into account the orientation of the nitroxide group, according to (32):

$$\tau_{R\parallel} = 1/6D_{R\parallel} = (\tau_R^{eff}/2)\sin^2\theta \qquad [8]$$

where θ is the angle between the principal nitroxide z-axis (*Figure 5*) and the membrane normal. It is possible to estimate θ using uniformly oriented membranes prepared on glass plates and fixed at a defined angle in the magnetic field, but the technical requirements for this approach are demanding. Alternatively the maximum value of $\tau_{R\parallel}$, given by $\tau_R^{eff}/2$ when $\theta = 90°$, which can be used to impose an upper limit on the protein molecular radius as estimated from the modified Stokes–Einstein relation in *Equation 6*.

5 Protein–ligand interactions

5.1 Experimental design

The previous sections describe some of the basic methodology that will enable the experimentalist to use EPR spectroscopy as a tool for examining the effects of ligands, agonists, and substrates on membrane protein structure and dynamics. How the tool is applied will depend, of course, on the questions being asked of the system, but the experimental design is likely to include some or all of these steps:

- protein purification
- protein spin labelling

- characterization of the spin label sites
- recording spectra
- addition of ligand
- observation of changes in the spectra
- interpretation of the experimental results

Changes seen in either the conventional EPR (V_1) or ST-EPR (V_2') spectral displays in response to the addition of ligand will ideally reflect changes in local environment of the spin label or in the global rotational dynamics of the protein.

The simple experiment described in *Protocol 10* is designed to record spectra of the membrane sample before and after incubation with ligand solution.

Protocol 10

Examining the effects of ligand on the conventional EPR spectrum

Equipment and reagents

- 1 mm i.d. capillary tube
- 5 mm quartz NMR tube
- EPR spectrometer
- Hamilton syringe
- Ligand
- Buffer

Method

1 Prepare the sample in a 1 mm i.d. capillary tube sealed at one end and cut to about 5 mm. Tie a length of cotton around the tube and place within a 5 mm quartz NMR tube so that the cotton extends over the top of the outer tube.

2 Place in the EPR cavity, allow the temperature to stabilize, and record a conventional spectrum.

3 Leave the spectrometer on standby and remove the capillary tube using the cotton, leaving the outer tube positioned in the cavity.

4 Add the ligand (and any cofactors) in a small concentration of buffer solution (1 μl) to the sample at the desired concentration. A 10 μl Hamilton syringe is ideal for this operation. Incubate at a suitable temperature for the desired length of time.

5 Replace the capillary in the NMR tube and record another spectrum.

6 Repeat steps 1–5 with a fresh sample, adding ligand-free buffer solution as a control experiment.

7 Analyse the spectra obtained before and after addition of ligand by measuring hyperfine anisotropy or deconvoluting multiple component spectra as in *Protocol 7*.

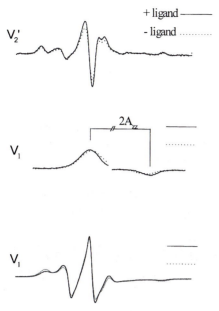

+ ligand ——————
- ligand ....

V_2'

$2A_{zz}$

V_1

V_1

Figure 18 Examples of saturation transfer (*top*) and conventional (*middle* and *bottom*) EPR spectra of nitroxide spin labelled membrane proteins before (solid line) and after (dotted line) addition of ligand. The ST-EPR spectra are consistent with an increase in protein rotational correlation time after addition of ligand, suggesting that the ligand promotes protein–protein interactions. The reduction of the outer hyperfine splitting (middle spectrum) suggests that the rotational mobility of the nitroxide is increased upon ligand binding. The change in the relative proportions of the two components in the bottom spectrum is consistent with a ligand-induced conformational change within the protein.

In principle, any changes in the spectrum after replacement of the sample are attributable to the ligand, provided the sample is subjected to minimal handling and is replaced in the cavity at the same position. Typical changes in the EPR spectrum that might be observed after addition of a ligand or substrate are illustrated in *Figure 18*.

ST-EPR spectra can be recorded using the general approach described above, but extra care should be taken not to increase the sample height in the capillary tube or alter the position of the tube in the cavity after addition of ligand.

It may not be desirable to add the ligand directly to the sample in the capillary tube, if it is sparingly soluble in buffer solution, for example. In such a case, the membrane sample can be removed from the capillary and suspended in buffer solution (1 ml) before adding the ligand in a small volume (1 µl) of a suitable solvent. Ethanol and dimethyl sulfoxide will dissolve many compounds and do not usually disrupt the membranes if they represent less than 5% of the total solvent volume. The membranes are then obtained as a pellet by centrifugation and transferred back to a capillary. The disadvantage of this approach is that transient effects of ligand binding over a short time scale will be missed.

5.2 Time-resolved spectroscopy

In cases where the ligand is found to have an appreciable effect on the conventional EPR spectrum (*Figure 18*), the rate of change of the spectral line shape may be slow enough to follow in real time to study the kinetics of the ligand binding response. The procedure for such an experiment involves first identifying the part of the spectrum that is most sensitive to ligand binding. The electromagnet is then held at the magnetic field B_0 corresponding to that part of the spectrum and the absorption of microwave radiation at the fixed B_0 recorded over a period of time.

Protocol 11

Time-resolved spectroscopy

Equipment and reagents

- EPR spectrometer
- Capillary tube
- Protein sample
- Ligand

Method

1   Spectra can be recorded as in-phase, first harmonic (V_1 display) or in-phase, second harmonic (V_2) to maximize sensitivity.

2   Increase the receiver gain to the maximum possible value without truncating the region of interest. Use the DC offset facility of the spectrometer to bring the signal within range of the detector if necessary.

3   Increase the modulation amplitude to the value given the maximum signal amplitude in the region of interest. If the modulation amplitude is too high, line broadening will occur which reduces the signal amplitude (*Figure 4*).

4   Set the centre magnetic field value to the chosen fixed field position and reduce the sweep width to 0 mT. Increase the field resolution to 4096 points per scan to allow the maximum sampling time on one trace.

5   Reduce the conversion time for each point to between 20 ms and 1 ms, and set the time constant to a quarter of this value.

6   Add the ligand to the sample in the capillary tube, place in the cavity, and tune.

7   Record the transient signal.

References

1. Chinkers, M. and Wilson, E. M. (1992). *J. Biol. Chem.*, **267**, 18589.
2. Trumpp-Kallmeyer, S., Hoflack, J., Bruinvels, A., and Hibert, M. (1992). *J. Med. Chem.*, **35**, 3448.
3. Esmann, M., Sar, P. C., Hideg, K., and Marsh, D. (1993). *Anal. Biochem.*, **213**, 336.
4. Marsh, D. (1989). In *Spin labelling theory and applications* (ed. L. J. Berliner and J. Reuben), pp. 255–303. Plenum Press, NY and London.

5. Hubbell, W. L., Mchaourab, H. S., Altenbach, C., and Lietzow, M. A. (1996). *Structure*, **4**, 779.

6. Middleton, D. A., Reid, D. G., and Watts, A. (1995). *Biochemistry*, **34**, 7420.

7. Rowntree, J. I. and Watts, A. (1988). *Biochem. Biophys. Res. Commun.*, **155**, 1412.

8. Jun, L., Rutz, J. M., Klebba, P. E., and Feix, J. B. (1994). *Biochemistry*, **33**, 13274.

9. Wu, J., Frillingos, S., Voss, J., and Kaback, H. R. (1994). *Protein Sci.*, **3**, 2294.

10. Ottemann, K. M., Thorgeirsson, T. E., Kolodziej, A. F., Shin, Y.-K., and Koshland, D. E. (1998). *Biochemistry*, **37**, 7062.

11. Hall, J. A., Thorgeirsson, T. E., Liu, J., Shin, Y. K., and Nikaido, H. (1997). *J. Biol. Chem.*, **272**, 17610.

12. Sterk, M., Hauser, H., Marsh, D., and Gehring, H. (1994). *Eur. J. Biochem.*, **219**, 993.

13. Rowand, J. K. and Berliner, L. J. (1992). *J. Protein Chem.*, **11**, 483.

14. Chen., Z., Coan, C., Fielding, L., and Cassafer, G. (1991). *J. Biol. Chem.*, **266**, 12386.

15. Altenbach, C., Marti, T., Khorana, G., and Hubbell, W. L. (1990). *Science*, **248**, 1088.

16. Farahbaksh, Z. T., Hideg, K., and Hubbell, W. L. (1993). *Science*, **262**, 1416.

17. Block, C., Lohmann, R. D., and Beyersmann, D. (1990). *Biol. Chem.*, **371**, 1145.

18. Megli, F. M., van Loon, D., Barbuti, A. A., Quagliariello, E., and Wirtz, K. W. A. (1985). *Eur. J. Biochem.*, **149**, 585.

19. Hubbell, W. L. and Altenbach, C. (1994). *Curr. Opin. Struct. Biol.*, **4**, 566.

20. Mahaney, J. E. and Grisham, C. M. (1992). *Biochemistry*, **31**, 2025.

21. Earnest, J. P., Limbacher, P., McNamee, M. G., and Wang, H. (1986). *Biochemistry*, **25**, 5809

22. Wojcicki, W. E. and Beth, A. H. (1993). *Biochemistry*, **32**, 9454.

23. Jacobsen, K. A., Ukena, D., Padgett, W., Kirk, K. L., and Daly, J. W. (1987). *Biochem. Pharmacol.*, **36**, 1697.

24. Palma, A. L. and Wang, H. H. (1991). *J. Membr. Biol.*, **122**, 143.

25. Lange, R., Maurin, L., Larroque, C., and Bienvenue, A. (1988). *Eur. J. Biochem.*, **172**, 189.

26. Esmann, M. (1982). *Biochim. Biophys. Acta*, **688**, 251.

27. Esmann, M., Karlish, S. J. D., Sottrup-Jensen, L., and Marsh, D. (1994). *Biochemistry*, **33**, 8044.

28. Horváth, L. I. and Marsh, D. (1988). *J. Magn. Reson.*, **80**, 314.

29. Hemminga, M. A. and De Jager, P. A. (1989). In *Spin labelling theory and applications* (ed. L. J. Berliner and J. Reuben), pp. 131–78. Plenum Press, NY and London.

30. Hemminga, M. A., De Jager, P. A., Marsh, D., and Fajer, P. (1984). *J. Magn. Reson.*, **59**, 160.

31. Ryba, N. J. and Marsh, D. (1992). *Biochemistry*, **31**, 7511.

32. Marsh, D. and Horváth, L. I. (1989). In *Advanced EPR. applications in biology and biochemistry* (ed. A. J. Hoff), p. 707. Elsevier, Amsterdam.

Chapter 10

NMR studies of protein–ligand interactions

Lu-Yun Lian

University of Leicester, Department of Biochemistry, University Road, Leicester LE1 7RH, UK.

1 Introduction

The interactions between pairs of molecules are crucial in many molecular recognition processes in biology, the main examples being found in complexes of proteins with other proteins (e.g. in signalling processes), with small molecules (e.g. enzyme–substrate or protein–drug interactions), and with nucleic acids (e.g. protein–RNA complexes, transcription factors with DNA). Like many other biophysical techniques, nuclear magnetic resonance (NMR) provides information about protein–ligand interactions. However, unlike many of the other techniques such a fluorescence spectroscopy, isothermal titration calorimetry, and surface plasmon resonance, NMR provides information on many different aspects of these interactions, ranging from structures to dynamics, kinetics and thermodynamics. Recent developments in NMR technology (1, 2), together with the routine use of stable isotopes (^{13}C, ^{15}N, ^{2}H), have enabled the use of a much greater variety of NMR-based techniques for studying protein–ligand interactions, and involving much larger protein systems; these techniques have been used in a many different basic and applied research environments (3, 4). Examples of the latter are evident, for example, in the use of NMR as a method for high-throughput screening in drug development programmes in the pharmaceutical industry (5–7).

The aim of this chapter is to highlight the basic practical considerations that must be borne in mind when performing protein–ligand studies by NMR, the variety of experiments that are possible, and the type of information that can be obtained. It does not aim to describe in detail the different experiments and ways to analyse the data since there are other reviews that achieve this (8). Many of the examples given in the chapter are taken from the author's laboratory. Methods to address the three main questions commonly asked when using NMR to investigate protein–ligand interactions—does the ligand bind, what is the binding constant, and what is the structure of the complex—are covered in this chapter. Certain areas are covered in greater detail in ref. 9.

2 Preparation of sample and preliminary binding studies

Most proteins are studied in the recombinant form and hence protein supply is not normally a limiting factor once the protein has been over-expressed in bacteria—and the subsequent protein production and purification procedures optimized. Growing the cells on ^{13}C-labelled sources (glucose, acetate, or pyruvate) and /or ^{15}N-labelled sources (ammonium chloride or sulfate) produces protein uniformly labelled with ^{13}C and/or ^{15}N (10). This means that in a protein–ligand complex, it is possible to either detect or filter-out the protein signals. Another method of removing the protein signals is to replace the protons of the protein with deuterons by growing the E. coli in a deuterated media (11).

The concentration of protein needed for the NMR experiment will vary depending on the type of experiment. Typically, 50–100 µM for 1D spectra to approximately 3 mM for some 2D experiments is required. For a protein of molecular weight (molar mass) M ~ 20 000 between 0.5 mg and 30 mg of protein would be sufficient to make a 500 µl NMR sample. The highest concentration is dependent on the solubility, the aggregation characteristics, and the long-term stability of the protein. The solution conditions can also govern the binding constant between a protein and ligand; these conditions can be adjusted depending on the type of NMR experiments to be performed (see later sections on chemical exchange).

Once purified samples are available, it is then necessary to establish the best condition for the long-term stability of the sample. Factors such as the pH, concentration, ionic strength, type of ions, type of buffer, and temperature, all influence the stability of the sample and need to be considered. Use of additives such as reductants (dithiothreitol) (together with flashing the sample with argon gas) and/or sodium azide to prevent bacterial growth, is recommended for long-term stability of the NMR samples; reductants are needed to prevent oxidation of the cysteine and methionine residues.

NMR studies of proteins are often carried out in aqueous solution, either in H_2O or D_2O. Some proteins will withstand lyophilization, others will not; hence, the method for buffer exchange would depend on this factor. If the protein can be lyophilized, the most desirable procedure for buffer exchange is to dialyse the protein against water or a very low buffer concentration (for example, phosphate buffer 1 mM). The NMR sample can later be made up by weighing out the approximate quantity of protein and redissolving in the desired buffer concentration. If the protein cannot be lyophilized, buffer exchange is best performed using a concentration device such as a stirred cell or a spin-column; using either method, the concentration step and the buffer exchange step can be carried out during the same procedure. Buffer exchange can also be performed on small concentrated sample using dialysis cassettes. Buffers are normally made up in either 100% D_2O, or 10% D_2O:90% H_2O. It is advisable to ensure that the NMR protein sample is a clear solution before loading into an NMR tube; it may be necessary to centrifuge (perhaps through a spin-filter) the sample to pellet any precipitate formed during the final stages of sample preparation. Protein concentrations in the NMR samples should be established using, for example, UV spectrophotometry.

3 Chemical exchange and analysing the NMR spectra for protein–ligand complexes

The first step in any study of protein–ligand interactions by NMR is to establish the rate of exchange between the free and complexed forms, since the type of experiments most appropriate for the system is dependent on the chemical exchange rates on the NMR time scale (12). NMR parameters that are used to provide meaningful data in studies of protein–ligand interactions—chemical shift, the relaxation rates (in particular, transverse, T_2, relaxation) of the ligand, and the nuclear Overhauser enhancement effects (NOEs)—also of the ligand—are affected by the rate of chemical exchange. Hence, the appearance in an NMR spectrum of a complex would depend on this rate of exchange. On the NMR time scale, exchange regimes are broadly classified as *slow*, *intermediate*, and *fast* with the regime being related to the equilibrium dissociation constant, K_d. The binding of a ligand L to an enzyme E can be considered a two-site second-order exchange:

$$E + L \underset{k_{on}}{\overset{k_{off}}{\rightleftharpoons}} EL \qquad [1]$$

The above equilibrium is described by the dissociation constant, $K_d = k_{off}/k_{on}$, where k_{off} is the rate of dissociation (off-rate) of ligand from the protein–ligand complex, and k_{on} is the rate constant for association (on-rate) of the ligand with the protein. The off- and on-rates define the lifetime, τ, of a particular state. The lifetime of a NMR nucleus in the ligand molecule in state EL is given by $\tau_{EL} = 1/k_{off}$, and the lifetime in state L by $\tau_L = 1/(k_{on}[E])$.

For a nucleus in the enzyme, the lifetime in state EL is given by $\tau_{EL} = 1/k_{off}$, and the lifetime in state E by $\tau_E = 1/(k_{on}[L])$. [E] and [L] are, respectively, the concentrations of the free enzyme and ligand, and [EL], the concentration of the complex. In most discussions of exchange effects on the NMR spectrum, a single lifetime is used to characterize the process defined by Equation 1; for the ligand: $1/\tau = 1/\tau_{EL} + 1/\tau_L = k_{off}(1 + p_{LE}/p_L)$, where p_{LE} $(= [EL]/[L_{total}])$ and p_L $(= [L]/[L_{total}])$ are, respectively, the fraction of the ligand in the bound and free states, and $[L_{total}] = [EL] + [L]$. It is important to note that the τ—and hence the actual appearance of the spectrum—depends on the ligand concentration.

As a first approximation, the rate of exchange can be estimated from the appearance of the 1D NMR spectrum. As many resonances of interest as possible should be monitored since different resonances can show different exchange behaviour. If separate resonances for the free and bound ligand are observed, then the *slow exchange* condition is met, that is the rate of equilibration, $1/\tau$, is slower than the chemical shift separation, $\Delta\delta$, of the resonances of the bound and free ligand:

$$1/\tau << \Delta\delta \text{ where } \Delta\delta = |(\delta_{EL} - \delta_L)|.$$

When the rate of equilibration $1/\tau$ is faster than the chemical shift separation and faster than the difference in relaxation rates:

$$1/\tau >> \Delta\delta$$

then the *fast exchange* condition exists, and a single resonance is observed for the ligand at a 'weighted average' chemical shift, δ_{obs}, given by:

$$\delta_{obs} = \delta_L p_L + \delta_{EL} p_{EL} \qquad [2]$$

where p_{EL} and p_L are the mole fractions of the bound and free ligand, δ_{EL} and δ_L are the chemical shifts of the bound and free ligand. Note: For a nucleus on the protein, under the same conditions, the resonance position is given by $\delta_{obs} = \delta_E p_E + \delta_{EL} p_{EL}$.

In the case of a ligand undergoing *moderately fast exchange* between the free (L) and bound (EL) states, the transverse relaxation rate $1/T_2$ (and hence the line width) of the resonances of a ligand is given by:

$$1/T_{2, obs} = p_{EL}/T_{2,EL} + p_L/T_{2,L} + [p_{EL}p_L{}^2 4\pi^2(\delta_{EL} - \delta_L)^2]/k_{-1} \qquad [3]$$

where $T_{2,EL}$ and $T_{2,L}$ are the transverse relaxation times of the bound and free ligand, $T_{2,obs}$ is the measured relaxation time, and k_{-1} the dissociation rate constant of the complex. In the case of very fast exchange (no exchange contribution to the line width) the third term in Equation 3 reduces to zero.

When the rate of equilibration $1/\tau$ is approximately equal to the chemical shift separation, $1/\tau > \Delta\delta$ then *intermediate exchange* conditions exist to give rise to broad complex spectra that are difficult to analyse.

It is clearly important to ascertain NMR chemical exchange regime of a protein–ligand complex before carrying out detailed NMR work. For most small ligand–protein interactions, the diffusion-limited on-rate is often estimated at 10^8 M^{-1}s^{-1}. With typical ^{1}H chemical shift differences observed between free and bound species in proton spectra ($\sim$ 1 ppm) a $K_d > 10^{-3}$ M will usually result in a spectrum where the resonances of the free and the bound species are in *fast exchange* while ligands that bind tightly (e.g. $K_d \sim 10^{-9}$ M; $k_{off} \sim 10^{-1}$ s^{-1}) can be considered to be in *slow exchange* on the NMR time scale. Very often, for many ligands where the protons have very different chemical shifts, some of the ligand signals can be in fast exchange (small shift changes) while others are in slow exchange (large shift changes). It is essential in all cases, as a first step in studying protein–ligand interaction to distinguish between fast, intermediate, and slow exchange. The methodology for obtaining this information is presented in *Protocol 1*.

With the current availability of ^{15}N-labelled proteins, it may be advantageous to perform some of these preliminary experiments using an isotopically labelled sample; this has the advantage that the protein signals can be conveniently monitored without the interference of the ligand signals. Independent of whether labelled or unlabelled proteins are used, the general principles for the establishment of exchange regime are the same.

Protocol 1

Determination of fast and slow exchange in protein–ligand complexes

Equipment and reagents
- NMR spectrometer
- NMR tube
- Syringe with a long needle
- Protein sample
- Ligand sample
- D_2O or H_2O buffer

Method
The protocol described here includes general guidelines for performing ligand titrations.

1 Sample preparation and ligand titration: Whether a D_2O or H_2O buffer should be used depends on the chemical nature of the ligand. If, for example, there are well-resolved aromatic resonances of the ligand outside the normal chemical shift range of protein aromatic resonances, a protein sample prepared in D_2O, where the protein amide protons are replaced by deuterons, would simplify the spectrum, allowing the ligand aromatic signals to be observed. Use the sample buffer to prepare both the protein and ligand samples. (It may be desirable to first solubilize a ligand sample using a small amount of organic solvent. If this is the case, use the minimum amount of organic solvent and make up the final stock ligand sample to be used for NMR titration by adding the protein buffer to the concentrated solubilized ligand; avoid titrating ligand samples in organic solvent directly into the protein sample.) The concentration of the stock ligand solution should be such that aliquots (e.g. 5–10 μl) can be conveniently added without significant dilution of the protein sample. Check the pH of the ligand stock sample. In most cases, the most efficient method is to add the ligand aliquots directly to the protein sample in the NMR tube using a syringe with a long needle; gently tap the NMR tube to mix the sample.

2 Record the ^1H spectrum of both the free protein and free ligand. The concentration-dependence of the ligand sample should also be checked by acquiring the ligand spectrum at a few different concentrations.

3 Titrate the enzyme with small aliquot (5 μl) of concentrated ligand solution. Record the spectra of the protein complexes with 0.2, 0.4, 0.6, 0.8, and 1 equivalents of the ligand. You can use either a simple 1D ^1H or, if a ^{15}N-labelled sample is available, a 2D ^{15}N-^1H HSQC spectrum. It is a good idea to check the pH of the sample at the end of the titration; if the protein and ligand samples were prepared using the same buffer, the pH change should only be minimal.

4 Observe whether the *resolved* protein (histidine C(2)H, C(4)H and/or high-field methyls

in unlabelled protein, amide signals in ^{15}N-labelled protein) and ligand signals are shifting (*fast exchange*) or changing in intensity, with concomitant appearance of new resonances (*slow exchange*).

5 Continue titrating ligand and monitor the spectrum in the following manner:

(a) **Ligand signals**. When the free ligand signals can be observed, look back at the spectra corresponding to lower concentrations to see if the free signals can be observed with hindsight. Determine whether the chemical shift is constant (*slow exchange*) or changing during the titration (*fast exchange*).

(b) **Protein signals**. Monitor the resolved protein resonances (either using the upfield methyl or downfield histidine resonances in a 1D spectrum, or the amide resonances in a 2D ^{15}N–^{1}H HSQC). A continuous change in chemical shift with increasing ligand concentration suggests that the protein resonance is in *fast exchange*. The appearance of a new signal with increasing intensity as more ligand is added, with concomitant decrease in intensity of the free protein signal, means that the protein resonance is in *slow exchange*.

6 The signals might be broadened due to exchange effects. Record the spectra at several suitable temperatures; ensure that the protein and ligand are stable within the temperature range used. Sharpening of the ligand signal with increasing temperature indicates *intermediate exchange* progressing to *fast exchange*; signal broadening indicates *slow exchange* progressing to *intermediate exchange*.

7 It is generally desirable to avoid the intermediate exchange regime; hence changing the temperature and/or solution conditions (e.g. increasing/decreasing the ionic strength) can alter the exchange rate to be either fast or slow. As low a temperature as possible (to give the desired exchange regime) should be used in order to improve the long-term stability of the protein complex.

4 Screening for ligand interactions

There are occasions when no optical biophysical methods can be conveniently used for screening protein–ligand interactions. Secondly, it is often the case—particularly in the development of a new drug—that a library of small compounds need to be screened for binding and it is necessary to characterize the identity and structure of these small compounds. The transferred NOE experiment can achieve both these objectives at the same time (8).

The amplitude and sign of NOEs depend on the rates of molecular motion (overall tumbling and internal mobility), which are usually characterized by an effective motional correlation time τ_c. Low molecular weight compounds in solution tumble rapidly compared with the Larmor frequency ($\omega\tau_c < 1$) and show weak positive NOEs (< 0.5). With increasing molecular weight, the overall molecular tumbling rate slows down, and if internal mobility is absent or of limited nature, then the condition $\omega\tau_c > 1$ applies, with the NOE becoming negative, reaching a theoretical maximum of -1. For intermediate motion ($\omega\tau_c$

~ 1), the NOE may be close to zero. When a small ligand binds to a macromolecule, it acquires the motional features of the large molecule particularly that part of the ligand that is directly involved in binding. Therefore, in the bound state, the ligand can develop a large negative NOE in a NOESY experiment. When exchange between the bound and free states is sufficiently fast (on the order of $1/T_1$ or faster), changes in magnetization of a bound ligand proton from intra- or intermolecular NOEs are transferred to the free ligand by exchange between the bound and free ligand. Therefore, when the bound and free ligand molecules are in equilibrium, the total NOE observed for ligand signals is predominantly controlled by the bound form, and the resulting effect is called *transferred NOE* (trNOE) (see Section 7). The trNOE experiment relies on indirect characterization of ligand–protein interactions and of the conformation of the bound ligand via the NMR spectrum of the free ligand. Typically, trNOEs are best observed with a 10–30-fold excess of ligand over the macromolecule to which it binds, using mixing times of 200–400 ms. Hence, the trNOE method is an excellent tool for establishing if a small ligand binds to a macromolecule, for distinguishing ligands that bind from those which do not, and for testing mixtures of compounds. In many cases, the 2D NOESY spectrum is expected to give the spectrum of the individual compounds regardless of whether they bind— only the signs of the cross-peaks in the 2D trNOE spectrum are different for the bound ligands versus unbound ligands—this method provides a nearly complete data set, allowing almost immediate identification of the chemical nature of the bound ligand without the need for spectral deconvolution.

Protocol 2

Screening for weak protein–ligand interactions

Equipment and reagents

- NMR spectrometer
- Ligand

- 0.1–0.5 mM protein sample in either D_2O or H_2O

Method

This protocol uses the transferred NOE approach and is most suitable for ligands of molecular weight below 1 kDa.

1   Prepare a protein sample of about 0.1–0.5 mM concentration (in either D_2O or H_2O, whichever is more convenient).

2   Collect a 1D proton spectrum of protein to ensure that the protein is folded, not aggregated, etc.

3   Add an aliquot of ligand to the protein sample to give a ligand:protein concentration ratio of about 30:1. Adjust the protein concentration accordingly if the ligand solubility is the limiting factor.

Protocol 2 continued

4 Acquire a 2D NOESY spectrum of the protein–ligand sample using a mixing time of 100–400 ms.

5 Process the 2D data. If the ligand resonances give cross-peaks which have the same phase as the protein diagonal peaks, then the ligand is binding to the protein in the *fast exchange* regime. If the ligand cross-peaks are of opposite phase to the protein diagonal peak then there are no protein–ligand interactions.

More recently, our laboratory (13, 14) and others (15) have demonstrated the use of NMR translation diffusion measurements for screening *weak protein–ligand interactions*. The method uses pulse field gradient to spatially encode molecules in solution. The technique is based on the principle that translational diffusion in solution is dependent on macromolecular size (as well as on macromolecular shape and the temperature and viscosity of the medium)—see Chapter 4 in this volume. The (translational) diffusion coefficient of a small molecule in solution can be altered (on a time-averaged basis) by complexation with another molecule. If the complex is, for example, between a small ligand and a protein, then the diffusion coefficient of the combined ligand–protein complex will be sufficiently different from the free ligand. Pulsed field gradient experiments can be performed to detect such a change in diffusion character-istics. However, there are practical and technical difficulties associated with this approach and hence is less routinely used although developmental work is currently being undertaken.

5 Overview of NMR techniques used to study protein–ligand interactions

Once ligand binding and the chemical exchange rate regime for the inter-actions have been established using the above procedure (together with other biochemical data), it is possible to proceed to the second stage of ligand binding studies using NMR. This second stage includes the quantitative analysis of either the line shape or magnetization transfer data when kinetic infor-mation is required, and/or one of a range of NOE experiments when structural information is needed, and/or the determination of more specific information, for example, the ionization states either of residues in a protein or of groups on the ligand. While the objectives of NMR studies of protein–ligand com-plexes are often *either* kinetic *or* structural, there are occasions when both types of information are desirable. It may therefore be necessary to manipulate the binding conditions (temperatures, ionic strengths, types of ligands, etc.) such that different exchange regimes occur to allow the different modes of analysis to be applied.

In the *fast* exchange regime, the transferred NOE experiment can be used to obtain structural information (from both intramolecular and intermolecular NOEs) and, in the very fast exchange case, the chemical shift changes as a function of ligand (or protein) concentrations can be used to estimate the equilibrium constant. A significant limitation to the interpretation of the spectra in the fast exchange regime should always be borne in mind: the spectra represent an average across all the states which the nuclei experience, but it is not known *a priori* how many states there are, nor what their parameters may be. A model is required for the interpretation of the fast exchange spectra and as many resonances as possible should be analysed to maximize the chances of detecting any inconsistencies which might indicate that the model being used is incorrect.

When the exchange is *slow*, it is best to perform the experiments using isotopically labelled protein. Most often the bound ligand conformation and the structure of the entire protein–ligand complex are determined at the same time, using strategies similar to the ones used for the calculation of the protein structures (16, 17): such structures are calculated using distance constraints obtained from intra- and intermolecular NOE data.

The relative ease with which most proteins can be labelled with stable isotopes in many different ways (uniformly, residue-specific, site-specific) and the development of multinuclear multidimensional NMR methods have made investigations of protein–ligand interactions simpler. With such techniques available, the main limitation to the use of NMR for studying protein–ligand binding is the resonance line widths of the system under consideration; even this limitation is gradually been overcome (1). Provided resonance line width is not the limiting factor, a whole plethora of experiments is available using proteins or ligands that have been enriched with ^{13}C, ^{15}N, and/or ^2H. The most important factor affecting the choice of experiments and strategies is to decide on the question to be addressed and the information required. Readers are referred to ref. 8 for a summary of the methods available.

Several factors give rise to broad line widths which are unfavourable for protein–ligand interaction studies. These include intrinsic line widths due to size of protein, mobile regions that are exchanging between multiple conformations at unfavourable rates, and protein aggregation. As a rule of thumb, it is necessary to perdeuterate a protein of molecular weight over 25 kDa in order to optimize the resonance line widths for detection of signals, regardless of the type of experiments to be performed; the main consideration that must however, be given is the level of deuteration; for example, if intermolecular proton–proton NOEs are required for structure calculations, a sufficient level of protonation ($\sim 50\%$) must be available to give the required signal-to-noise ratio. Perdeuteration can sometimes be used to alleviate problems of protein aggregation. If a protein is known or suspected to contain mobile or flexible region(s), is it often possible to reduce (or eliminate) this unfavourable dynamic behaviour by complexing the protein to a known ligand; in this case, the protein can only be studied as a protein–ligand complex.

Protocol 3

Optimizing the NMR spectral line widths in the presence of protein aggregation

Equipment and reagents

- NMR spectrometer
- Protein solution
- Ligand solution
- Range of buffers and solvents

Method

1 Check for protein aggregation using several biophysical methods such as ultra-centrifugation, light-scattering techniques, or using a gel-filtration column. As far as possible, use the same buffer condition as for a NMR sample.

2 For the NMR sample itself, perform a quick experiment to determine the diffusion coefficient (13) or the transverse relaxation, T_2, of the protein. Estimate the size of the protein from the T_2 or the diffusion coefficient.

3 If the protein appears aggregated try several strategies (either individually or in combination) to reduce or remove aggregation:

(a) Vary the temperature and/or pH (ensure that the protein does not unfold within the temperature range).

(b) Vary the ionic strength of the sample. Use the Hoffmeister series as a guide.

(c) Use an aqueous chaotropic solvent (e.g. 0.1 M urea) but ensure that the protein remains folded.

(d) If the protein has a defined ligand, add some ligand to the protein sample (at a concentration ratio respective binding sites of 1:1).

6 Exchange rate measurements

The example described here, using the binding between coenzyme A (CoA) (*Figure 1*) and chloramphenicol acetyl transferase (CAT) (18), serves to illustrate many of the points described in *Protocol 1*.

To estimate the rate of exchange of the ligand CoA between the bound and free states, CoA is titrated in small increments into a sample of perdeuterated CAT and a 1D ^1H spectrum was obtained after each addition; the changes in the spectrum as the ligand concentration is increased reveal the region of exchange. *Figure 2a* shows examples of spectra for the upfield region of the protein. At low CoA concentrations (< 1 molar equivalent) two new signals appear, at 0.91 ppm and 0.43 ppm. These signals correspond to the two methyl groups $C^{P9}H_3$ and $C^{P8}H_3$ of the pantetheine moiety of CoA. As the ligand concentration is increased, the resonance at 0.91 ppm increases in intensity and rapidly becomes sharper, while the chemical shift changes only slightly; this is characteristic of *fast exchange* behaviour. This signal was assigned, based on the crystal structure of the CoA–CAT complex to the $C^{P9}H_3$ group (18).

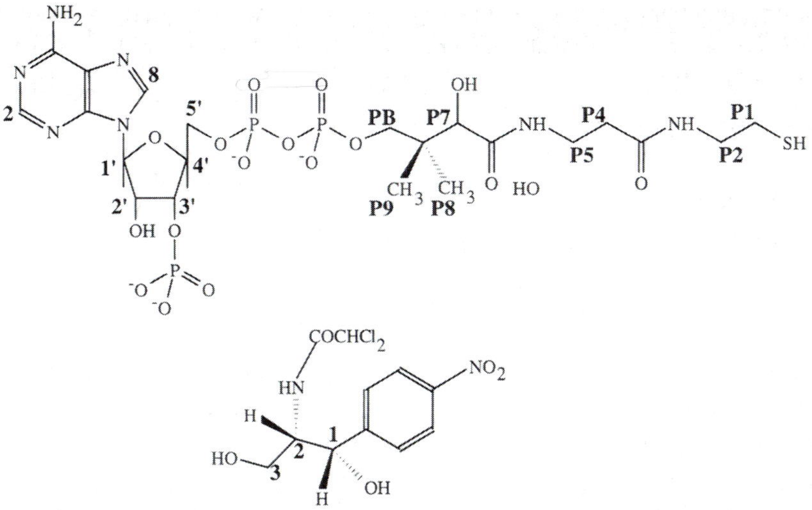

Figure 1 Structures of CoA (*left*) and chloramphenicol (*right*) indicating the atomic nomenclature used. (Reproduced with permission from ref. 18.)

The signal appearing at 0.43 ppm, assigned to the $C^{P8}H_3$ group of bound CoA, shows a much larger progressive change in chemical shift as the CoA concentration is increased, approaching the chemical shift of free CoA only at a CoA:CAT ratio of 10. This signal also remains broader than that at 0.91 ppm as the CoA concentration is increased (*Figure 2a*). This effect can be attributed to an exchange contribution to the line width, indicating *moderately fast exchange* behaviour.

Most of the other CoA signals show *fast exchange* behaviour with minimal changes in chemical shift, line narrowing only being observed as the ligand concentration is increased in the presence of the enzyme. Two notable exceptions are the C^2H and $C^{P7}H$ protons. The C^2H resonance is broadened beyond detection at CoA: CAT ratios below 3:1, suggesting an *intermediate* exchange regime.

The behaviour of the $C^{P7}H$ signal is similar to the $C^{P8}H_3$ discussed earlier, with an exchange contribution to the line width. For $C^{P8}H_3$ and $C^{P7}H$ resonances, quantitative estimates of the rate of exchange can be obtained from line shape analysis of the resonances at 298 K. At a lower temperature of 280 K, however, the broadening of the signals of the bound state prevents the use of line shape analysis; this illustrates the importance of choosing the correct temperature for the appropriate experiment and method of analysis (see *Protocol 1*, step 7). The equilibrium dissociation constant, K_d, for CoA binding to CAT is 4×10^{-5} M, as determined using reverse reaction steady-state kinetics. This was used to calculate the fraction of CoA bound to the enzyme at different CoA: CAT ratios, and the observed chemical shifts of the $C^{P8}H_3$ and $C^{P7}H$ resonances were found to be linearly dependent on the fraction of the free ligand (*Figures 2b and 2c*), as expected from *Equation 2*. Extrapolation of the straight lines in *Figure 2b*

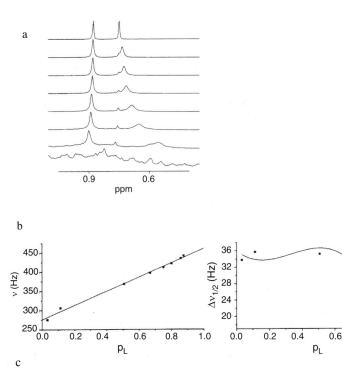

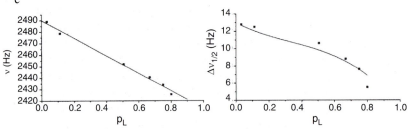

Figure 2 (a) The high-field region of ^{1}H spectra of perdeuterated CAT recorded at 298 K in the presence of increasing concentrations of CoA, showing the changes in the methyl resonances of the ligand. From bottom to top the spectra are those of the free protein, 1:1, 1:2, 1:3, 1:5, 1:7, and 1:10 molar ratios of protein:CoA, and free CoA. (b, c) The chemical shift, υ (*left*) and line width, $\Delta\upsilon_{1/2}$ (*right*) of the $C^{P8}H_3$ (b) resonances and $C^{P7}H$ (c) resonances as a function of p_L, the fraction of free CoA, as the ratio of CoA to perdeuterated CAT is varied at 298 K. Continuous curves represent least-squares fits of the model equations for chemical shift (Equation 3) and for the line width (Equation 4). (Reproduced with permission from ref. 18.)

and *2c* was used to calculate the difference in chemical shift between the bound and free states, $(\delta_{EL} - \delta_L)$, yielding estimates of 185 and -74 Hz for $C^{P8}H_3$ and $C^{P7}H$ protons respectively. The off-rate was then estimated by a least-squares fit of *Equation 3* to the experimental values of the line width as a function of CoA concentration (*Figures 2b* and *2c*). The k_{off} values thus obtained were 3.2×10^3 s^{-1} from the $C^{P8}H_3$ resonance and 4.2×10^3 s^{-1} from the $C^{P7}H$ resonance at 298 K.

7 Assignment of resonances

Before detailed information (such as conformation, ionization states, dynamics) about a protein–ligand complex can be obtained, it is first necessary to assign the protein and ligand resonances. The extent of the assignments required would depend on the nature of the studies undertaken. If the complete structure of the protein–ligand complex is required, a complete assignment of resonances of the protein and the ligand in the complex will be required. If only the interactions between specific protein residues and the ligand are of interest, the resonance assignments can be confined to those residues and regions of a protein and ligand.

7.1 Protein resonance assignments

Protein resonance assignments are most frequently achieved using multidimensional NMR (2D, 3D, and/or 4D) in combination with ^{2}H-, ^{13}C-, and ^{15}N-labelled proteins. Many of these methods and the protocol for resonance assignments are described in detail in ref. 19 and references therein. When the protein can be studied in the free form, the resonance assignments in this form is made first. These resonance assignments can be transferred to the complex form of the protein as it is often that, except for residues around the binding site and under conditions of gross conformational changes, the chemical shifts between the free and complexed protein are not vastly different. There are two common methods of transferring the assignments between the free and the bound form of the protein: by tracking the change in chemical shifts from the 'free' shift to the 'bound' shift (fast exchange), and by using 2D exchange spectroscopy to connect the resonances in the two forms (slow exchange). In the latter case, the free and bound protein must undergo exchange that is sufficiently slow to give two separate signals but fast enough to show exchange cross-peaks. In the extreme case of a tightly bound complex where the chemical shifts in the free and the complex protein gives sufficiently different spectra, it is advisable to assign the complexed protein independently, treating it as a 'new' protein sample.

If only a limited region or specific residues in a protein are of interest in the complex, selective resonance assignments can be made using several strategies. The most costly approach is known as the CN approach where the protein is labelled in a pairwise manner, with the carbonyl carbon of residue (i) labelled with ^{13}C and the succeeding residue (i + 1) labelled with ^{15}N at the amide nitrogen (20). This method was used to assign the Fc fragment of IgG (M ~ 50 000), as shown in *Figure 3*; sequence-specific resonance assignments of all the histidine and methionine residues was achieved (21). The second method is to use a combination of residue-type selective labelling in combination with site-directed mutagenesis (22). This approach has been successfully used in many systems. However, as with many experiments involving mutant proteins, care must be taken when using these mutants. Since in most cases, the mutated residue for which resonance assignments are required are catalytically and/or structurally

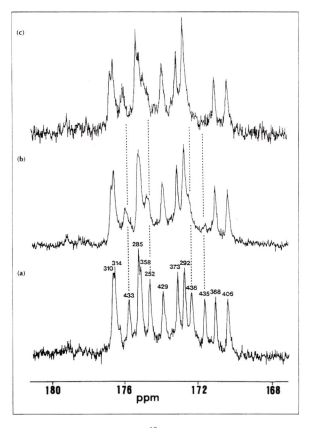

Figure 3 Part of the 100 MHz ^{13}C-NMR spectrum of the Fc fragment of mouse IgG (γ2a) labelled with [1-^{13}C]histidine and [1-^{13}C]methionine in the presence of (a) 0, (b) 0.5, and (c) 1.0 molar equivalents of domain II of protein G. (Reproduced with permission from ref. 21.)

important residues, a mutation of this residue could cause adverse structural effects, making interpretation of the NMR data unsafe.

7.2 Ligand resonance assignments

For weak binding ligands, the fast exchange behaviour is observed and the bound ligand shifts are obtained using the plot of chemical shift against concentration of ligand (binding curve) (*Figure 2b*). Ligands with $K_d \sim 10^{-6}$ M will sometimes show moderately slow exchange behaviour, giving spectra with separate peaks for the free and bound ligand. Because the dissociation rate constants are sufficiently large to allow transfer of saturation, the 2D exchange experiment can be used to connect the resonance of the bound and free species. For very tight binding ligands ($K_d < 10^{-9}$ M), the resonance assignments of bound ligand are more challenging. Since most ligands are not isotopically labelled with the standard stable isotopes used for protein labelling (^{13}C, ^{15}N, ^{2}H), advantage of this is taken. To assign the bound ligand signals, it is first necessary to detect the ligand proton signals only. This is achieved by removing the protein proton resonances from the spectrum using:

(a) ^{13}C-, ^{15}N-labelled proteins in combination with isotope-filtered experiments in which only proton bound to the unlabelled centres (that is, ligand protons) are detected.

(b) Removal of the protein protons by deuterium labelling (8). The disadvantages of this latter method are that as close to 100% deuterium labelling is required and that back-exchange of labile protons means that the NMR experiment needs to be done in D_2O; exchangeable protons on the ligand would not be detected.

It is often the case that ligands contain fluorine and phosphorus nucleii, both of which occur at 100% natural abundance. These resonances provide a valuable structural and mechanistic probe to monitor protein–ligand interactions; information such as the pKa around the binding site, and the presence of multiple conformations, etc. can be conveniently obtained without resorting to expensive stable-isotope labelling of the protein.

8 Structural information

Several main approaches are commonly used to extract structural information from protein–ligand systems; these are summarized below, with detailed description of the commonly used ones discussed in detail in the sections that follow.

(a) **Chemical shift changes**. These are used to broadly locate ligand binding sites on proteins (23) or to determine the regions of ligands interacting with the protein. These experiments are best carried out using isotope-editing methods, where only protons attached to either ^{13}C or ^{15}N are detected. This method can be used irrespective of whether the protein structure is known, so long as sequence-specific resonance assignments are available.

(b) **Nuclear Overhauser effects**. Again, stable-isotope labelling is required. The NOE experiment provides more precise quantitative structural information. The structure of the protein–ligand complex can be calculated once sufficient NOE information is available. In the case of weak binding ligands, detailed structural studies are normally performed on systems where the structures of proteins are known; the NMR studies would focus on the structure of the bound ligand (18).

(c) **Hydrogen exchange rates**. Reduced solvent exchange rates are often observed for protein or ligand groups that are located at the intermolecular interface in the complex. Provided the protein–ligand interface is well-defined, shielded from solvent molecules and not conformationally mobile, the amide proton–solvent exchange rate of these 'interface groups' will be slower compared to their rates in the uncomplexed molecules.

LU-YUN LIAN

(d) **Surface mapping by paramagnetic agents**. This method can be used to determine solvent exposure. Nitroxyl radicals such as 4-hydroxy-2,2,6,6-tetra-methylpiperidinyl-1-oxy (HyTEMPO) can broaden the resonances arising from residues on the surface of proteins or resonances from ligand atoms that are solvent accessible without affecting residues in the interior of protein or ligand atoms that are buried at the protein–ligand interface. It may be necessary to measure the concentration effects of HyTEMPO on the T_1 values rather than to rely simply on an inspection of the line width changes.

(e) **Determination of protein or ligand ionization states**. Characterization of the charge state of ionizable groups in a protein and ligand help to elucidate enzyme mechanisms and molecular recognition.

(f) **Protein dynamics**. Changes in the dynamics of a protein, particularly of the active site, can be used to study ligand-induced perturbation in protein mobility (24).

8.1 Chemical shift changes

The simplest experiments involve direct observation using ^{13}C, ^{15}N, ^{31}P, or ^{19}F NMR. An example of this is given in *Figure 3*; chemical shift changes in the carbonyl carbon resonances of the immunoglobulin Fc fragment allowed identification of a protein G binding site involving methionine 252, 433, 435, and 436, all of which are located at the hinge between the CH2 and CH3 domains.

The isotope-edited experiments detect protons attached to the ^{13}C and ^{15}N. *Figure 4* (top) shows a ^{1}H–^{13}C HSQC spectrum recorded on the same protein G-Fc complex as described above but using ^{13}C-labelled protein G and unlabelled Fc (23, 21). The resonances of many residues undergo significant chemical shift changes (greater than the line width of the cross-peak) on complex formation or are so broadened as to be undetectable. These specific changes may reflect direct protein–protein contacts and/or small changes in conformation on complex formation. The unaffected resonances, on the other hand, can be unambiguously assigned and must arise from residues that are not involved in intermolecular contacts. Since the ^{15}N–^{1}H spectrum give information primarily about the backbone of the protein, further information on the side chain contacts was obtained using ^{13}C–^{1}H correlation spectroscopy, using ^{13}C-enriched domain II. These changes in the resonances of protein G on complex formation allow identification of the regions of protein G that are involved in Fab binding, as shown in *Figure 4* (bottom, *a*).

The chemical shift changes in 1H–^{15}N HSQC experiments, as described above, form the basis of the SAR (structure–activity relationships) by NMR approach (7) for high-throughput screening of ligand interactions; it is desirable that sequence-specific assignments of the protein spectrum is available. Titrating the small ligand into a ^{15}N-labelled protein and monitoring the change in chemical shifts of the ^{15}N amide resonance using a 2D HSQC spectrum first identifies ligands with weak binding affinity. Two weak binding ligands, each of which perturb different sets of amide chemical shifts when added to the ^{15}N-labelled

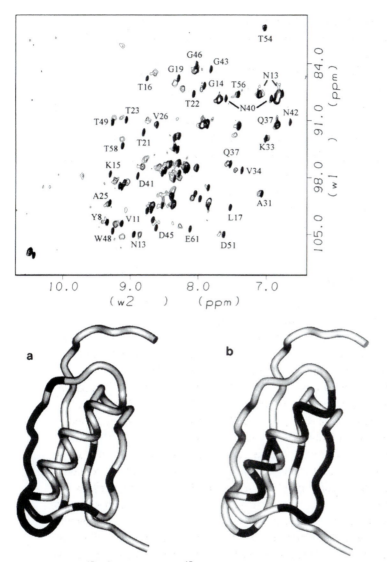

Figure 4 (*Top*) [15]N–[1]H correlation of ^{15}N-labelled protein G domain II in the presence and absence of Fab. Cross-peaks from domain II are shown in grey and those from the complex in unshaded contours. (*Bottom*) Comparison of the residues of the IgG binding domains of protein G affected by binding to (a) Fab and (b) Fc fragments of IgG. Those whose ^{1}H/^{15}N/^{13}C chemical shifts and/or line widths are altered on formation complex are coloured in black. (Reproduced with permission from Academic Press.)

protein, are first identified; their location and orientation in the ternary complex is next determined experimentally using isotope-filtered or transferred NOE methods. The information from these latter experiments are used to synthetically link the two lead, weak-binding compounds together in such a way that the relative spatial orientation (with respect to each other and to the target protein) are preserved in order to produce the final high-affinity ligand.

8.2 Nuclear Overhauser effects and structure calculations

The strategy used for structure calculations of a tightly bound protein–ligand complex is similar to that for the calculation of proteins (16, 25); the only difference is that the distance constraints for structure calculations include intra-ligand distances and intermolecular distances between the protein and the ligand. Methods for obtaining the intermolecular distances centre on the use of isotope-filtered experiments ((26) and references therein).

In the case of weakly bound ligands, where *fast exchange* rates between the free and the bound ligand apply, the transferred NOE (trNOE) experiment will provide data for intra-ligand, inter-ligand (in the case of ternary complexes with more than one bound ligand), and protein–ligand distances. It should be noted that trNOEs can be observed whether exchange is slow or fast with respect to chemical shift, that is whether separate signals or a single averaged signal are seen for the ligand in the bound and free states: the condition is simply that the relaxation rates are averaged. Under the right conditions, the trNOE approach is also used to obtain bound ligand structures in systems where the protein is too large for complete structure determination by NMR. For these weak binding systems, it is the structure of the ligand and the protein site(s), to which the ligand binds that are of interest. The structure of the protein–ligand complex is normally calculated using the known structure of the free protein as the starting structure.

The use of transferred NOE experiments to determine in detail the conformation of a ligand bound to a protein requires careful quantitative analysis of the NOE data, in particular:

(a) The determination of the exchange rate of the ligand between the bound and free states.

(b) The determination of the rotational correlation time.

(c) A careful distinction between direct and indirect magnetization transfer (spin diffusion).

We use the example of the binding of CAT to coenzyme A (CoA) to discuss in some detail the analysis of transferred NOE data. A more detailed description of how to analyse transferred NOE data is given by Lian *et al.* (1994). The determination of the exchange rate was dealt with in Section 6.

To estimate the rotational correlation time (τ_c) of the free ligand (CoA) the ratio of the cross to diagonal peak intensity for the $C^{PB}H_2$ protons was taken; this gave a τ_c of 0.3 ns. For the bound ligand, τ_c was estimated to be 40 ns from the dependence of the transferred NOE intensities upon the mixing time (of the cross-peaks between the $C^{PB}H_2$ protons and between the protons of the ribose ring) (*Figure 1*) (18).

One of the main problems encountered in trying to obtain proton–proton distance in large proteins is the phenomenon of spin-diffusion; this occurs when protons not close in space show 'NOE' interactions as if they were close together, hence giving erroneous distance information. The spin-diffusion has effects on the intensities of cross-peaks and its contributions to the intensities of cross-peaks corresponding to proton pairs at large internuclear distances become more pronounced as the ligand concentration increases. Spin diffusion can occur through protons of both the ligand and the protein and these must be dealt with in different ways. The only satisfactory way of dealing with spin diffusion through the protein is to eliminate it by using perdeuterated protein. Spin diffusion via ligand protons can be accounted for using the full relaxation matrix calculations (8).

One good way to distinguish between direct magnetization transfer and spin diffusion is to use rotating frame NOE spectroscopy (ROESY) experiments. The positive NOE effects in the rotating frame leads to cross-peaks arising from direct effects and from two-step spin diffusion having opposite signs. Thus cross-peaks in the ROESY spectrum which have opposite phase to the diagonal peaks correspond to direct magnetization transfer, while those which are in-phase with diagonal peaks arise from indirect transfer. Zero cross-peak intensity indicates either that there is no magnetization transfer, due to large interproton distance, or that direct and indirect transfer is equally effective. However, it must be borne in mind that the ROESY experiment does not eliminate spin-diffusion; it simply gives it the opposite sign to that of direct transfer. The danger of using ROESY experiments alone is that vital proton–proton distances may not be observed in the ROESY spectra simply because of the cancellation of direct and indirect effects. Thus, it is recommend that structural information be derived from NOESY spectra with the help of relaxation matrix calculations, ROESY being used for stereospecific assignment and to estimate spin diffusion contributions.

In summary, the optimal conditions for obtaining the bound ligand structure is to use a combination of NOESY and ROESY data. The data should be collected at a ligand:protein ratio of 15:1, guaranteeing strong cross-peaks while the contribution from the free state remains small, and using perdeuterated ($> 85\%$ ^{2}H) protein to prevent magnetization transfer through protein protons. As low a temperature to give fast exchange should be used to improve the stability of the sample. *Figure 5* shows a superimposition of ten CoA–chloramphenicol structures calculated using trNOE data to obtain the intra-ligand and inter-ligand distances.

8.3 Determination of ionization states

The charge state of a ligand and protein can both provide information on enzyme mechanisms and molecular recognition, in addition to affecting the affinity between two molecules. Changing the charge state of an ionizable group in a protein or ligand is usually accompanied by large and characteristic changes in the chemical shifts of nuclei in close proximity to the ionizable group (through-bond or through-space) due to changes in shielding of the nuclei. NMR has been used to report on the ionization states of specific groups in proteins and ligands. The ability to isotopically label proteins with ^{13}C and ^{15}N either uniformly or selectively has enable detailed pKa values of important protein side

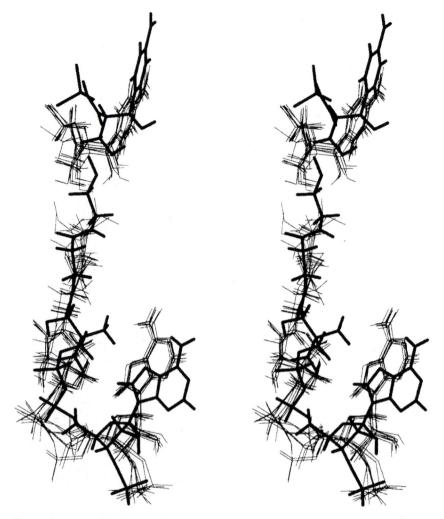

Figure 5 Stereo view of ten structures to show the relative orientation of CoA and chloramphenicol (Cm) for the ternary CAT–CoA–Cm complex at 290 K, superimposed on the X-ray structure (in bold) of the two binary complexes. (Reproduced with permission from ref. 18.)

chains to be obtained provided the ionizable group(s) titrate within the pH range in which the protein is stable. ^{31}P NMR has been a useful probe for studying the ionization states of bound ligand molecules.

8.4 Protein ionization states

Ionization effects on glutamate, aspartate, histidine, cysteine, and lysine residues have all been reported; a majority of these data have been obtained using isotopically labelled proteins. For the glutamate and aspartate, protonation of the carboxyl group have been monitored using direct detection ^{13}C NMR of the carboxyl carbon. Protonation of the carboxyl carbon causes an upfield shift of between 2.0–5.0 ppm. The C2 proton of the imidazole rings of histidines are easy to detect using with 1D proton spectra of a protein sample dissolved in D_2O or a proton-detected 1D or 2D spectrum using a $^{13}C(2)$-his-labelled sample. The pKa values of histidines in protein are normally in the range 4.5–8.5; deprotonation of a histidine residue causes an upfield shift of the C(2) and C(4) protons; the overlap of the C(4) carbon of histidine residues make these carbons less suitable as probes for obtaining the pKa values of histidines.

The deprotonation of cysteine side chains cause a downfield shift of up to 0.3 ppm and 4 ppm respectively for the H^β proton and C^β carbon signals. Less well characterized are the pKa values of the lysine, arginine, and tyrosine side chains in proteins. The normal pKa values of these residues are above 9; hence experiments to obtain pKa values are limited by the instability of most proteins over the high pH range over which a protein has to be titrated in order to obtained reliable pKa values.

During a pH titration, there is usually intermediate to fast exchange between the protonated and unprotonated species. We shall consider the proton resonances, as these are the ones most often used. In the case of *intermediate exchange*, the proton signals are normally observable in the fully protonated and non-protonated states; however, over the pH range when the group is titrating, the signals are normally too broad to be interpretable. In the presence of *fast exchange* between the protonated and unprotonated species, the signals have averaged chemical shifts weighted according to the populations of the two states. A plot of the chemical shift against pH values can be fitted to the Henderson–Hasselbach equation and thus provide values of the pKa_a ('pKa_a' for an amino acid) and the chemical shifts of the protonated and non-protonated species. In some cases the titration curves of the titratable amino acid residue do not conform to the Henderson–Hasselbach curve or titrate at a pKa which is very different from the expected value: in such cases more detailed analyses are required in order to derive useful information.

The example chosen here is the determination of the pKa_a of a lysine side chain in the enzyme TEM-1 β-lactamase, a member of a group of enzymes that catalyses the hydrolysis of β-lactam antibiotics (27). The design of efficient antibacterial compounds requires an improved understanding of the catalytic mechanism of this class of enzymes. *Figure 6* shows the ^1H–^{13}C heteronuclear single quantum coherence (HSQC) spectra of [ε-^{13}C] lysine-labelled β-lactamase at two pH values, 5.8 and 11.4, and the pH dependence of the ^1H and ^{13}C chemical shifts of peaks arising from Lys 73. It is important to note that the enzyme is still in its native conformation at pH 11.4, as judged from the NMR spectrum and the catalytic activity of the enzyme. Two types of cross-peaks can be discerned from *Figure 6a*; sharp intense peaks, which imply magnetic equivalence of the Cε-methylene protons, and broader peaks appearing as pairs of signals with the same ^{13}C chemical shift but two different ^1H shifts. These paired broad signals (B1, B2 and B3, B4) each corresponds to a single lysine residue in which the Cε-methylene protons are non-equivalent. Peaks B1, B2 were assigned to

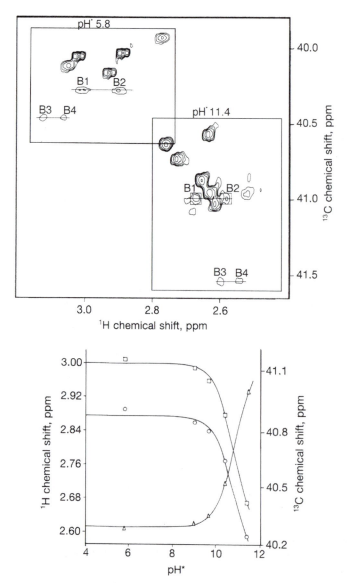

Figure 6 (*Upper*) $^1H–^{13}C$ HSQC spectra of [ε-^{13}C]lysine-labelled β-lactamase at pH 5.8 and pH 11.4. (*Lower*) pH dependence of ^1H and ^{13}C chemical shifts of peaks B1 and B2 arising from Lys 73. □ and ○, ^1H chemical shifts of B1 and B2, respectively, Δ, ^{13}C chemical shifts for B1 and B2. (Reproduced with permission from ref. 27.)

active site Lys 73 and the pH dependence of the ^1H and ^{13}C chemical shift of these peaks show that the pKa_a value for Lys 73 is greater than 10, that is a normal pKa_a value. Hence, it is unlikely that this residue could act as a proton acceptor in catalysis and that an alternative mechanism of catalysis is required.

References

1. Wider, G. and Wuthrich, K. (1999). *Curr. Opin. Struct. Biol.*, **9**, 594.

2. Wuthrich, K. (1998). *Nature Struct. Biol. Suppl. II*, **5**, 492.

3. Roberts, G. C. K. (1999). *Curr. Opin. Biotech.*, **10**, 42.

4. Stockman, B. (1998). *Prog. NMR Spectrosc.*, **33**, 109.

5. Moore, J. M. (1999). *Biopolymers*, **51**, 221.

6. Chen, A. and Shapiro, M. J. (1998). *J. Am. Chem. Soc.*, **120**, 10258.

7. Shuker, S. B., Hajduk, P. J., Meados, R. P., and Fesik, S. W. (1996). *Science*, **274**, 1531.

8. Lian, L. Y., Barsukov, I. L., Sutcliffe, M. J., Sze, K. H., and Roberts, G. C. K. (1994). In (ed. T. L. James) *Methods in enzymology* Vol. 239, pp. 657–700. Academic Press, London and N.Y.

9. Feeney, J. L. and Birdsall, B. (1993). *Practical approaches in NMR of biological macromolecules* (ed. G. C. K. Roberts), pp. 183–212. Oxford University Press.

10. Kainosho, M. (1997). *Nature Struct. Biol.*, **4**, 858.

11. Sattler, M. and Fesik, S. W. (1996). *Curr. Biol.*, **4**, 1245.

12. Lian, L. Y. and Roberts, G. C. K. (1993). *Practical approaches in NMR of biological macromolecules*, pp. 153–82. Oxford University Press.

13. Tillett, M. L., Lian, L. Y., and Norwood, T. J. (1998). *J. Magn. Reson.*, **133**, 379.

14. Tillett, M. L., Horsfield, M. A., Lian, L. Y., and Norwood, T. J. (1999). *J. Biomol. NMR*, **13**, 223.

15. Bleicher, K., Lin, M., Shapiro, M. J., and Wareing, J. R. (1998). *J. Org. Chem.*, **63**, 8486.

16. Clore, G. M. (1998). *Proc. Natl. Acad. Sci. USA*, **95**, 5891.

17. Guntert, P. (1998). *Q. Rev. Biophys.*, **31**, 145.

18. Barsukov, I. L., Lian, L. Y., Ellis, J., Sze, K. H., Shaw, W. V., and Roberts, G. C. K. (1996). *J. Mol. Biol.*, **262**, 543.

19. Clore, G. M. and Gronenborn, A. M. (1997). *Nature Struct. Biol. Supp 1*, **4**, 849.

20. Kainosho, M. and Tsuji, T. (1982). *Biochemistry*, **21**, 6273.

21. Kato, K., Lian, L. Y., Barsukov, I. L., Derrick, J. P., Kim, H., Tanaka, R., *et al.* (1995). *Structure*, **3**, 79.

22. Derrick, J. P., Lian, L. Y., Roberts, G. C. K., and Shaw, W. V. (1992). *Biochemistry*, **31**, 8191.

23. Lian, L. Y., Barsukov, I. L., Derrick, J. P., and Roberts, G. C. K. (1994). *Nature Struct. Biol.*, **1**, 355.

24. Kay, L. E. (1998). *Nature Struct. Biol. Suppl. II*, **5**, 513.

25. Nilges, M. and O'Donoghue, S. I. (1998). *Prog. NMR Spectrosc.*, **32**, 107.

26. Garrett, D. S., Seok, Y.-J., Peterkofsky, A., Gronenborn, A. M., and Clore, G. M. (1999). *Nature Struct. Biol.*, **6**, 166.

27. Damblon, C., Raquet, X., Lian, L. Y., Lamotte Brasseur, J., Fonze, E., Charlier, P., *et al.* (1996). *Proc. Natl. Acad. Sci. USA*, **93**, 1747.

Chapter 11

Quantification and mapping of protein–ligand interactions at the single molecule level by atomic force microscopy

Clive J. Roberts, Stephanie Allen, Martyn C. Davies, Saul J. B. Tendler, and Philip M. Williams
University of Nottingham, School of Pharmacy, University Park, Nottingham NG7 2RD, UK.

1 Introduction

This chapter illustrates the recent application of atomic force microscopy (AFM) to the measurement of the forces that bind protein–ligand complexes. The experiments highlighted will demonstrate the potential of AFM to quantify the adhesion force between individual molecular pairs and how this can lead to a deeper understanding of protein structure and function. The possibility of spatial mapping active surface bound proteins though their ability to bind ligands will also be illustrated.

An understanding of the recognition mechanisms is not only of importance in understanding protein structure, folding, and function but has practical implications for the development of novel devices in areas such as sensors, screening, and drug discovery.

AFM has long been recognized as having significant potential for the imaging of biological systems due to its nanometre spatial resolution and ability to operate in ambient/aqueous environments. For example, AFM has been employed to image sub-molecular structure of membrane proteins (1), which cannot be resolved using X-ray crystallography or electron microscopy. When employed as an imaging tool a very sharp probe, supported on the end of a flexible lever, is made to scan over a sample surface whilst monitoring the resultant deflection of the lever (*Figure 1*) (2). During this process the probe can be either in contact with the sample surface, in intermittent contact with the surface, or just off-contact. Since AFM probes typically have terminal radii of 2–10 nm, the spatial resolution of AFM is of the order of nanometres. However, just as important is the lever, and typically the levers are 100–400 μm in length and 2–10 μm in

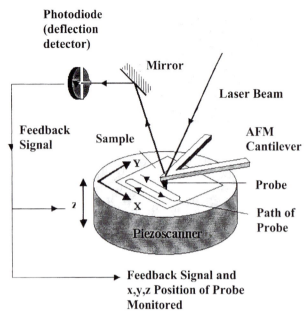

Figure 1 Schematic of the principle components of an atomic force microscope. A sharp probe supported upon a flexible cantilever is moved over a sample surface by a piezo-ceramic scanner. Feedback of the probe's (and hence cantilever's) motion over the sample is gained via an optical lever set-up. This signal is typically utilized in a feedback system which maintains constant imaging conditions.

thickness, and hence very little force is required to cause a detectable deflection. This allows the probe to ride over (and hence image) even very soft objects such as hydrated proteins (1, 3). To utilize this potential resolution the spatial position of the probe relative to the sample must be controlled to sub-nanometre resolution. This is achieved using piezo-ceramic actuators. The deflection of the lever itself is typically monitored using an 'optical lever' set-up based upon the reflection of a laser beam from the back of the cantilever.

The potential of AFM to image, and interact with, dynamic molecular processes has also been exploited. For example AFM has been employed to visualize *Escherichia coli* RNA polymerase transcribing two different linear double-stranded (ds) DNA templates (4) and has been used to remove extrinsic proteins from the photosystem I (PSI) reaction centre from *Synechococcus* exposing the stromal side of the PSI core, whose surface structure could then be imaged at a resolution better than 1.4 nm (5). This last application hints at the ability of AFM to not only image but to sense and interrogate forces at a level which permits the inter-molecular forces between individual molecular complexes to be explored. Whilst AFM is not alone in this ability; for example optical tweezers have also been recently employed to investigate such interactions (6), it does however, offer a unique combination of spatial resolution and force sensitivity. Hence, if an AFM tip can be coated with a protein of interest, its interaction with a potential ligand immobilized to a substrate can be studied and measured.

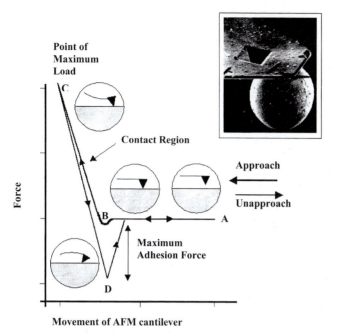

Figure 2 Idealized view of force–distance data gained from an AFM. The inset images in the circles indicate the relative position of the probe and sample surface during the approach and unapproach phase of the data acquisition. The inset electron microscope image shows a bead placed on the end of an AFM cantilever as can be used for spring constant calibration and also functionalization of the probe (7).

Such force measurements are recorded by monitoring the deflections of the cantilever as the tip and sample are brought into contact (approach trace) at a constant velocity, and then separated (retract trace) (*Figure 2*), the result is a plot of cantilever deflection force versus distance moved by cantilever-base or scanner (7). Initially (A in *Figure 2*), the tip and sample are separated by a large distance with the cantilever in the rest position (zero deflection). As the tip–sample separation is reduced (B in *Figure 2*), the cantilever can be deflected by long-range forces acting on the probe. Attractive forces (for example, van der Waals or attractive electrostatic forces) bend the cantilever towards the probe. Repulsive forces (for example, electrostatic repulsive forces) deflect the probe away from the surface. At a point close to the surface, the probe can 'jump into contact' if it experiences an attractive force greater than the stiffness of the cantilever spring. Once the probe is in contact with the surface, cantilever deflection will increase (C in *Figure 2*) as the fixed end of the cantilever is moved closer to the sample. This continues until a predetermined point of maximum load is reached (which should be as low as possible to avoid damaging any immobilized protein). The process is then reversed. As the cantilever is withdrawn (D in *Figure 2*), the probe may adhere to the surface due to protein–ligand interactions, formed during contact, between the probe and sample. Further retraction overcomes this interaction and the cantilever breaks free from the surface (E in *Figure 2*).

409

The magnitude of the interaction formed during contact can be calculated from the difference between the maximal cantilever deflection during retraction and the point of zero deflection of the cantilever.

The following sections of this chapter detail recent applications of AFM to the quantification of protein–ligand interactions and how practically such experiments are executed. To illustrate these points the measurement of specific antibody–antigen interactions will be discussed in detail. The points highlighted in these studies are general to the use of AFM to measure biomolecular interactions.

2 Applications

The first reports of the measurement of the forces binding a protein–ligand complex were in 1994 for (strept)avidin–biotin (8, 9). Since then a number of groups have investigated specific interactions between various ligand–receptor pairs, such as IgG–protein G (10, 11), heavy meromyosin and actin (12), intermolecular and intramolecular forces of dextran molecules (13). Similar approaches have also facilitated the measurement of stretching forces of individual biopolymers such as titin (14) and DNA. For example, the end-to-end stretching of a duplex DNA oligonucleotide has been studied using a combination of molecular dynamics simulations and AFM experiments (15). The authors show that the barrier to strand separation occurs when unfavourable DNA interstrand repulsion cannot be compensated for by favourable DNA–solvent interactions.

3 Quantification of the adhesion forces of individual protein–ligand complexes

3.1 General guidelines

When attempting to quantify forces of adhesion between individual protein and ligand molecules using AFM, a number of key practical points need to be addressed. Namely:

(a) How many molecules are interacting (ideally a single pair).

(b) What is the level and nature of any non-specific interaction between the AFM probe and sample surface.

(c) The effect of any molecular immobilization strategies on conformational freedom of the protein and/or ligand.

(d) The preservation of molecular conformation/activity during multiple AFM measurements.

(e) The rate of loading on molecular complexes once formed (i.e. the combined effect of AFM cantilever spring constant and rate of motion of the sample from the AFM tip).

(f) How is the data to be processed and analysed to produce useful thermodynamic and structural insights.

Each of these points will be considered separately in the following text.

3.1.1 Number of interactions measured

Control over the number of interacting molecules is in fact not as difficult as it may first seem. The majority of AFM experiments employ the very sharp probe used for imaging materials at nanometre resolution. These tips typically have terminal radii in the 2–10 nm region. Hence the contact area between the tip and a substrate is rarely able to accommodate more than a few protein molecules to potentially interact with a substrate. In addition by controlling the surface density of functional proteins and/or ligands this number can be optimized to one-pair. This is particularly relevant if a large contact area is expected, as in the case where protein are immobilized to micron-sized beads (usually silica or agarose) (9), and these beads are then mounted on the AFM lever for molecular interaction studies. If a protein is to be immobilized onto a self-assembled monolayer coated surface or probe the use of mixed functionality monolayers allows the dilution of protein coverage. Similarly if proteins are immobilized to functionalized polymer coated probes, the protein coverage may be diluted using non-functional polymer units.

It is worth noting that the forces involved in rupturing single protein–ligand complexes can also be obtained from statistical study of the Poisson distribution of the forces required to break very many such complexes (16). Potentially this approach does offer advantages, since it requires no assumptions about the surface energies and is not limited by the force resolution of the AFM employed. However, this approach does preclude studying the effects of external factors such as competing binders at a single molecule level.

3.1.2 Non-specific forces and controls

To identify the forces of rupture of individual or small number of protein–ligand complexes it is vital to reduce the level of the non-specific forces between the probe and surface to near zero. A number of approaches to this problem have been employed, including the use of biotinylated bovine serum albumin as both an immobilization anchor (for avidin and subsequently biotinylated proteins) and a screen, the coating of AFM probes in polymer brushes and suitably functionalized self-assembled monolayers. Following the measurement of proposed specific protein–ligand adhesion forces it is necessary to carry out a control which blocks protein–ligand binding (e.g. through the introduction of excess ligand). In such a case the forces of adhesion observed should be reduced to near zero, so here too the absence of non-specific forces is important.

3.1.3 Immobilization of the protein and ligand

The choice of immobilization method of the protein and ligand is an important consideration. During a rupture experiment any complex formed between immobilized species will be subject to a loading sufficient to overcome the bonding between the protein and ligand, however, it must not be sufficient to break the bonds involved in the immobilization. Typically this means that some form of covalent attachment is employed. It has been reported that the force required to break a covalent bond is if the order of 1 to 10 nN (17, 18) which is larger than the largest force measured in protein–ligand rupture experiments (typically 50–200 pN). Other important considerations are the orientation of the protein and ligand and their freedom of movement following immobilization. Sections 3.2.1 and 3.2.2 deals with these matters in some detail.

3.1.4 Effect of loading rate on rupture force

Currently there is some discussion in the literature as to the relationship between the forces observed in AFM protein–ligand rupture experiments and thermodynamic consideration of molecular binding/unbinding (19, 20). For example, site-directed mutagenesis of the streptavidin system, indicates a relationship between the force and the enthalpic barrier of dissociation (19). Other experimental data and computational modelling support this view. It is worth noting that any protein–ligand complex if left alone for a suitable period is likely to dissociate, and therefore, there exists a dependence upon the amount of work required to break an interaction and the time over which the interaction is forced to rupture. In practical terms, this means that the rate at which the AFM is loaded is a key experimental factor. The loading rate is directly related to the speed of retraction of the AFM cantilever and its spring constant, a high loading rate causing a relatively large force of rupture to be observed and a low rate a small force of rupture. To exploit AFM measurements to the full it is therefore, necessary to execute the rupture experiment with a large range of loading rates, and at least across three orders of magnitude.

3.1.5 Processing raw AFM force data

Having recorded the force data proposed to correspond to protein–ligand interactions it is important to process the data correctly, since in its raw form the data since this is not a true representation of force applied to the protein–ligand complex. Raw data is obtained as cantilever deflection (nA) versus displacement of the z-piezo (nm). Cantilever deflection (nA) is first converted to a deflection distance (nm) using the gradient of the linear portion of the retract trace. Using the cantilever spring constant (k), and Hooke's Law (F = k.d), the cantilever deflection is converted to a force (pN). The spring constants of individual cantilevers may be experimentally determined using the resonant frequency method of Cleveland *et al.* (21), where small masses are added to the cantilever and the change in resonant frequency recorded. The horizontal distance axis (nm) is converted from z-piezo displacement to probe-sample separation by subtracting the cantilever deflection distance (nm) from the z-piezo displacement for every data point in each force curve.

3.2 Measurement of antibody–antigen interactions

A number of groups have exploited the force sensitivity of AFM to quantify forces of adhesion between individual antibody–antigen molecular pairs. For example we have quantified interaction between anti-ferritin (IgG) and ferritin (11) and anti-hCG (IgG) and β-subunit of human chorionic gonadotrophin (βhCG) (22). Others and we have adopted essentially two separate approaches to the immobilization of proteins for such experiments. The first involves the using a 'tight' covalent binding of the protein to the AFM probe and sample surface, the second and more recent approach involves the use of long-chain linkers to maximise orientational and conformational freedom of the proteins.

3.2.1 Direct covalent binding of proteins to the AFM probe

Protocol 1 illustrates a typical immobilization approach of this type where we have employed amino silanized surfaces activated using glutaraldehyde to produce protein coated probes with tightly bound molecular layers (see also *Figure 3*).

Force measurements were recorded using a Thermomicroscopes Explorer AFM (Thermomicroscopes Corporation, Bicester, Oxford, UK). Experiments were performed in freshly prepared potassium phosphate buffer (100 mM, pH 7) in a glass liquid cell which was cleaned thoroughly prior to use. Force measurements were obtained between ferritin functionalized AFM probes and silicon wafers to which anti-ferritin antibody had been immobilized (*Figure 4a*). To confirm the presence of specific ferritin binding, ferritin binding sites on the antibody functionalized surface were blocked by flooding the experimental system with ferritin (0.15 mg/ml) (*Figure 4b*). After 1 h the AFM liquid cell was rinsed thoroughly with potassium phosphate buffer to remove excess ferritin. Force measurements were then obtained between the derivatized probe and the blocked surface. Additional force measurements were also recorded between ferritin derivatized probes and wafers coated with anti-HSA as a control experi-

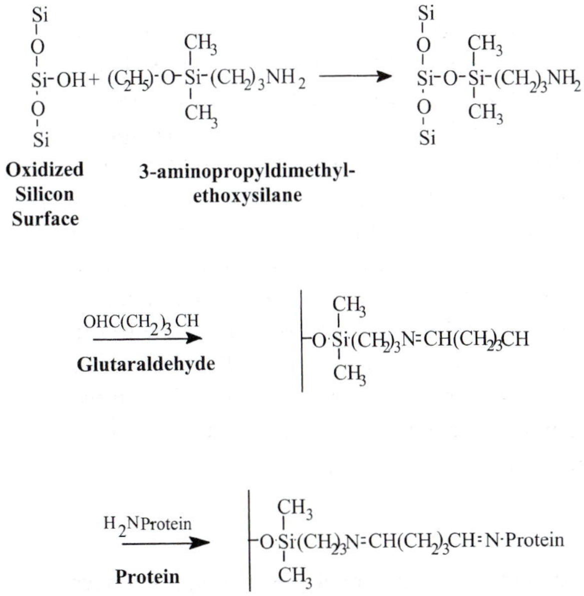

Figure 3 Summary of the procedure detailed in *Protocol 1* for the immobilization of proteins to a silicon AFM probe or sample surface.

413

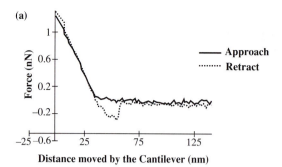

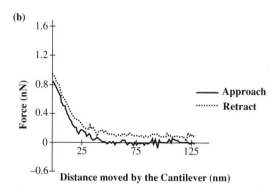

Figure 4 (a) AFM force–distance data showing the additional force required to break an individual anti-ferritin–ferritin complex on retraction of a suitably functionalized probe from a anti-ferritin functionalized surface. The average force observed for this event was 50 pN. (b) An example of a control force–distance measurement used to confirm the presence of specific ferritin binding; ferritin binding sites on the antibody functionalized surface were blocked by flooding the experimental system with ferritin (11).

ment. The spring constants of individual cantilevers were determined using the resonant frequency method outlined by Cleveland *et al.* (21), and were found to be within the range 0.043–0.057 Nm^{-1}.

Quantized forces of approximately 50 pN were observed in the adhesive force distribution for ferritin–anti-ferritin, which we interpreted as single unbinding events between individual antigen and antibody molecules (*Figure 5a*) (11). It was also noted, that some force measurements displayed multiple adhesions in their retract traces, and we postulated that such force curves could arise from multiple interactions between the ferritin molecules on the probe and surface immobilized antibodies. However, as ferritin is a protein that is comprised of 24 homologous subunits, it contains many repeats of the same binding site or epitope for the monoclonal antibody employed. Thus, the surface bound antibodies could bind in numerous ways with the ferritin molecules on the AFM probe (*Figure 6a*). The further interpretation of the force data was therefore, hindered by the complexity of the ferritin–anti-ferritin interaction. Utilizing *Protocol 1* for the monoepitopic antigen βhCG and anti-βhCG, illustrates this point. The force distribution observed for βhCG yields a periodic force observed

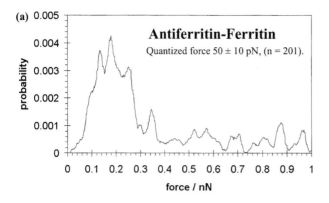

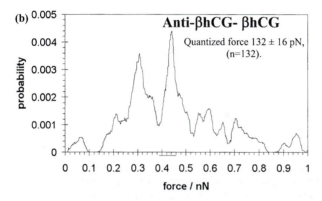

Figure 5 The distribution of adhesive forces obtained between (a) ferritin functionalized AFM probes and silicon surfaces coated with anti-ferritin antibody (b) βhCG functionalized AFM probes and silicon substrates coated with anti-βhCG antibody. The data is normalized to indicate the probability of observing an adhesive force of a particular magnitude. The periodicity of the quantized forces in each distribution is utilized to obtain a value for the single antigen–antibody rupture force (22).

in the adhesive force distribution, due to the rupture of single antigen–antibody interactions, which is found to be larger and more clearly observed than for the ferritin system (132 ± 16 pN compared to 49 ± 10 pN) (*Figure 5*). These findings indicate the potential of the AFM to distinguish between multivalent and mono-valent antibody-antigen interactions, and demonstrate the influence of the number of expressed epitopes upon such binding studies.

Adhesion forces ranging from 60–244 pN (23–25) have been reported for similar experiments employing different antigen–antibody systems. In addition, it has been estimated that antigen–antibody complexes with affinity constants ranging from 10^2 to 10^{10} M^{-1} require rupture forces of 35–135 pN (23). This sug-gests that the force of rupture measured for the anti-ferritin-ferrin and the anti-βhCG–βhCG are in the range expected for the rupture of a single antibody-antigen complex.

(a)

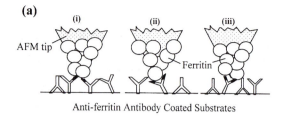

(b)

Figure 6 (a) Due to the distribution of epitopes on the surface of ferritin, their availability for epitope binding is not highly dependent on molecular conformation/orientation. Thus many unbinding scenarios exist for the ferritin–anti-ferritin interaction (bound portions of antibodies are coloured black). (b) Epitope availability will be much more influenced by molecular conformation/orientation, for proteins expressing only single epitopes, e.g. βhCG. It is therefore, unlikely that an antibody molecule would be able to bind to epitopes on neighbouring molecules, and hence fewer unbinding scenarios exist (22).

Protocol 1

Immobilization of antibody for AFM force measurements

Equipment and reagents

- SDS–polyacrylamide gel electrophoresis equipment
- Polished silicon wafers
- Anti-ferritin mouse monoclonal IgG2a
- Chloroform, isopropanol, methanol
- 3-aminopropyldimethylethoxysilane in toluene
- Glutaraldehyde (Grade II, 25% in aqueous solution)
- Potassium phosphate buffer (100 mM, pH 7)

Method

1 Anti-ferritin mouse monoclonal IgG2a (Johnson and Johnson Clinical Diagnostics, Chalfont St. Giles, Buckinghamshire, UK) was purified from mouse ascites fluid by protein A chromatography.

2 The purity of the antibody preparation was confirmed using SDS–polyacrylamide gel electrophoresis.

3 Purified anti-ferritin antibody was immobilized onto polished silicon wafers using a method adapted from Vinckier *et al.* (26).

4 Silicon wafers were cleaned by sonication in a series of solvents (chloroform, iso-

propanol, methanol, and then water). The surfaces of the wafers were then oxidized using a oxygen plasma (200 W, 1 min), after which they were immediately transferred to a 1.5% (v/v) solution of 3-aminopropyldimethylethoxysilane in toluene. After 2 h the silanized substrates were sonicated in the series of solvents to remove any unbound silane. The amino functionalized substrates were then activated by incubating the silanized wafers in a 10% (v/v) solution of glutaraldehyde in potassium phosphate buffer (100 mM, pH 7) for 1 h at room temperature.

5   Wafers were thoroughly cleaned with deionized water to remove any unreacted glutaraldehyde, the antibody solution (0.9 mg/ml in phosphate buffer) was dropped onto the surface of each wafer and incubated at room temperature for 1 h. After this time the wafers were thoroughly rinsed with deionized water and buffer to remove any loosely bound biological material. Wafers were stored in phosphate buffer, at 4 °C, until used. X-ray photoelectron spectroscopy analysis of each surface after the last stage in the immobilization process confirmed the presence of protein on the surface of the silicon wafer.

6   The surfaces of the AFM probe can be functionalized in a similar manner. Cantilevers (ThermoMicroscopes, UK) with silicon nitride (Si_3N_4) probes were functionalized with human spleen ferritin (Johnson and Johnson Clinical Diagnostics) using the immobilization protocol (steps 1–5). Probes should be cleaned/oxidized prior to protein immobilization using an oxygen plasma (10 W, 30 sec). After activating the surface of silanized probes with glutaraldehyde, they were incubated in a solution of ferritin (0.15 mg/ml in phosphate buffer) for 1 h. The probes were then rinsed with deionized water and buffer to remove loosely bound ferritin, and stored in buffer, at 4 °C, until used.

The immobilization strategy employed in these experiments attaches proteins to surfaces by the ε-amino groups of lysine residues, and hence the antigen molecules will be immobilized onto the probe surfaces in a variety of orientations (similarly antibodies on the silicon surface). Due to the distribution of epitopes on the surface of the ferritin molecule, their availability for antibody binding is not highly dependent on the molecular conformation/ orientation. In contrast, for proteins expressing only single epitopes for the employed antibodies, e.g. βhCG, the availability will be much more influenced by molecular conformation and orientation. It is therefore not surprising that we observed that ferritin coated probes were able to form more antigen–antibody pairs than a corresponding βhCG coated probe of the same size.

To overcome this problem with the lack of control over orientation and also to maximize conformational freedom immobilization protocols based upon the use of long-chain linkers have been developed (24).

3.2.2 Immobilization via long-chain linkers

The provision of long-chain linkers between the AFM tip surface and protein under study offers a number of potential advantages. Namely, the possibility of

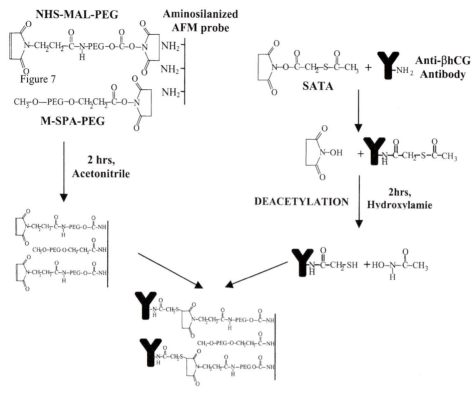

Figure 7 Summary of the procedure detailed in *Protocol 2* for the immobilization of proteins to a silicon AFM probe or sample surface via a long-chain PEG linker.

controlling molecular orientation and conformational freedom and the removal of the molecular adhesion events under study outside the range of non-specific forces associated with interaction between the AFM and sample surfaces. We and others have employed hetero-bifunctional poly(ethylene glycol) (PEG) linkers to immobilize antibodies and their antigens to AFM probes (see *Protocol 2* and *Figure 7*). Besides, successful protein immobilization an important consideration is the surface density of protein required.

We have employed PEG spacers for the measurement of βhCG–anti-βhCG interactions. The experimental protocol for data acquisition is the same as detailed in Section 3.2.1. The average force of adhesion for a single βhCG–anti-βhCG was consistent with the 50 pN previously observed (11). Before complex rupture occurs some stretching of the PEG linkers on the tip and surface occurs and this is highlighted in the force data in *Figure 8a*. As in the previous experiments a control was performed whereby excess ligand (βhCG) was injected into the AFM liquid cell and a complete reduction of interaction force observed (*Figure 8b*). Depending upon the dissociation time of the complex under study it is of course still possible to observe interaction events during this control phase, however, the occurrence of such events should be considerably less than is

(a)

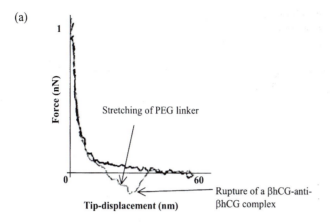

(b)

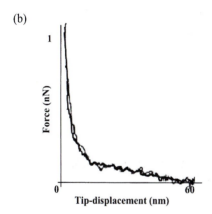

Figure 8 (a) AFM force–distance data showing the additional force required to break and individual βhCG–anti-βhCG antibody complex on retraction of probe and surface to which the proteins have been immobilized via long-chain PEG linkers. The average force observed for this event was 50 pN. (b) An example of a control force–distance measurement used to confirm the presence of specific binding; binding sites on the antibody functionalized AFM probe surface were blocked by flooding the experimental system with βhCG.

observed in the non-excess ligand case. In general, when utilizing the PEG spacer approach with an acceptable level of protein coverage on tip and surface inter-action events should be observed in approximately one in five attempts on differ-ent sample areas. If the tip is maintained in a position where an interaction has been previously observed then all subsequent measurements on that point should display interactions; if not then instrumental drift may have inadvertently moved the contact point, or worse the protein coverage on the tip and/or surface may-be damaged. With care in excess of 1000 force–distance measurements may be recorded before degradation of the protein coverage on the AFM tip, the most important point being not to approach the AFM tip any further than is necessary to form the protein–ligand complex.

Protocol 2

Immobilization of proteins to an AFM probe via a PEG linker

Equipment and reagents

- Excellulose™ GF-5 Desalting Column (Pierce, Product No. 22582)
- 50 mM sodium phosphate, 1 mM EDTA pH 7.5
- *N*-succinimidyl *S*-acetylthioacetate (SATA, Pierce)
- Dimethyl sulfoxide (DMSO)
- Poly(ethylene glycol), alpha-maleimide-omega-NHS ester (NHS-MAL-PEG), molecular weight 2000 Da (Shearwater Polymers Inc., Huntsville, USA)

A Introduction of protected thiol groups into proteins for reaction with succinimide functionalized polyethylene linkers

1 Dissolve protein in a 50 mM sodium phosphate, 1 mM EDTA pH 7.5 to a concentration of 60 μM.

2 Dissolve 13–15 mg of *N*-succinimidyl *S*-acetylthioacetate (SATA) in 1.0 ml DMSO.

3 Combine 1.0 ml of the 60 μM protein solution (step 1) with 10 μl of the SATA solution (step 2). Cover with Parafilm, mix, and allow to react at room temperature for 30 min.

4 Purify the protein using an Excellulose™ GF-5 Desalting Column. Equilibrate one column for each reaction tube with 10 ml of 50 mM sodium phosphate, 1 mM EDTA pH 7.5. Apply 1.01 ml of the reacted protein solution to each column. Collect 1.0 ml fractions using 50 mM sodium phosphate, 1 mM EDTA pH 7.5 as the eluent. Monitor A_{280} of the fractions. Fractions 2 and 3 contain the bulk of the protein.

5 Prepare deacetylation solution by dissolving 1.74 g hydroxyalamine•HCl and 0365 g EDTA in 40 ml of 62.5 mM sodium phosphate pH 7.5 buffer. Readjust pH to 7.5 using sodium hydroxide and bring final volume to 50l ml. Combine 1.0 ml of the protein solution with 100 μl of the deacetylation solution in 13 × 100 mm test-tubes. Cover and mix. Allow to react for 2 h.

B Immobilization of heterobifunctionalized PEG spacer group and antibody to AFM probe

1 Functionalize surface of AFM probe with amino groups. Various methods are available. For example silanization using 3-aminopropyldimmethylethoxysilane (as in *Protocol 1*, steps 1–4).

2 Poly(ethylene glycol), alpha-maleimide-omega-NHS ester (NHS-MAL-PEG) dissolved in acetonitrile to 50 mg/ml. The aminosilanized AFM probes are incubated in this solution for 2 h.

3 Rinse AFM probes in IPA and water and incubate in deacetylated antibody solution (part A, step 5) for at least 2 h. Rinse prior to use in water.

[a] See also Pierce protocol 26102, Pierce, Rockford, Illinois, USA.

4 Spatial mapping of protein–ligand interactions

The ability to sense and measuring forces due to specific molecular recognition between proteins and ligands with AFM allows the possibility of utilizing the spatial resolution of AFM to map such interactions at the molecular level. However, whilst currently a number examples of the mapping of non-specific molecular interactions exist (27, 28), fewer for specific interactions have been reported (29, 31). The principle reason for this is the challenging requirement of having to bring a functionalized probe to a sample surface the many thousands of time required to a map a surface whilst maintaining the integrity of the molecules on that probe. As an extension of our work on ferritin and anti-ferritin we have achieved the mapping of ferritin–anti-ferritin interactions (31). In particular it was possible to flood the system with excess ferritin (as in the standard control), and then to remove this ferritin whilst mapping the surface. This facilitated the observation of an increase in probe–sample interactions as ferritin–anti-ferritin complexes dissociated naturally. Such a direct approach allows an estimation of the dissociation constant for protein–ligand complex. The immobilization of antibodies via a PEG linker has also facilitated the spatial localization of molecular binding for the anti-human serum albumin–human serum albumin (HSA) system (24). Analysis of this data revealed that the forces originated from the dissociation of individual Fab fragments from a HSA molecule, leading the conclusion that the two Fab fragments of the anti-human serum albumin antibody bind independently with equal probability.

5 Conclusions

The experiments described in this chapter represent early steps towards the development of AFM as a powerful analytical technique not only for molecular resolution imaging but also the study of biomolecular recognition processes. The direct measurement of the rupture of single protein–ligand complexes, when combined with complementary biophysical data and molecular modelling offers unique insights in molecular structure and function. Whilst current state-of-the-art has allowed the study of a range of the receptor–ligand systems, many of these measurements have been made at the limit of force sensitivity (approximately 5–10 pN). However, the pace of development is rapid. For example, recently the first commercial AFM system refined for and dedicated to molecular interaction studies has been launched (32), carbon nanotube AFM tips (which maybe functionalized) (33) are offering unparalleled spatial resolution and new modes of data acquisition offer more sensitivity with less chance of damage to immobilized molecules (34). These and developments on immobilization chemistries and a deeper understanding of the rupture process through theoretical studies will establish AFM as a routine tool for the study or protein–ligand interactions ate the molecular level.

References

1. Muller, D. J., Fotiadis, D., Scheuring, S., Muller, S. A., and Engel, A. (1999). *Biophys. J.*, **76**, 1101.
2. Hansma, H. G. and Pietrasanta, L. (1998). *Curr. Opin. Chem. Biol.*, **2**, 579.
3. Blakely, H. K. L., Patel, N., Davies, M. C., Roberts, C. J., Tendler, S. J. B., Wilkinson, M. J., *et al.* (1999). *Exp. Neurol.*, **158**, 437.
4. Kasas, S., Thomson, N. H., Smith, B. L., Hansma, H. G., Zhu, X. S., Guthold, M., *et al.* (1997). *Biochemistry*, **36**, 461.
5. Fotiadis, D., Muller, D. J., Tsiotis, G., Hasler, L., Tittmann, P., Mini, T., *et al.* (1998). *J. Mol. Biol.*, **283**, 83.
6. Grier, D. G. (1997). *Curr. Opin. Colloid Interface Sci.*, **2**, 264.
7. Allen, S., Davies, M. C., Roberts, C. J., Tendler, S. J. B., and Williams, P. M. (1997). *Trends Biotechnol.*, **15**, 101.
8. Lee, G. U., Kidwell, D. A., and Colton, R. J. (1994). *Langmuir*, **10**, 354.
9. Florin, E.-L., Moy, V. T., and Gaub, H. E. (1994). *Science*, **264**, 415.
10. Moy, V. T., Florin, E.-L., and Gaub, H. E. (1994). *Colloids Surfaces A: Physicochem. Eng. Aspects*, **93**, 343.
11. Allen, S., Chen, X., Davies, J., Davies, M. C., Dawkes, A. C., Edwards, J. C., *et al.* (1997). *Biochemistry*, **36**, 7457.
12. Nakajima, H., Kunioka, Y., Nakano, K., Shimizu, K., Seto, M., and Ando, T. (1997). *Biochem. Biophys. Res. Commun.*, **234**, 178.
13. Rief, M., Oesterhelt, F., Heymann, B., and Gaub, H. E. (1997). *Science*, **275**, 1295.
14. Rief, M., Gautel, M., Oesterhelt, F., Fernandez, J. M., and Gaub, H. E. (1997). *Science*, **276**, 1109.
15. MacKerell, A. D. and Lee, G. U. (1998). *Eur. Biophys. J. Biophys. Lett.*, **28**, 415.
16. Lo, Y. S., Huefner, N. D., Chan, W. S., Stevens, F., Harris, J. M., and Beebe, T. P. (1999). *Langmuir*, **15**, 1373.
17. Dammer, U., Popescu, O., Wagner, P., Anselmetti, D., Guntherodt, H. J., and Misevic, G. N. (1995). *Science*, **267**, 1173.
18. Grandbois, M., Beyer, M., Rief, M., Clausen Schaumann, H., and Gaub, H. E. (1999). *Science*, **283**, 1727.
19. Chilkoti, A., Boland, T., Ratner, B. D., and Stayton, P. S. (1995). *Biophys. J.*, **69**, 2125.
20. Moore, A., Williams, P. M., Davies, M. C., Jackson, D. E., Roberts, C. J., and Tendler, S. J. B. (1999). *J. Chem. Soc. Perkin Trans.*, **2**, 419.
21. Cleveland, J. P., Manne, S., Bocek D., and Hansma, P. K. (1993). *Rev. Sci. Instrum.*, **64**, 403.
22. Allen, S., Chen, X., Davies, J., Davies, M. C., Dawkes, A. C., Roberts, C. J., *et al.* (1999). *Biochem. J.*, **341**, 173.
23. Dammer, U., Hegner, M., Anselmetti, D., Wagner, P., Dreier, M., Huber, W., *et al.* (1996). *Biophys. J.*, **70**, 2437.
24. Hinterdorfer, P., Baumgartner, W., Gruber, H. J., Schilcher, K., and Schindler, H. (1996). *Proc. Natl. Acad. Sci. USA*, **93**, 3477.
25. Ros, R., Schwesinger, F., Anselmetti, D., Kubon, M., Schäfer, R., Plückthun, A., *et al.* (1998). *Proc. Natl. Acad. Sci. USA*, **95**, 7402.
26. Vinckier, A., Heyvaert, I., Dhoore, A., McKittrick, T., VanHaesendonck, C., Engelborghs, Y., *et al.* (1995). *Ultramicroscopy*, **57**, 337.
27. Utriainen, M., Leijala, A., Niinisto, L., and Matero, R. (1999). *Anal. Chem.*, **71**, 2452.
28. Chen, X., Davies, J., Davies, M. C., Dawkes, A. C., Edwards, J. C., Roberts, C. J., *et al.* (1998). *Appl. Phys. A*, **66**, S631.
29. Ludwig, M., Dettmann, W., and Gaub, H. E. (1997). *Biophys. J.*, **72**, 445.

30. Roberts, C. J., Allen, S., Chen, X., Davies, M. C., Tendler, S. J. B., and Williams, P. M. (1998). *Nanobiology*, **4**, 163.

31. Allen, S., Chen, X., Davies, J., Davies, M. C., Dawkes, A. C., Roberts, C. J., *et al.* (1998). *Appl. Phys. A*, **66**, S255.

32. Molecular Force Probe, Asylum Research, Santa Barbara, CA 93117, USA.

33. Wong, S. S., Joselevich, E., Woolley, A. T., Cheung, C. L., and Lieber, C. M. (1998). *Nature*, **394**, 52.

34. Schindler, H., Badt, D., Hinterdorfer, P., Kienberger, F., Raab, A., and Pastushenko, V. P. (1999). *Biophys. J.*, **76**, A354.

List of suppliers

Amika Corp., 8980 F Route 108, Columbia, MD 21045, USA.

Analytica of Branford, Inc. (Electrospray ionization source), Branford, CT, USA.

Anderman and Co. Ltd., 145 London Road, Kingston-upon-Thames, Surrey KT2 6NH, UK.
Tel: 0181 541 0035   Fax: 0181 541 0623

Applied Biosystems, 7 Kingsland Grange, Woolston, Warrington WA1 4SR, UK.
Applied Biosystems, 850 Lincoln Center Drive, Foster City, CA 94404, USA.

Applied Photophysics (Stopped flow equipment), 203/205 Kingston Road, Leatherhead KT22 7PB, UK.

Attwater and Sons Ltd., PO Box 39, Hopwood St Mills, Preston PR1 1TA, UK.

AVIV Instruments Inc. (Stopped flow and circular dichroism suppliers), 750 Vassar Avenue, Lakewood, NJ 08701-6907, USA.

Beckman Coulter (UK) Ltd., Oakley Court, Kingsmead Business Park, London Road, High Wycombe, Buckinghamshire HP11 1JU, UK.
Tel: 01494 441181
Fax: 01494 447558
URL: http://www.beckman.com

Beckman Coulter Inc., 4300 N Harbor Boulevard, PO Box 3100, Fullerton, CA 92834-3100, USA.
Tel: 001 714 871 4848
Fax: 001 714 773 8283
URL: http://www.beckman.com

Beckman Instruments (Analytical ultracentrifuge equipment), Progress Road, Sands Industrial Estate, High Wycombe, Buckinghamshire HP12 4JL, UK.
Beckman Instruments Inc., PO Box 3100, 2500 Harbor Boulevard, Fullerton, CA 92634, USA.
Beckman Instruments, Frankfurter Ring 115, Postfach 1416, D-80807 München, FR Germany.

Becton Dickinson and Co., 21 Between Towns Road, Cowley, Oxford OX4 3LY, UK.
Tel: 01865 748844
Fax: 01865 781627
URL: http://www.bd.com
Becton Dickinson and Co., 1 Becton Drive, Franklin Lakes, NJ 07417-1883, USA.
Tel: 001 201 847 6800
URL: http://www.bd.com

BIAcore (Surface plasmon resonance equipment), 2 Meadway Court, Meadway Technology Park, Stevenage, Hertfordshire SG1 2EF, UK.
Biacore Inc., 200 Centennial Avenue, Suite 100, Piscataway, NJ 08854, USA.

Biacore AB, Rapsgatan 7, SE-754 50 Uppsala, Sweden (Head Office).

Bio 101 Inc., c/o Anachem Ltd., Anachem House, 20 Charles Street, Luton, Bedfordshire LU2 0EB, UK.
Tel: 01582 456666   Fax: 01582 391768
URL: http://www.anachem.co.uk
Bio 101 Inc., PO Box 2284, La Jolla, CA 92038-2284, USA.
Tel: 001 760 598 7299
Fax: 001 760 598 0116
URL: http://www.bio101.com

Bio-Rad Laboratories Ltd. (Confocal laser scanning microscope, FTIR), Bio-Rad House, Maylands Avenue, Hemel Hempstead, Hertfordshire HP2 7TD, UK.
Tel: 0181 328 2000   Fax: 0181 328 2550
URL: http://www.bio-rad.com
Bio-Rad Laboratories Ltd., Division Headquarters, 1000 Alfred Noble Drive, Hercules, CA 94547, USA.
Tel: 001 510 724 7000
Fax: 001 510 741 5817
URL: http://www.bio-rad.com
Bio-Rad Spectroscopy Group, 237 Putnam Avenue, Cambridge, Massachusetts 02139, USA.

BOMEM, Inc. (FTIR equipment), 450 St-Jean Baptiste Avenue, Quebec PQ G2E 5S5, Canada.   E-mail: metal@bomem.co
URL: http://www.bomem.com

BDH Laboratory Supplies, Poole, Dorset BH15 1TD, UK.

Bruker UK Limited (NMR, diffraction, and FTIR equipment), Banner Lane, Coventry CV4 9GH, UK.
URL: http://www.bruker-axs.com/
Bruker Instruments, Inc., 44 Manning Road, Manning Park, Billerica, MA 01821-3991, USA.
E-mail: optics@bruker.com

Bruker Daltonics (Mass spectrometry equipment), Manning Park, Billerica, MA 01821, USA.

Calorimetry Sciences Corp., 155 West 2050 South, Spanish Fork, Utah 84660, USA.

Clairet Scientific Ltd. (Raman equipment), 17 Scirocco Close, Moulton Park Industrial Estate, Northampton NN3 6AP, UK.
Fax: 01604 494499
E-mail: clairet@compuserve.com
URL: http://www.clairet.co.uk

Composite Metal Services Ltd. (sources for capillaries), The Chase, Hallow, Worcester WR2 6LD, UK.

CP Instrument Co. Ltd., PO Box 22, Bishop Stortford, Hertfordshire CM23 3DX, UK.
Tel: 01279 757711   Fax: 01279 755785
URL: http://www.cpinstrument.co.uk

DG Electronics (Benchtop and high resolution spectrofluorimeters), 16/20 Camp Road, Farnborough GU14 6EW, UK.

Dianorm GmbH, Stöcklstrasse 5a, D-81247 München 65, FR Germany.

Dupont (UK) Ltd., Industrial Products Division, Wedgwood Way, Stevenage, Hertfordshire SG1 4QN, UK.
Tel: 01438 734000   Fax: 01438 734382
URL: http://www.dupont.com
Dupont Co. (Biotechnology Systems Division), PO Box 80024, Wilmington, DE 19880-002, USA.
Tel: 001 302 774 1000
Fax: 001 302 774 7321
URL: http://www.dupont.com

Eastman Chemical Co., 100 North Eastman Road, PO Box 511, Kingsport, TN 37662-5075, USA.
Tel: 001 423 229 2000
URL: http://www.eastman.com

Fisher Scientific UK Ltd., Bishop Meadow Road, Loughborough, Leicestershire LE11 5RG, UK.
Tel: 01509 231166   Fax: 01509 231893
URL: http://www.fisher.co.uk
Fisher Scientific, Fisher Research, 2761 Walnut Avenue, Tustin, CA 92780, USA.
Tel: 001 714 669 4600
Fax: 001 714 669 1613
URL: http://www.fishersci.com

Flow Laboratories, Woodcock Hill, Harefield Road, Rickmansworth, Hertfordshire WD3 1PQ, UK.

Fluka, PO Box 2060, Milwaukee, WI 53201, USA.
Tel: 001 414 273 5013
Fax: 001 414 2734979
URL: http://www.sigma-aldrich.com
Fluka Chemical Co. Ltd., PO Box 260, CH-9471, Buchs, Switzerland.
Tel: 0041 81 745 2828
Fax: 0041 81 756 5449
URL: http://www.sigma-aldrich.com

Hampton Research (Crystallization materials), 25431 Cabot Road, Suite 205, Laguna Hills, CA 92653-5527, USA.
URL: http://www.hamptonresearch.com
xtalrox@aol.com

Hewlett Packard, Cain Road, Bracknell, Berkshire RG12 1HN, UK.
Hewlett Packard, 3495 Deer Creek Road, Palo Alto, CA 94304, USA.
Hewlett Packard, 150 Route du Nant-d'Avril, CH-1217 Meyrin-Geneva 2, Switzerland.

Hi-Tech Scientific Ltd. (Stopped flow equipment), Hi-Tech House, Brunel Road, Church Fields, Salisbury, Wiltshire SP2 7PU, UK.
Tedk@Hi-tech.demon.co.uk

Hybaid Ltd., Action Court, Ashford Road, Ashford, Middlesex TW15 1XB, UK.
Tel: 01784 425000
Fax: 01784 248085
URL: http://www.hybaid.com
Hybaid US, 8 East Forge Parkway, Franklin, MA 02038, USA.
Tel: 001 508 541 6918
Fax: 001 508 541 3041
URL: http://www.hybaid.com

HyClone Laboratories, 1725 South HyClone Road, Logan, UT 84321, USA.
Tel: 001 435 753 4584
Fax: 001 435 753 4589
URL: http://www.hyclone.com

Invitrogen Corp., 1600 Faraday Avenue, Carlsbad, CA 92008, USA.
Tel: 001 760 603 7200
Fax: 001 760 603 7201
URL: http://www.invitrogen.com
Invitrogen BV, PO Box 2312, 9704 CH Groningen, The Netherlands.
Tel: 00800 5345 5345
Fax: 00800 7890 7890
URL: http://www.invitrogen.com

JASCO (Circular dichroism suppliers), 18 Oak Industrial Park, Great Dunmow, CM6 1XN, UK.

Jobin Yvon / Dilor Raman (Raman suppliers), 2-4 Wigton Gardens, Stanmore, Middlesex HA7 1BG, UK.
Fax: 020 8204 6142
E-mail: jy@jyhoriba.co.uk
Jobin Yvon SA (Head Office), 16-18 Rue du Canal, BP 118, 1165 Longjumeau, Cedex, France.
Fax: (33) 1.69.09.93.19
www.jobinyvon.com

Kaiser Optical Systems, Inc. (Raman equipment), PO Box 983, 371 Parkland Plaza, Ann Arbor, MI 48106, USA.
Fax: 734-665-8199
E-mail: sales@kosi.com
URL: http://www.kosi.com

KinTek Corporation (Stopped flow equipment), 7604 Sandia Loop, Suite C, Austin TX 78735, USA.

Life Technologies Ltd., PO Box 35, Free Fountain Drive, Incsinnan Business Park, Paisley PA4 9RF, UK.
Tel: 0800 269210   Fax: 0800 838380
URL: http://www.lifetech.com
Life Technologies Inc., 9800 Medical Center Drive, Rockville, MD 20850, USA.
Tel: 001 301 610 8000
URL: http://www.lifetech.com

Lipex BioMembranes (Pressurized extruder apparatus), 3550 West 11th Avenue, Vancouver, British Columbia V6R 2K2, Canada.

Merck Sharp & Dohme, Research Laboratories, Neuroscience Research Centre, Terlings Park, Harlow, Essex CM20 2QR, UK.
URL: http://www.msd-nrc.co.uk
MSD Sharp and Dohme GmbH, Lindenplatz 1, D-85540, Haar, Germany.
URL: http://www.msd-deutschland.com

Microcal Inc., 22 Industrial Drive East, Northampton, MA 01060-2327, USA.

Micromass UK Ltd. (Mass spectrometry equipment), Floats Road, Wynthenshaw, Manchester M23 9LZ, UK.

Millipore (UK) Ltd., The Boulevard, Blackmoor Lane, Watford, Hertfordshire WD1 8YW, UK.
Tel: 01923 816375
Fax: 01923 818297
URL: http://www.millipore.com/local/UK.htm

Millipore Corp., 80 Ashby Road, Bedford, MA 01730, USA.
Tel: 001 800 645 5476
Fax: 001 800 645 5439
URL: http://www.millipore.com

Molecular Dynamics, 4 Chaucer Business Park, Kemsing, Sevenoaks, Kent TN15 6PL, UK.
Molecular Dynamics, 880 East Arques Avenue, Sunnyvale, CA 94086, USA.
Molecular Dynamics, Elisabethstrasse 103-105, D-47799 Krefeld, FR Germany.

Molecular Kinetics/Bio-Logic (Stopped flow equipment), PO Box 2475 C.S, Pullman, WA 99165, USA.

Molecular Probes, PO Box 22010, 4849 Pitchford Avenue, Eugene, Oregon 97402-9144, USA.
Molecular Probes, Poort Gebouw, Rijnsburgerweg 10, NL-2333 AA Leiden, The Netherlands.

MSC (Diffraction equipment)
USA: http@///www/msc.com/
Japan: http@//www/msc.com/

New England Biolabs, 32 Tozer Road, Beverley, MA 01915-5510, USA.
Tel: 001 978 927 5054

Nicolet Instrument Ltd. (FTIR and Raman equipment), Nicolet House, Budbrooke Road, Warwick CV34 5XH, UK.
Nicolet Instruments Corporation, 5225 Verona Road, Madison, WI 53711, USA.
E-mail: nicinfo@nicolet.com

Nikon Inc., 1300 Walt Whitman Road, Melville, NY 11747-3064, USA.
Tel: 001 516 547 4200
Fax: 001 516 547 0299
URL: http://www.nikonusa.com

Nikon Corp., Fuji Building, 2-3, 3-chome, Marunouchi, Chiyoda-ku, Tokyo 100, Japan.
Tel: 00813 3214 5311
Fax: 00813 3201 5856
URL: http://www.nikon.co.jp/main/index_e.htm

Nikon Diaphot (Inverted fluorescence microscope), Nikon Inc., Melville, New York, USA.

Nonius Ltd. (Diffraction equipment), Delft, The Netherlands.
URL: http://www.nonius.com/

Nycomed Amersham plc, Amersham Place, Little Chalfont, Buckinghamshire HP7 9NA, UK.
Tel: 01494 544000   Fax: 01494 542266
URL: http://www.amersham.co.uk
Nycomed Amersham, 101 Carnegie Center, Princeton, NJ 08540, USA.
Tel: 001 609 514 6000
URL: http://www.amersham.co.uk

Oxford Cryosystems (Crosystems for crystallography), Oxford, UK.
URL: http://www.oxfordcryosystems.co.uk

Perkin Elmer Ltd., Post Office Lane, Beaconsfield, Buckinghamshire HP9 1QA, UK.   Tel: 01494 676161
URL: http://www.perkin-elmer.com
Perkin Elmer Cetus (The Perkin-Elmer Corporation), 761 Main Avenue, Norwalk, CT 06859, USA.
E-mail: info@pe-corp.com
Perkin Elmer Ltd., Postfach 101164, Askaniaweg 1, D-88647, Überlingen, FR Germany.

Pharmacia Biotech (Biochrom) Ltd., Unit 22, Cambridge Science Park, Milton Road, Cambridge CB4 0FJ, UK.
Tel: 01223 423723   Fax: 01223 420164
URL: http://www.biochrom.co.uk

Pharmacia and Upjohn Ltd., Davy Avenue, Knowlhill, Milton Keynes, Buckinghamshire MK5 8PH, UK.
Tel: 01908 661101   Fax: 01908 690091
URL: http://www.eu.pnu.com
Pharmacia LKB Biotechnology AB, Bjorngatan 30, S-75182 Uppsala, Sweden.

Pierce & Warriner (UK) Ltd., 44 Upper Northgate Street, Chester CH1 4EF, UK.
Pierce Chemical Company, 3747 Meridian Road, PO Box 117, Rockford, IL 61105, USA.

Polymicro Technologies, Inc. (sources for capillaries), 18019 N 25th Avenue, Phoenix, AZ, USA.

Promega UK Ltd., Delta House, Chilworth Research Centre, Southampton SO16 7NS, UK.
Tel: 0800 378994
Fax: 0800 181037
URL: http://www.promega.com
Promega Corp., 2800 Woods Hollow Road, Madison, WI 53711-5399, USA.
Tel: 001 608 274 4330
Fax: 001 608 277 2516
URL: http://www.promega.com

Protein Solutions, Unit 5, The Hillside Centre, Upper Green Street, High Wycombe, Buckinghamshire, UK.
Protein Solutions, 2300 Commonwealth Drive, Suite 102, Charlottesville, VA 22901, USA.

Qiagen UK Ltd., Boundary Court, Gatwick Road, Crawley, West Sussex RH10 2AX, UK.
Tel: 01293 422911
Fax: 01293 422922
URL: http://www.qiagen.com
Qiagen Inc., 28159 Avenue Stanford, Valencia, CA 91355, USA.
Tel: 001 800 426 8157
Fax: 001 800 718 2056
URL: http://www.qiagen.com

Renishaw (Raman equipment), New Mills, Wotton-under-Edge, Gloucestershire GL12 8JR, UK.
Fax: +44 1453 524901
E-mail: genenq@renishaw.co.uk
Renishaw Inc., 623 Cooper Court, Schaumburg, Illinois 60173, USA.
Fax: +1 847 843 1744
E-mail: usa@renishaw.com

Rheodyne, PO Box 996, Cotatim, CA 94928, USA.

Roche Diagnostics Ltd., Bell Lane, Lewes, East Sussex BN7 1LG, UK.
Tel: 01273 484644
Fax: 01273 480266
URL: http://www.roche.com
Roche Diagnostics Corp., 9115 Hague Road, PO Box 50457, Indianapolis, IN 46256, USA.
Tel: 001 317 845 2358
Fax: 001 317 576 2126
URL: http://www.roche.com
Roche Diagnostics GmbH, Sandhoferstrasse 116, 68305 Mannheim, Germany.
Tel: 0049 621 759 4747
Fax: 0049 621 759 4002
URL: http://www.roche.com

Savant Instruments, 110-103 Bi-County Boulevard, Farmingdale, NY 11735, USA.

Schleicher and Schuell Inc., Keene, NH 03431A, USA.
Tel: 001 603 357 2398

Separations Group, 1734 Mojave Street, PO Box 867, Hesperia, CA 92345, USA.

SETARAM, 7 rue de l'Oratoire, F-69300, Caluire, France.

Shandon Scientific Ltd., 93-96 Chadwick Road, Astmoor, Runcorn, Cheshire WA7 1PR, UK.
Tel: 01928 566611
URL: http://www.shandon.com

Sigma-Aldrich Co. Ltd., The Old Brickyard, New Road, Gillingham, Dorset XP8 4XT, UK.
Tel: 01747 822211
Fax: 01747 823779
URL: http://www.sigma-aldrich.com
Sigma-Aldrich Co. Ltd., Fancy Road, Poole, Dorset BH12 4QH, UK.
Tel: 01202 722114
Fax: 01202 715460
URL: http://www.sigma-aldrich.com
Sigma Chemical Co., PO Box 14508, St Louis, MO 63178, USA.
Tel: 001 314 771 5765
Fax: 001 314 771 5757
URL: http://www.sigma-aldrich.com

SLT Labinstruments, PO Box 13953, Research Triangle Park, NC 27709, USA.
SLT Labinstruments, Unterbergstrasse 1A, A-5082, Grodig, Austria.

Spectrum Medical Industries, Inc., 1100 Rankin Road, Houston, Texas 77073-4716, USA.
Spectrum Europe BV, European Headquaters, PO Box 3262, 4800 DG Breda, The Netherlands.

Stratagene Inc., 11011 North Torrey Pines Road, La Jolla, CA 92037, USA.
Tel: 001 858 535 5400
URL: http://www.stratagene.com
Stratagene Europe, Gebouw California, Hogehilweg 15, 1101 CB Amsterdam Zuidoost, The Netherlands.
Tel: 00800 9100 9100
URL: http://www.stratagene.com

Supelco, Inc. (sources for capillaries), Supelco Park, Bellefonte, PA 16823, USA.

ThermoMetric Ltd., 10 Dalby Court, Gadbrook Business Park, Northwich, Cheshire CW9 7TN, UK.
Thermometric AB, Spjutvägen 5A, S-175 61 Järfälla, Sweden.

Thermomicroscopes Corporation (Atomic force microscopy equipment), 1171 Burregas Avenue, Sunnyvale CA 94089, USA.

United States Biochemical, PO Box 22400, Cleveland, OH 44122, USA.
Tel: 001 216 464 9277

Varian Instruments (FTIR equipment), 2700 Mitchell Drive, Walnut Creek, CA 94598, USA.
E-mail: osiusa@varianinc.com

Waters Chromatography, 324 Chester Road, Hartford, Northwich, Cheshire CW8 2AH, UK.
Waters Chromatography, 34 Maple Street, Milford, MA 01757, USA.
Waters Chromatography, 6 rue JP Timbaud, F-78180 Montigny le Breton-neux, France.

Index